Methods and Tools in Biosciences and Medicine

Microinjection,
edited by Juan Carlos Lacal, Rosario Perona and James Feramisco, 1999
DNA Profiling and DNA Fingerprinting,
edited by Jörg Epplen and Thomas Lubjuhn, 1999
Animal Toxins - Facts and Protocols,
edited by Hervé Rochat and Marie-France Martin-Eauclaire, 2000
Methods in Non-Aqueous Enzymology,
edited by Munishwar Nath Gupta, 2000

Animal Toxins

Facts and Protocols

Edited by

Hervé Rochat
Marie-France Martin-Eauclaire

Springer Basel AG

Editors

Professor Dr. Hervé Rochat
Laboratoire de Biochimie
Faculté de Médecine
Bld Pierre Dramard
F-13916 Marseille Cedex 20
France

Dr. Marie-France Martin-Eauclaire
Laboratoire de Biochimie
Faculté de Médecine
Bld Pierre Dramard
F-13916 Marseille Cedex 20
France

A CIP catalogue record for this book is available from the library of Congress, Washington D.C., USA

Deutsche Bibliothek Cataloging-in-Publication Data
Animal toxins : facts and protocols / ed. by Hervé Rochat ;
Marie-France Martin-Eauclaire. – Basel ; Boston ; Berlin :
Birkhäuser, 2000
(Methods and tools in biosciences and medicine)
ISBN 978-3-7643-6020-7 ISBN 978-3-0348-8466-2 (eBook)
DOI 10.1007/978-3-0348-8466-2

Originally published by Birkhäuser Verlag in 2000
Softcover reprint of the hardcover 1st edition 2000
Printed on acid-free paper produced from chlorine-free pulp. TCF ∞
Cover Illustration: Aah II molecule (page 175)

ISBN 978-3-7643-6020-7 3-7643-5983-8

9 8 7 6 5 4 3 2 1

Contents

List of Contributors

LÁSZLO BÉRESS, Institute of Toxikology, Christian-Albrechts-Universität, Kiel, Brunswikerstr. 10 D-24105 Kiel, Germany, Fax.: #49-431-597 4937, Email beress@toxi.uni-kiel.de

NOUREDDINE BOUAÏCHA, Laboratoire Santé Publique-Environnement Faculté de Pharmacie, Université Paris-Sud, Rue J.B. Clément, 99296 Chatenay-Malabry, France

PIERRE E. BOUGIS, CNRS UMR 6560 Ingénierie des Protéines, Institut Fédératif de Recherche Jean Roche, Université de la Méditerranée, Marseille, France

JEAN-CLAUDE BOULAIN, Département d'Ingénierie et d'Études des Protéines, CEA/Saclay, 91191, Gif-sur-Yvette, France

THOMAS BRUHN, Institute of Toxikology, Christian-Albrechts-Universität, Kiel, Brunswikerstr. 10 D-24105 Kiel, Germany, Fax.: #49-431-597 4937, Email bruhn@toxi.uni-kiel.de

MARCEL CREST, Laboratoire de Neurobiologie LNB4, UPR CNRS 9024, 31 chemin Joseph Aiguier, 13402 Marseille, Cedex 20, France, Tel. 04 91 16 43 26, (33) 4 91 16 43 26, Fax. 04 91 16 46 15, (33) 4 91 16 46 15. Email gola@irlnb.cnrs-mrs.fr

LOURDES J. CRUZ, Marine Science Institute, U. P. Diliman, Quezon City 1101, Philippines and Department of Biology, University of Utah, Salt Lake City, UT 84112, USA

CHRISTIANE DEVAUX, Laboratoire de Biochimie-Ingénierie des Protéines, CNRS UMR 6560, Faculté de Médecine-Nord, Bd P. Dramard – 13916 Marseille Cedex 20, France

GEORGES DIOGENE, CEMATMA, Laboratoire de Cryptogramie, Muséum National d'Histoire Naturelle, 12 rue Buffon 75005, Paris, France

PASCAL DREVET, Département d'Ingénierie et d'Études des Protéines, CEA/Saclay, 91191, Gif-sur-Yvette, France

FRÉDÉRIC DUCANCEL, Département d'Ingénierie et d'Études des Protéines, CEA/Saclay, 91191, Gif-sur-Yvette, France

MOHAMED EL AYEB, Laboratoire des Venins et Toxines, Institut Pasteur de Tunis, B.P. 74, 1002 Tunis, Tunisie

MARIA L. GARCIA, Department of Membrane Biochemistry and Biophysics, Merck Research Laboratories, P.O. Box 2000, Rahway, NJ 07065, USA

MAURICE GOLA, Laboratoire de Neurobiologie LNB4, UPR CNRS 9024, 31 chemin Joseph Aiguier, 13402 Marseille, Cedex 20, France, Tel. (33) 4 91 16 43 26, (33) 4 91 16 43 26, Fax. 04 91 16 46 15, (33) 4 91 16 46 15. Email gola@irlnb.cnrs-mrs.fr

WILLIAM R. GRAY, Department of Biology, University of Utah, Salt Lake City, UT 84112, USA

EVGENIY GRISHIN, Shemyakin and Ovchinnikov Institute of Bioorganic Chemistry RAS. 16/10 Miklukho-Maklaya, 117871, GSP-7, Moscow V-437, Russia

Markus Hanner, Department of Membrane Biochemistry and Biophysics, Merck Research Laboratories, P.O. Box 2000, Rahway, NJ 07065, USA

Alan L. Harvey, Department of Physiology and Pharmacology, and Strathclyde Insitute for Drug Research, University of Strathclyde, Glasgow G4 0NR, UK, Tel. (44) 141 553 4155, Fax (44) 141 552 8376

Julita S. Imperial, Department of Biology, University of Utah, Salt Lake City, UT 84112, USA

Gregory J. Kaczorowski, Department of Membrane Biochemistry and Biophysics, Merck Research Laboratories, P.O. Box 2000, Rahway, NJ 07065, USA

William R. Kem, Department of Pharmacology and Therapeutics, University of Florida College of Medicine, Gainesville, Florida 32610-0267, USA, Phone 352-392-3541, FAX 352-392-9696, E-mail: Kem@pharmacology.ufl.edu

Hans-Günther Knaus, Institute for Biochemical Pharmacology, Peter Mayr Strasse 1, A-6020 Innsbruck, Austria

R. Manjunatha Kini, Bioscience Centre, Faculty of Science, National University of Singapore, 10 Kent Ridge Crescent, Singapore 119260

Pascale Marchot, CNRS UMR 6560 Ingénierie des Protéines, Institut Fédératif de Recherche Jean Roche, Université de la Méditerranée, Marseille, France

Marie-France Martin-Eauclaire, Laboratoire de Biochimie, Unité Mixte de Recherche 6560 du Centre National de la Recherche Scientifique, Institut Fédératif de Recherche Jean Roche, Université de la Méditerranée, Faculté de Médecine Secteur Nord, 15 Bd Pierre Dramard, F-13916 Marseille Cedex 20, France

J. Michael McIntosh, Department of Biology, University of Utah, Salt Lake City, UT 84112, USA and Department of Psychiatry, University of Utah, Salt Lake City, UT 84112, USA

André Ménez, Département d'Ingénierie et d'Études des Protéines, CEA/Saclay, 91191, Gif-sur-Yvette, France

Gilles Mourier, Département d'Ingénierie et d'Études des Protéines, CEA/Saclay, 91191, Gif-sur-Yvette, France

Hidehito Mukai, Laboratory of Life Science & Biomolecular Engineering, Japan Tobacco Inc., 6–2, Umegaoka, Aoba-ku, Yokohama, Kanagawa 227–8512, Japan, Tel. +81-45-972-5901, Fax +81-45-972-6205

Terumi Nakajima, Suntory Institute for Bioorganic Research, 1-1-1 Wakayamadai, Shimamoto-cho, Mishima-gun, Osaka 618-8503, Tel. +81-75-962-1660, Fax +81-75-962-2115

Tom Piek, Ebrudreef 42, 3561 JN Utrecht, Netherlands

Kirill A. Pluzhnikov, Shemyakin and Ovchinnikov Institute of Bioorganic Chemistry RAS. 16/10 Miklukho-Maklaya, 117871, GSP-7, Moscow V-437, Russia

Mark A. Poli, U.S. Army Medical Research Institute of Infectious Diseases, Toxinology Division, Ft. Detrick, Frederick, MD 21702 U.S.A.

Simone Puiseux-Dao, CEMATMA, Laboratoire de Cryptogramie, Muséum National d'Histoire Naturelle, 12 rue Buffon 75005, Paris, France

Hervé Rochat, Laboratoire de Biochimie, Unité Mixte de Recherche 6560 du Centre National de la Recherche Scientifique, Institut Fédératif de Recherche Jean Roche, Université de la Méditerranée, Faculté de Médecine Secteur Nord, 15 Bd Pierre Dramard, F-13916 Marseille Cedex 20, France

Edward G. Rowan, Department of Physiology and Pharmacology, University of Strathclyde, Glasgow G4 0NR, UK, Tel. (44) 141 548 2689, Fax (44) 141 552 2562

Jean-Marc Sabatier, Laboratoire de Biochimie, Ingénierie des Protéines, CNRS UMR 6560, IFR Jean Roche, BD Pierre Dramard, F-13916, Marseille Cedex 20, France. Tel. (33) 491698852, Fax (33) 491 657595 Email: sabatier.jm@jean-roche.univ-mrs.fr

Judith Sánchez-Rodríguez, Institute of Toxikology, Christian-Albrechts-Universität, Kiel, Brunswikerstr. 10 D-24105 Kiel, Germany

Hugues Schweitz, Institut de Pharmacologie Moléculaire et Cellulaire, 660 route des Lucioles, Sophia Antipolis, 06560 Valbonne, France

Ludmilla V. Shevchenko, Shemyakin and Ovchinnikov Institute of Bioorganic Chemistry RAS. 16/10 Miklukho-Maklaya, 117871, GSP-7, Moscow V-437, Russia

Jüri Siigur, Laboratory of Bioorganic Chemistry, National Institute of Chemical Physics and Biophysics, Akadeemiatee 23, 12618 Tallinn, Estonia

Ene Siigur, Laboratory of Bioorganic Chemistry, National Institute of Chemical Physics and Biophysics, Akadeemiatee 23, 12618 Tallinn, Estonia

Robert S. Slaughter, Department of Membrane Biochemistry and Biophysics, Merck Research Laboratories, P.O. Box 2000, Rahway, NJ 07065, USA

Peter N. Strong, Division of Biomedical Sciences, Sheffield Hallam University, Sheffield S1 1WB, UK

Vera L. Trainer, National Marine Fisheries Service, Northwest Fisheries Science Center, Environmental Conservation Division, 2725 Montlake Blvd. E., Seattle, WA 98105 USA

Elmar Wachter, Institut für Physiologische Chemie der Ludwig-Maximilians Universität München, Goethestr. 33, D-80336 München, Germany

Jonathan D.F. Wadsworth, Department of Neurogenetics, Imperial College School of Medicine at St. Mary's, Norfolk Place, London WZ IPG, UK

Kaori Wakamatsu, Department of Biochemical Sciences, Faculty of Engineering, Gunma University, 1-5-1 Tenjin-cho, Kiryu, Gunma 376-8515, Japan, Tel/Fax +81-277-30-1439

Abbreviations

$[^3H]$PbTx-3	tritiated brevetoxin
$[^3H]$STX	tritiated saxitoxin
ACh	acetylcholine
AChE	acetylcholinesterase
AL-B	alboaggregin-B
ATCh	acetylthiocholine
BAEE	N(α)-benzoyl-L-arginine ethyl ester
BAPNA	N(α)-benzoyl-DL-arginine p-nitroanilide
BSA	bovine serum albumin
CM	carboxymethyl
DCC	dextran-coated charcoal
DEAE	diethylaminoethyl
DFP	(or DIFP) diisopropyl fluorophosphate
DMSO	dimethyl sulfoxide
DSP	Diarrhetic Shellfish Poisoning
DTNB	5,5'-dithiobis-2-nitrobenzoic acid
DTT	dithiothreitol
DTX	dinophysistoxin
EDTA	ethylene diamine tetraacetic acid
EGTA	ethylene glycol bis(β-aminoethyl ether)-N,N,N',N'-tetraacetic acid
Fas 1,2,3	fasciculins 1,2,3
G protein	trimeric GTP-binding regulatory protein
GP	glycoprotein
GPIb	glycoprotein 1b
GPIIb/IIIa, $\alpha_{II}b\beta_3$	glycoprotein integrin receptor
HPLC	high performance liquid chromatography
IC_{50}	the concentration at which 50% of total ligand binding is inhibited
LDH	lactate dehydrogenase
mAChE	recombinant AChE from mouse
MCYST	microcystin
MEPQ	7-[[(methylethoxy)phosphonyl]-oxy]-1-methylquinolinium
MTX	maitotoxin
MW	molecular weight
$NaPO_4$	Na_2HPO_4/NaH_2PO_4
NH_4Ac	ammonium acetate
NH_4-formate	ammonium formate
NH_4HCO_3	ammonium bicarbonate
NMR	nuclear magnetic resonance
OA	okadaic acid
PAGE	polyacrylamide gel electrophoresis
PAI	platelet aggregation inhibitor
PBS	phosphate-buffered saline
PbTx-3	brevetoxin from *P. brevis* (*G. breve*)
Pip	Piperidyl
PL	phospholipids
PMSF	phenylmethyl-sulfonyl fluoride
pNA	p-nitroaniline
PPP	platelet-poor plasma
PRP	platelet-rich plasma
PTX	pertussis toxin
rFas2	recombinant fasciculin 2
RIA	radioimmunoassay
RVV-V	Russell's viper venom factor V activator
RVV-X	Russell's viper venom factor X activator
SDS	sodium dodecyl sulfate
SDS-PAGE	sodium dodecyl sulfate-polyacrylamide gel electrophoresis
SPE	solid phase extraction
SRA	specific radioactivity
SRTX	saraotoxin
STX	saxitoxin
TAME	N(α)-p-tosyl-L-arginine methyl ester
TcAChE	*Torpedo californica* AChE
TFA	trifluoroacetic acid
ToxC	toxin C
Tris	tris-(hydroxymethyl)aminoethane
VBXAE	*V. berus* venom factor X activating enzyme
vWF	von Willebrand factor

Preface

It is now something of a truism to say that natural toxins form a major component of the molecular tools used increasingly frequently by the ever growing number of laboratories of various kinds. Evidence for this is provided not only by the increasing number of firms including such toxins in their catalogues but also by the large number of demands received by those who discover new toxins.

This book is designed to facilitate the work of scientists interested in exploring a new domain of toxicology or in using toxins as tools for research. The length of this book has been kept reasonable by concentrating on animal toxins, most of which are polypeptides.

The venom of each animal contains a large number of toxins, which may be similar in terms of their physical and chemical properties and are therefore difficult to purify. However, their pharmacological activities may be very different. These toxins are extremely active, and it is therefore necessary to demonstrate conclusively that the activity observed really does correspond to the newly characterized molecule.

The toxins active against a particular target often constitute a group of molecules that are very similar in structure, although the properties of the interaction of each toxin with the receptor site may be characteristic. The diversity of these toxins corresponds, as we are beginning to understand with increasing clarity, to the diversity of the intended target. It is thus necessary to be very precise in the choice of "toxin" tool used to study a specific target within a given tissue.

The articles in this book have been written bearing these limitations in mind. The reader will find authorized trials studying the mechanism of action of animal toxins at the cellular and molecular level. Such information is already available, so information about the use of these molecular "scalpels" for dissecting the molecular mechanisms of their pharmacological targets is also included. These polypeptide toxins may also be treated as biological entities, miniproteins with very strong binding activities making it possible, for a wide diversity of affinities, to correlate structure and activity. These miniproteins are frequently extensively cross-linked by disulphide bridges. Comparison of their sequences may provide information about the relationship between the primary and folded structures of these molecules.

This book does not deal with all animal toxins, but each of its chapters aims to help everyone concerned optimize the use they make of these molecules.

H. Rochat
M.F. Martin-Eauclaire

December 1999

1 Assays for Dinoflagellate Toxins, Specifically Brevetoxin, Ciguatoxin, and Saxitoxin

Vera L. Trainer and Mark A. Poli

Contents

1 Introduction

Dinoflagellate toxins have been widely recognized as the compounds responsible for the human intoxications named paralytic shellfish poisoning (due to ingestion of saxitoxin and related compounds), neurotoxic shellfish poisoning (due to ingestion of brevetoxins), and ciguatera poisoning (due to ingestion of

Methods and Tools in Biosciences and Medicine
Animal Toxins, ed. by H. Rochat and M.-F. Martin-Eauclaire

ciguatoxins). These toxins can accumulate in molluscs through filter feeding (saxitoxins and brevetoxins) or in fish through grazing and/or food-chain amplification (ciguatoxins), and are then passed on to humans through the ingestion of seafood. Each toxin binds specifically to a binding site (receptor) on the sodium channel of nerves, thereby either blocking or opening the channel (for structures of these toxins and their action on nerves, see Fig. 1). A result of this binding event is the impairment of normal nerve function, resulting in death in extreme cases. For this reason, it is important that accurate and sensitive assays for these toxins be made available. Two related types of assays, the radioimmunoassay (RIA) and the receptor binding assay, are very useful for detecting these toxins in a variety of matrices.

A. *Toxin:* Saxitoxin
Abbreviation: **STX**
Origin: Marine dinoflagellates of the genera, *Alexandrium (Gonyaulax), Pyrodinium, and Gymnodinium*
Structure:

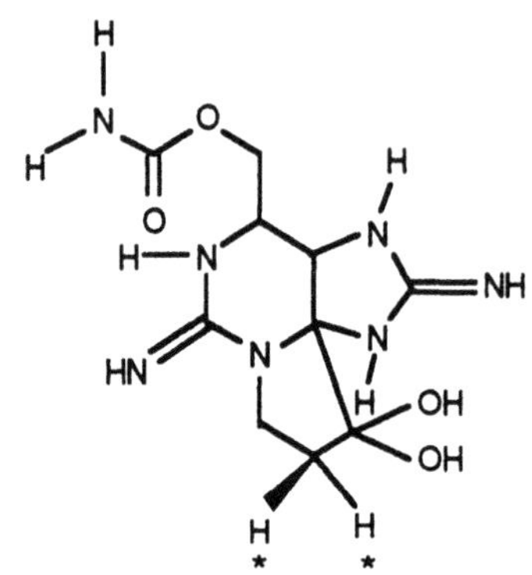

Effect: Blocks ion transport into sodium channels
MW: 419

B. *Toxin* Tetrodotoxin
Abbreviation: **TTX**
Origin: pufferfish, bacteria, members of the classes *Turbellaria, Gastropoda*, and *Amphibia*
Structure:

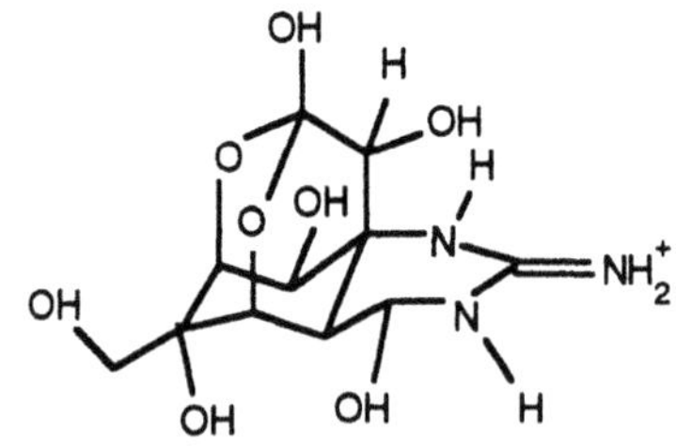

Effect: Blocks ion transport into sodium channels
MW: 319

Figure 1 Toxins which can be analyzed using receptor binding assays or radioimmunoassay. The asteriks denote the position of tritium (^3H) in the labeled toxin.

Both of these assays rely upon the specific recognition of the toxins in solution by a binding protein (antibody) or a membrane protein complex (receptor). The affinity of the toxins for the receptor or antibody used in each assay is theoretically based on different structural components of the toxins themselves. In the RIA, affinity is based on the antibody recognition of the antigenic determinants on the immunizing compound, in this case, a toxin. In the receptor binding assay, affinity is based on receptor recognition of the portions of each toxin molecule that bind to their specific site on the sodium channel. Comparing these two assays, the advantages of the RIA are its exceptional sensitivity, lower expense, and ease of use. The major advantage of the receptor binding assay is its ability to measure the composite potency of a sample.

C. *Toxin:* Brevetoxin
Abbreviation: **PbTx-3**
Origin: The marine dinoflagellate, *Gymnodinium breve* (*Ptychodiscus brevis*)
Structure:

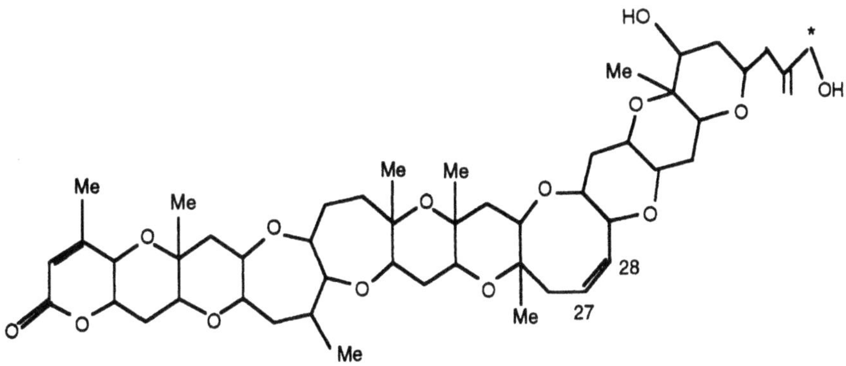

Effect: Repetitive firing, persistent opening of sodium channels
MW: 895

D. *Toxin:* Ciguatoxin
Abbreviation: **CTX**
Origin: Marine dinoflagellate, *Gambierdiscus toxicus*
Structure:

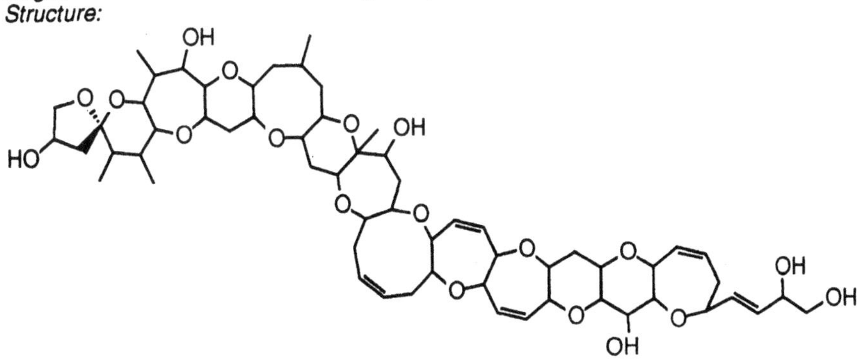

Effect: Repetitive firing, persistent opening of sodium channels
MW: 1113

In this chapter we present the protocols for

- RIA
  (1) Optimization of antibody dilution for use in the RIA.
  (2) Generation of the RIA standard curve and its use in determining toxin concentrations in samples.
- Receptor binding assay
  (1) Optimization of membrane dilution for use in the receptor binding assay.
  (2) Generation of the receptor binding assay standard curve and its use in determining toxin concentrations in samples.

2 Radioimmunoassay

2.1 Materials and chemicals

Buffers and Reagents

- **Phosphate-buffered saline (PBS) 10 mM, pH 7.4**
  Dissolving 1 packet in 1 l distilled water gives a solution comprised of 10 mM phosphate buffered salts (Na_2HPO_4, NaH_2PO_4), 0.138 M NaCl, 2.7 mM KCl.
- **Assay Buffer (Solution 1)**
  PBS containing 0.01% Alkamuls EL-620
- **Dextran-coated charcoal (DCC)**
  10 g activated charcoal suspended in 100 ml PBS
  Centrifuge at 1500 x g and discard supernatant containing floating particles.
  Repeat three times.
  After the third wash, resuspend charcoal pellet in PBS containing 0.25% dextran T-500 and 0.01% thimerosal **(Solution 2)**.

Chemicals

Acetic acid, glacial	Fisher	BP1185-500
Alkamuls EL-620	Rhone-Poulenc	331616
PbTx polyclonal antibodies	Center for Marine Research*	PCG-PB03
Brevetoxin, PbTx-3	Center for Marine Research*	PB03
Brevetoxin, [^{3}H]PbTx-3	Center for Marine Research*	TRI-PB03
Charcoal, activated	Sigma	C-9157
Dextran T-500	Pharmacia	17.0320.01
EnviroSafe scintillation fluid, or	Anoroc Scientific	LSC-060
Ecolume	ICN	882470
Glacial acetic acid	Sigma	A6283
Methanol	Fisher	A935-4
Phoshate-buffered saline	Sigma	P-3813
Thimerosal	Sigma	T-5125

* Center for Marine Research, WINTIR, UNC Wilmingston, 601 S. College Road, Wilmington, NC 28403

Equipment

- pipettors and test tube racks
- vortex mixer
- vacuum oven
- benchtop centrifuge
- scintillation counter
- refrigerator
- kitchen blender or equivalent

2.2 Methods

At least nine distinct brevetoxin derivatives which are based upon two separate backbone structures have been purified from both field blooms and laboratory cultures of *G. breve* (for complete structures of the brevetoxins, see Baden and Trainer, 1993). The following radioimmunoassay is specific for brevetoxins sharing the PbTx-3 backbone structure shown in Figure 1 (for method of antibody production, see Trainer and Baden, 1991). Brevetoxins possessing the PbTx-1 backbone structure are minimally cross-reactive with these antibodies. Antibody stock solutions should be kept refrigerated in PBS containing a bacteriostatic agent such as thimerosal or sodium azide (0.01%). Under these conditions, antibody stocks should last for several months.

Optimization of antibody and radiolabeled toxin concentrations

The principle of RIA is shown in Figure 2. A key step in performing this assay is optimization of the concentrations of the antibody and the labeled toxin. The detection limit of any competitive immunoassay is dependent upon the affinity of the antibody for the ligand, the radiolabeled toxin concentration at which the competition occurs, and the specific activity of the radiolabeled toxin. The affinity of the antibody for the ligand is constant. However, the labeled toxin concentration can be varied within the linear portion of the curve. The detection limit and 50% inhibition value (IC_{50}; the toxin concentration at which 50% of total binding is inhibited) will decrease (sensitivity will increase) with decreasing concentrations of labeled toxin. The practical lower limits of labeled toxin concentration within the linear portion of the curve is the number of detectable bound counts (disintegrations per minute, DPMs), which is dependent upon the specific activity of the labeled toxin and the dilution of antibody. The optimal antibody dilution must always be determined empirically for each new lot of antibody or [^{3}H]PbTx-3. In addition, the optimal dilution may change slightly if the activity of either the labeled toxin or the antibody preparation decreases over time. The goal of the optimization procedure is to arrive at concentrations of labeled toxin and antibody yielding the greatest sensitivity and

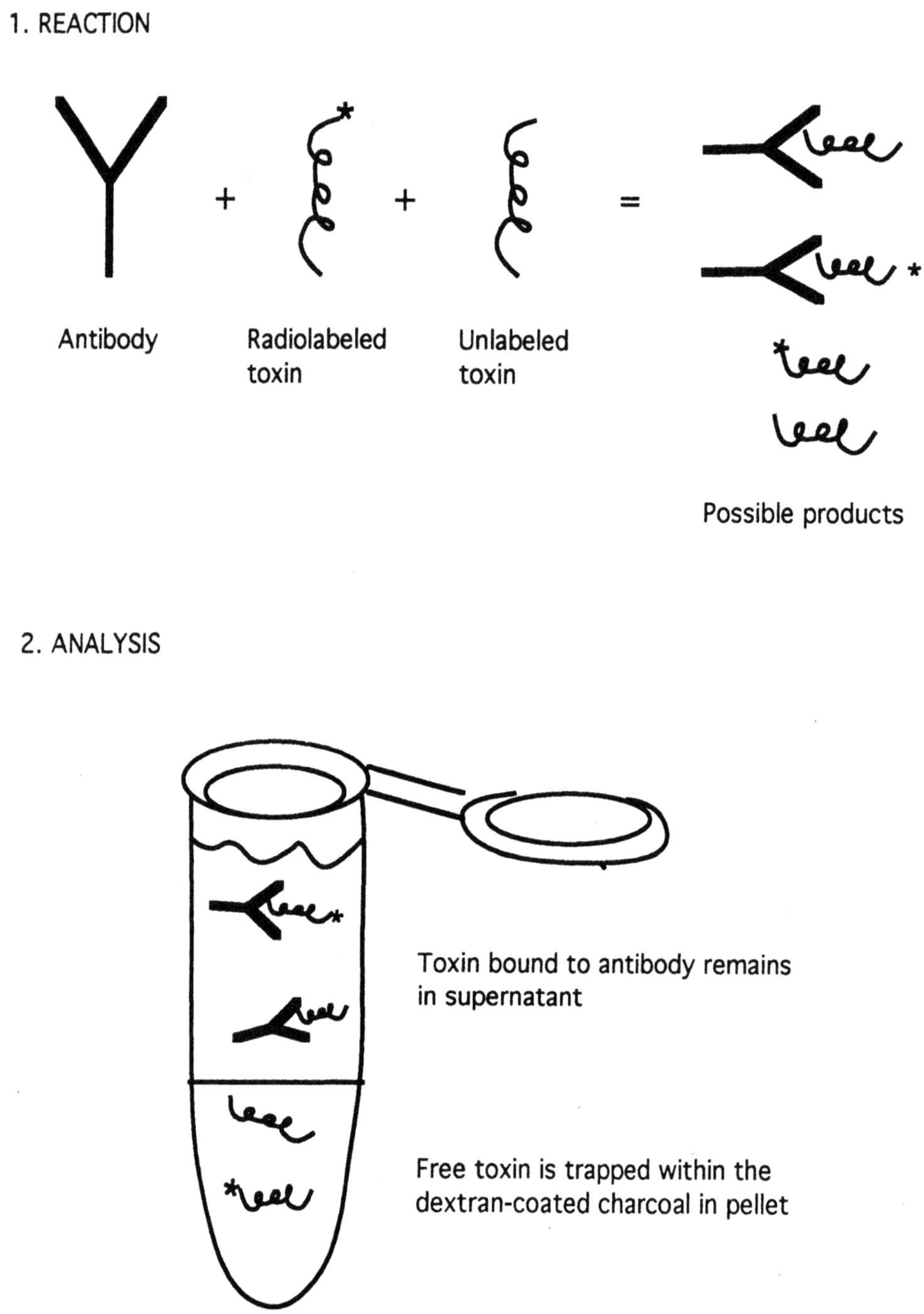

Figure 2 The schematic diagram of a radioimmunoassay.

economy of reagents. To optimize the dilution of antibody, begin with a label concentration that is at or slightly below the apparent affinity constant (Kd) of the antibody. The apparent Kd can be derived from a saturation isotherm that is obtained by measuring the biological response as a function of the toxin concentration. The affinity constants for toxins used in this chapter have been previously determined.

Protocol 1 Optimization of antibody dilution

1. Prepare needed volume of 1.8 ng/ml [^{3}H]PbTx-3 in assay buffer **(Solution 1)** (the volumes of each reagent used in this optimization experiment are the same as in the RIA method diagrammed in Figure 4).
2. Make a series of half-log dilutions of PbTx antibody in assay buffer covering the range of 1:100–1:100,000.
3. Set up a series of test tubes where each set of triplicates contains 0.5 ml of [^{3}H]PbTx-3 (made in step 1) and 0.5 ml of each antibody dilution (made in step 2). This dilution brings the final [^{3}H]PbTx-3 concentration to 1 nM. Vortex and incubate at least 1–2 h at 4°C. Samples can be prepared at the end of the day and incubated overnight in the refrigerator for sedimentation and analysis the next day.
4. To each tube, add 0.5 ml of a 1:50 dilution of dextran-coated charcoal (DCC) in PBS **(Solution 2)** (recipe found in the section on buffers and reagents). Vortex and incubate 15 min at 4°C.
5. Immediately sediment the DCC by centrifugation at 2000 x g for 15 min at 4°C.
6. Carefully pipette 1 ml of each supernatant into scintillation vials containing 50 µl of glacial acetic acid. Add scintillation fluid, shake until solution clears, and count each sample for 5 min in a scintillation counter.

Notes:
The optimal antibody dilution will result in bound DPMs 8- to 10-fold above background that should be in the range of 2000–2500 DPM. In this case, background binding is measured in the absence of antibody.

Sample preparation

Samples are typically organic solvent extracts of shellfish. The basic method of toxin extraction from tissue is as follows.

Protocol 2 Toxin extraction from shellfish tissue

1. Remove tissue from the surrounding shell, weigh, and homogenize in a minimal volume of distilled water.

   Note:
   Homogenization should be done on ice using a motor-driven tissue homogenizer such as a kitchen blender. The sample should be blended thoroughly but not allowed to foam significantly (proteolysis will occur).

2. The slurry is then extracted three times with two volumes of acetone and then filtered.
3. The filtrate is evaporated to dryness and redissolved in 90% methanol.
4. Neutral lipids are removed by liquid/liquid partitioning against equal volumes of light petroleum ether in a separatory funnel.
5. The methanol (heavier) fraction is again evaporated to dryness and redissolved in 100% methanol at 1 ml/g of original tissue weight.
6. Dilutions of these samples are then prepared in the same solvent, and a constant volume of each dilution is tested in triplicate.

   Note:
   Whenever possible, standard curves should be prepared in a similar dilution of a nontoxic control extract. For example, if an oyster extract of 1 g/ml is being tested at dilutions of 1:100, 1:500, and 1:1000, then the standard curve should be prepared in a 1:500 dilution of a parallel extract of nontoxic control oysters.

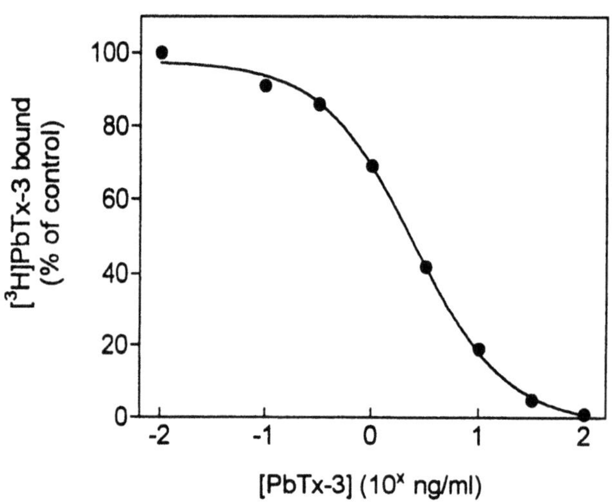

Figure 3 A typical standard curve.

The nonspecific (background) value, the level of binding at 100 ng/ml PbTx-3, is substracted from all measurements. The IC_{50} is calculated as the value on the abscissa which corresponds to 50% of control binding.

In our experience, shellfish extracts present no matrix interference at dilutions of 1:100 or greater, but matrix effects should always be evaluated with non-toxic control extracts prior to analyzing samples.

Standard curve and analysis of samples

In this competitive RIA, standard curves are constructed by incubating antibody with increasing concentrations of unlabelled competitor (PbTx-3) in the presence of a constant concentration of [^{3}H]PbTx-3. After allowing binding to come to equilibrium, a suspension of DCC is added. DCC scavenges free toxin from solution. Centrifugation pellets the charcoal and effectively removes unbound toxin from solution. Increasing competitor (unlabeled toxin) concentration results in a dose-dependent decrease in binding of radiolabeled toxin to the available antibody. The percentage of [^{3}H]PbTx-3 bound at each concentration of competitor toxin compared to that bound in the absence of competitor toxin (control) is plotted against the log of competitor toxin concentration to generate a standard curve. Toxin in unknown samples is quantitated by plotting sample values as a percentage of control binding on this standard curve (Fig. 3). The following RIA protocol is diagrammed in Figure 4.

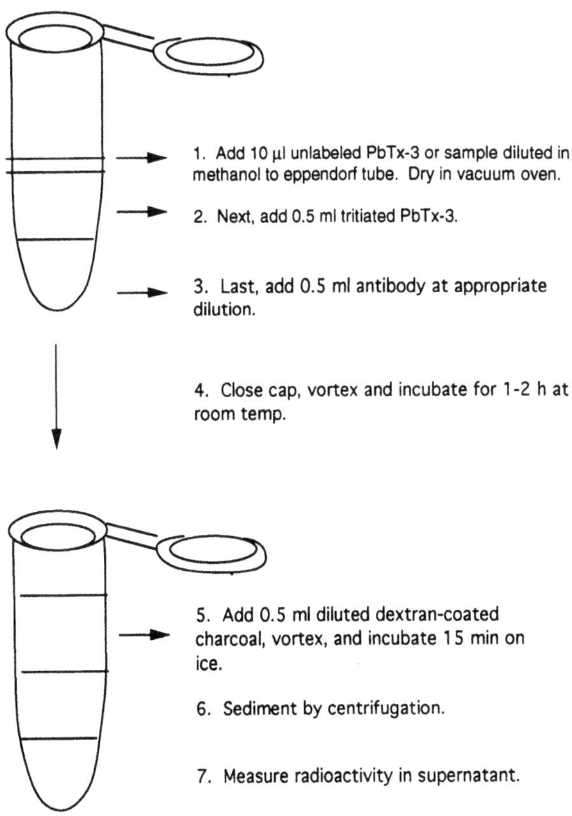

Figure 4 Radioimmunoassay method.

Protocol 3 Standard curve and analysis of toxin samples

1. Prepare the standard curve by diluting a stock solution of unlabeled PbTx-3 (100 µg/ml in methanol or in nontoxic methanol extract of the tissue being tested for toxin content) in half log dilutions over the range of 0.01–30 ng/ml final concentration. Aliquots (10 µl) of each dilution are added to triplicate tubes and the tubes dried in a vacuum oven set at about 60°C. If a vacuum oven is not available, air drying at room temperature or drying under a stream of nitrogen is acceptable.
2. Aliquots (10 µl) of unknown samples at three separate concentrations (e.g., 1:100, 1:500, and 1:1000) diluted in 100% methanol are placed in triplicate tubes and dried as above.
3. Prepare a solution of the optimal antibody dilution (previously determined) in the necessary final amount of assay buffer. To this add 0.5 ml [^{3}H]PbTx-3 (1 nM final concentration), vortex, and incubate at least 1–2 h at room temperature or overnight in the refrigerator. Overnight incubation does not compromise assay results, but refrigeration is recommended.
4. Add 0.5 ml of a 1:50 dilution of DCC in PBS **(Solution 2)** to each tube, vortex, and incubate 15 min at 4°C.
5. Sediment the DCC by centrifugation at 2000 x g for 15 min at 4°C.
6. Carefully pipette 1 ml of the supernatant from each tube into scintillation vials containing 50 µl glacial acetic acid. Add scintillation fluid, shake until the solution clears, and measure bound radioactivity in a scintillation counter.
7. For the standard curve, plot the percentage of total bound DPMs in the standard against the log of PbTx-3 concentration. Quantify unknowns by comparing percent bound sample values to the standard curve. Express results as "ng/ml PbTx-3 equivalents".

2.3 Troubleshooting

Standard Curve: The standard curve should be sigmoidal in shape, with the slope of the linear portion ranging from 0.9 to 1. If it is significantly less than 0.9, then repeat the experiment paying close attention to the dilutions for the standard curve. Background (nonspecific) binding (i.e., DPMs at the highest concentration of unlabeled toxin, usually about 100 times the concentration of radiolabeled toxin) should be no more than 10–15% of bound DPMs at the lowest competitor concentration. If not, raise the antibody concentration slightly to increase the number of bound DPMs in solution in the absence of competitor. The difference between background and total DPMs should increase.

Unknown Samples: For accurate quantitation, unknown samples should be tested at three concentrations within the linear portion of the standard curve. This usually requires a range-finding experiment over a wide concentration range, followed by a more directed experiment over a narrow range. All three concentrations, after adjustment for differing dilutions, should calculate to approximately the same final value. If not, interfering matrix effects (the effects of the material in which toxin is embedded) should be suspected and the proper control experiments should be done to evaluate the problem.

Matrix Effects: In our experience, matrix effects are not a problem in shellfish extracts when diluted at least 1:100, but significant matrix effects may occur in other matrices (i.e., hemolyzed blood), especially when analyzed at very low dilutions (Poli, 1996). Control curves for matrix effects should always be generated by increasing concentrations of the control matrix in the absence of competitors. Most matrices will show some level of interference at very high concentrations. If the interference is minor (10–30%), then generating a standard curve using the same dilution of control matrix as was used for samples is sufficient. If the interference is severe, then a pre-assay sample cleanup step may be required.

Dextran-Coated Charcoal Concentration: If the unknown sample contains a great deal of extraneous material, it is possible for the DCC to become saturated and no longer scavenge the unbound toxin effectively. This should be suspected when either of the following situations occurs:

1) When running a standard curve in a control matrix, bound DPM levels are elevated in comparison to the standard curve run in assay buffer alone, especially at the lower end of the curve (background levels are elevated).
2) When evaluating an unknown at a range of dilutions, the bound DPMs increase disproportionately with higher sample concentration. This is especially evident when the samples contain competitor, because the expected sigmoidal decrease in bound counts may become abberant and even show a slight increase at high competitor concentration. In the event of mild interference, the slope of the sigmoidal competition curve may decrease relative to the expected slope, and back-calculation of concentration to the original sample may give widely divergent values.

In these situations, increasing the DCC concentration usually solves the problem. In the past, we have found changing the DCC dilution from 1:50 to 1:40 or even 1:25 is usually sufficient. However, DCC at these dilutions typically begins to bring down a portion of the bound label as well, so a concomitant increase in antibody concentration may be necessary to maintain the desired signal-to-noise ratio.

3 Receptor Binding Assay

3.1 Materials and chemicals

Buffers and reagents

- **PMSF stock solution**
  10 mM phenylmethanesulfonyl fluoride (PMSF) in methanol
- **Sucrose buffer (Solution 3)**
  5 mM sodium phosphate
  0.32 M sucrose, pH 7.4
- **Binding medium (Solution 4)**
  50 mM HEPES (pH adjusted to 7.4 with HCl)
  130 mM choline chloride
  5.5 mM glucose
  0.8 mM magnesium sulfate
  5.4 mM potassium chloride
  1 mg/ml bovine serum albumin
  0.01% Alkamuls EL-620
- **Wash medium (Solution 5)**
  5 mM HEPES (pH adjusted to 7.4 with HCl)
  163 mM choline chloride
  1.8 mM calcium chloride
  0.8 mM magnesium sulfate,
  1 mg/ml bovine serum albumin

Chemicals

Acetic acid, glacial	Fisher	BP1185-500
Alkamuls EL-620	Rhone-Poulenc	331616
Bovine serum albumin	Sigma	A-7888
Brevetoxin PbTx-3, unlabeled	Center for Marine Research, WINTIR	PB03
Brevetoxin [^{3}H] PbTx-3	Center for Marine Research, WINTIR	TRI-PB03
Calcium chloride	Sigma	C-3881
Choline chloride	Sigma	C-1879
EnviroSafe scintillation fluid, or	Anoroc Scientific	LSC-060
Ecolume	ICN	882470
Glucose	Sigma	G-5000
HEPES buffer	Sigma	H-3375
Magnesium sulfate	Sigma	M-7506
Methanol	Fisher	A935-4
Phenylmethanesulfonyl fluoride (PMSF)	Sigma	P-7626
Potassium chloride	Sigma	P-4504
Rat brains, frozen	Pel-Freez Biologicals	56004-2

Saxitoxin, unlabeled	Sigma	S1417
Saxitoxin, [^{3}H]STX	Amersham	TRK877
Sodium phosphate, dibasic	Sigma	S-0876
Sucrose	Baker	V-133-8
Thimerosal	Sigma	T-5125

Equipment

- pipettors and test tube racks
- vortex mixer
- vacuum oven
- filtration manifold (Millipore)
- scintillation counter
- glass fiber filters (Whatman GF/F, 25 mm)
- tissue homogenizer, Teflon/glass, motorized
- centrifuge
- refrigerator
- kichen blender

3.2 Methods

Unlike an immunoassay, which detects toxin based upon molecular structure and is blind to pharmacological activity, the receptor binding assay detects toxin based upon a pharmacologically relevant event, namely, its binding to an endogenous receptor site on the voltage-dependent sodium channel. Thus, where an immunoassay measures the amount of *toxin* present, the receptor assay effectively measures the amount of *biological activity* present in the sample. Like the brevetoxin RIA described above, the receptor binding assay is a competition assay, and takes advantage of the competition between unlabeled toxin in the unknown sample and the radiolabeled toxin for binding to the receptor site in solution. Since saxitoxin and tetrodotoxin act at the same sodium channel receptor site by blocking the influx of sodium ions, the receptor assay used to detect unknown quantities of saxitoxin will also detect tetrodotoxin. Likewise, since brevetoxin and ciguatoxin both act at the same sodium channel receptor site by causing persistent activation, unknown amounts of ciguatoxin will also displace radiolabeled brevetoxin from its specific receptor (for similarities between brevetoxin and ciguatoxin as well as STX and TTX structures, see Fig. 1). It is important to note that since these toxins have slightly different affinities for their sodium channel receptor, results will be expressed as "ng PbTx-3 equivalents" or "ng STX equivalents".

Preparation of the rat brain membranes

Nerve cells are a rich source of voltage-dependent sodium channels. We use a preparation of rat brain membranes for the "receptors" in this assay. Rat brains can be purchased frozen from commercial sources, thus eliminating the need for animal care facilities and the handling of live laboratory animals.

Protocol 4 Preparation of rat brain membranes

Note:
Throughout the preparation procedure, it is very important to keep the solutions ice cold to minimize receptor degradation due to proteolysis.

1. Immediately prior to use, add PMSF stock solution (1 µl/ml buffer) to an appropriate volume of sucrose buffer.
2. Homogenize each brain (3–4 g) in approximately 20 ml of ice-cold sucrose buffer **(Solution 3)** containing PMSF using 10 strokes of a Teflon/glass tissue homogenizer at 800 RPM.

   Note:
   Hold the glass dounce containing brains in a container of ice during this step and during all subsequent steps that utilize the homogenizer.
3. Sediment the homogenate at 1000 x g for 10 min at 4°C. Remove and place the supernatant on ice.
4. Resuspend the pellet in 20 ml cold sucrose/PMSF solution and repeat step 3. Pool the two supernatant fractions and discard the pellet.
5. Centrifuge the combined supernatants at 17,000 x g for 60 min at 4°C. Discard the supernatant.
6. Resuspend the pellet in cold sucrose/PMSF solution at 4 ml/brain, homogenizing with 5 strokes of the tissue homogenizer (200 RPM) to ensure homogeneity.
7. Place 1 ml aliquots into airtight vials and immediately freeze in liquid nitrogen or dry ice. Store frozen aliquots at –70°C.

Membranes prepared in this manner and stored at –70°C are stable for several years. However, once thawed the receptors are extremely susceptible to proteolysis. Frozen aliquots should *never* be allowed to thaw at room temperature. Keep frozen until the last moment, then thaw quickly by placing the tube under hot running water and immediately suspend them in ice-cold binding medium. For examples of other methods used to prepare neuronal membranes, see Dodd *et al.* (1981) and Ray *et al.* (1978).

Optimization of membrane and radiolabeled toxin concentrations

As for the RIA, optimization of the receptor and radiolabeled toxin concentrations is a key step in assay development. Again, the concentration of [^{3}H]PbTx-3 should ideally be at or somewhat below the Kd of the receptor. Under the described conditions, the Kd for this receptor is approximately 3 nM PbTx-3 (Poli *et al.*, 1986). We therefore set our [^{3}H]PbTx-3 concentration at 1 nM and optimize our membrane dilution to give us the desired level of bound DPMs. The principle of receptor binding is shown in Figure 5.

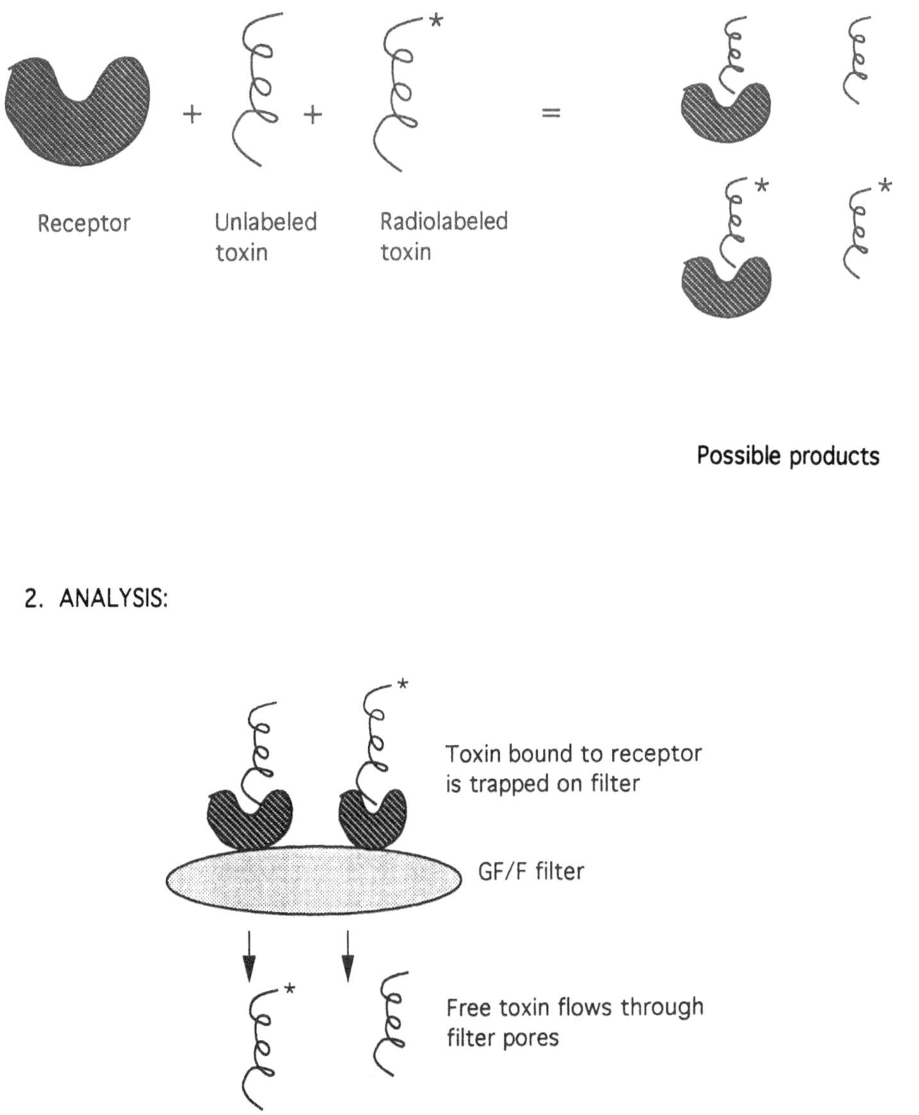

Figure 5 The schematic diagram of a receptor binding assay.

Protocol 5 Optimization of membrane and radiolabeled toxin concentrations

1. Prepare needed volume of 1.25 nM [^{3}H]PbTx-3 in binding medium **(Solution 4)**.
2. Rapidly thaw 1 tube (1 ml) of frozen rat brain membranes under hot running water. Immediately dilute with 1 ml ice-cold binding medium and homogenize by hand with 10 strokes using a Teflon/glass homogenizer in an ice bath. Make a series of dilutions of this membrane preparation in ice-cold binding medium from 1:2 to 1:32 (vol:vol). Add 100 µl of each dilution into triplicate test tubes containing 0.4 ml of the [^{3}H]PbTx-3 solution (made in step 1). This dilution brings the final [^{3}H]PbTx-3 concentration to 1 nM. Vortex and incubate 1 h in an ice bath. Always add the membrane preparation last.
3. Dilute each sample with 3 ml ice-cold wash medium **(Solution 5)** and immediately filter through glass fiber filters using a vacuum filter manifold. Wash the filtered membranes twice with additional 3 ml aliquots of ice-cold wash medium **(Solution 5)**. Filter and rinse each tube before beginning with the next tube.
4. Place filters in scintillation vials and add 10 ml scintillation fluid. Vortex and let sit overnight at room temperature to allow the scintillant to fully permeate the filters.
5. Measure bound radioactivity in a scintillation counter by counting each sample for 5 min.
6. Plot bound DPM versus increasing rat brain membrane concentration.

Choose a tissue dilution which is on the linear portion of the "bound DPM versus tissue concentration" curve. This optimal dilution should give approximately 2000–2500 DPM bound.

Sample preparation, standard curve, and sample analysis

Unknown samples should be prepared as for radioimmunoassay. The following receptor binding method is diagrammed in Figure 6.

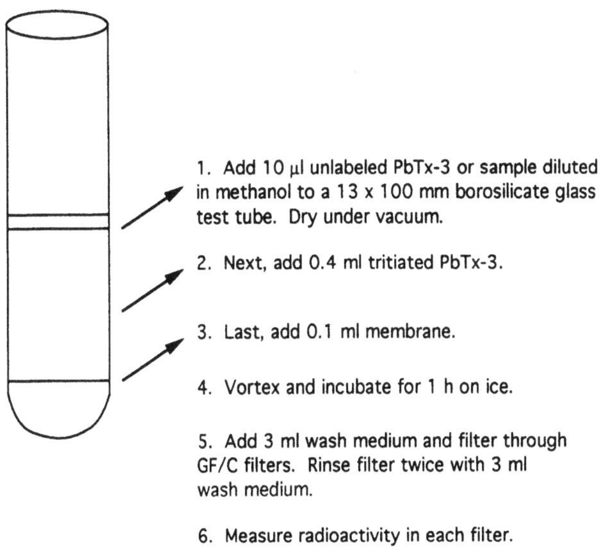

Figure 6 Receptor Binding Method.

Protocol 6 Receptor binding method

1. Prepare the standard curve by diluting a stock solution of unlabeled PbTx-3 (100 μg/ml in 100% methanol) in half log dilutions over the range of 0.01–100 ng/ml final concentration. Aliquots (10 μl) of each dilution are added to triplicate tubes and dried in a vacuum oven set at 60°C or using a stream of nitrogen.
2. Aliquots (10 μl) of unknown samples in methanol at three separate dilutions are prepared in triplicate tubes and dried as above.
3. Prepare needed volume of 1.25 nM [^{3}H]PbTx-3 in binding medium. Add 0.4 ml of this solution to the tubes prepared in steps 1 and 2. Then prepare a solution of ice-cold binding medium containing rat brain membrane at the previously determined optimal dilution. Add 0.1 ml of this preparation to each assay tube, vortex, and incubate in an ice bath for 1 h.
4. After the 1-h incubation, dilute each sample with 3 ml of ice-cold wash medium **(Solution 5)** and immediately filter through glass-fiber filters in a vacuum filtration manifold. Rinse the filtered membranes twice with additional 3-ml aliquots of ice-cold wash medium **(Solution 5)**.

5. Place filters in scintillation vials, add 10 ml of scintillation fluid, vortex, and let sit overnight at room temperature to allow the scintillant to fully permeate the filters.
6. Measure bound radioactivity in a scintillation counter by counting each sample for 5 min.
7. For the standard curve, plot percentage bound DPM versus the log of PbTx-3 concentration. Toxin in the unknown samples is determined by comparison to this standard curve. Express results as "ng/ml PbTx-3 equivalents", since either ciguatoxin or brevetoxin can be present in samples.

   Notes:
   The shape of the resulting standard curve will be similar to that shown in Figure 3.

Saxitoxin assay

General Comments: For the saxitoxin assay, [^{3}H]STX will be used as the radiolabeled toxin and STX as the unlabeled competitor toxin. Rat brain membrane will be prepared as described previously as the source of specific STX receptor. The optimization of membrane concentration and utilization of a standard curve to determine STX concentrations in unknown samples are identical to the procedures detailed above for brevetoxin. Dilutions of unlabeled STX used to derive competition curves can be made in distilled water rather than methanol since saxitoxin is a hydrophilic molecule. If dilutions are made in water, drying in the vacuum oven is not necessary, and final reaction volumes in each tube should be adjusted to 1 ml. Saxitoxin has a Kd of 2 nM (Catterall *et al.*, 1979) for its sodium channel receptor; therefore, a final concentration of 1 nM is appropriate for development of a standard curve using radiolabeled saxitoxin. Results will be expressed as "ng STX equivalents", since saxitoxin or tetrodotoxin can be present in samples. Once the membrane concentration is optimized using either radiolabeled brevetoxin or saxitoxin, that same concentration can be used with either radiolabeled toxin in future assays.

Sample preparation: The standard method of sample preparation prior to analysis for saxitoxin content which is outlined here is described fully in the AOAC official methods of analysis (AOAC, 1990). Remove tissue from the surrounding shell, clean with fresh H_2O, and weigh. Homogenize in a kitchen blender while keeping sample ice cold. Add an equal volume of 0.1 N HCl, stir thoroughly, and check pH. If necessary, adjust pH to 3. Heat mixture, boil gently for 5 min, and allow to cool to room temp. Adjust this mixture to pH 2–4 using 5 N HCl or 0.1 N NaOH dropwise with constant stirring. Dilutions of these samples are then prepared in H_2O, and a constant volume of each dilution is tested in triplicate. Whenever possible, standard curves should be prepared in a similar dilution of a nontoxic control extract.

3.3 Troubleshooting

General Comments: Troubleshooting is the same as described for RIA except for the section pertaining to DCC concentrations.

Saxitoxin Assay: The tritium label on the [^{3}H]STX probe is exchangeable; therefore, incubation for much longer than 1 h in an ice bath (or a few minutes at room temp) can lead to back-exchange of the ^{3}H and a decrease in bound DPM. If low DPM are observed in the assays using [^{3}H]STX, decrease incubation times and make an effort to keep solutions at 4°C at all times.

Mention of trade names is for information only and does not constitute an endorsement by the U.S. Department of Commerce.

References

AOAC (1990) 959.08 Paralytic shellfish poison biological method final action. In: K Helrich (ed): Official Methods of Analysis of the Association of Official Analytical Chemists, Food Composition; Additives; Natural Contaminants, 15th edn, pp 881–882

Baden DG, Trainer VL (1993) Mode of action of toxins of seafood poisoning. In: IR Falconer (ed): Algal Toxins in Seafood and Drinking Water, pp 49–74, Academic Press, New York

Baden DG, Mende TJ, Szmant AM et al (1988) Brevetoxin binding: molecular pharmacology versus immunology. *Toxicon* 26: 97–103

Baden DG, Mende TJ, Lichter W et al (1981) Crystallization and toxicology of T34: a major toxin from Florida's red tide organism (Ptychodiscus brevis). *Toxicon* 19: 455–462

Dodd PJ, Hardy JA, Oakley AE et al (1981) A rapid method for preparing synaptosomes: comparison with alternative procedures. Brain Res 226: 107–118

Poli MA (1996) Three dimensional binding assays for the detection of marine toxins. In: DG Baden (ed): *Proceedings of the Workshop Conference on Seafood Intoxication: Pan American Implications of Natural Toxins in Seafood*, University of Miami, Miami

Poli MA, Mende TJ, Baden DG (1986) Brevetoxins, unique activators of voltage-sensitive sodium channels, bind to specific sites in rat brain synaptosomes. Mol. Pharmacol. 30: 129–135

Ray R, Morrow CS, Catterall WA (1978) Binding of scorpion toxin to receptor sites associated with voltage-sensitive sodium channels in synaptic nerve ending particles. J. Biol. Chem. 253: 7307–13

Trainer VL, Baden DG (1991) An enzyme immunoassay for the detection of Florida red tide brevetoxins. *Toxicon* 29: 1387–94

2 Maitotoxin, Okadaic Acid, and Microcystins: Toxins That Disturb Signal Transduction and Phosphorylations

Simone Puiseux-Dao, Noureddine Bouaïcha and Georges Diogene

Contents

1 Introduction

Seafood poisoning surveys point out the increasingly wide occurrence of illness risk due to toxins found in molluscs or fish consumed by humans. Some poisonings such as paralytic shellfish poisoning induce neurologic symptoms; in others, digestive disorders are predominant, as in diarrhetic shellfish poisoning (DSP). The toxins isolated from these animals are, in fact, synthesized by microalgae, in general dinoflagellates but also diatoms or cyanobacteria, and they accumulate along the food chain.

This chapter deals with three types of toxins which have been more or less extensively used to analyse fundamental cell processes such as signal transduction and phosphorylations.

Methods and Tools in Biosciences and Medicine
Animal Toxins, ed. by H. Rochat and M.-F. Martin-Eauclaire

2 Toxin Chemistry

2.1 Maitotoxins

Maitotoxin-1 was first isolated from the viscera of a herbivorous surgeon fish, *Ctenochaetus striatus,* but is produced by a benthic marine dinoflagellate, *Gambierdiscus toxicus.* It is a large disulphated polyether compound (3422 Da for the disodium salt) as shown in Figure 1 (Murata *et al.*, 1993), the activity of which is associated to the sulphate groups (Murata *et al.*, 1991).

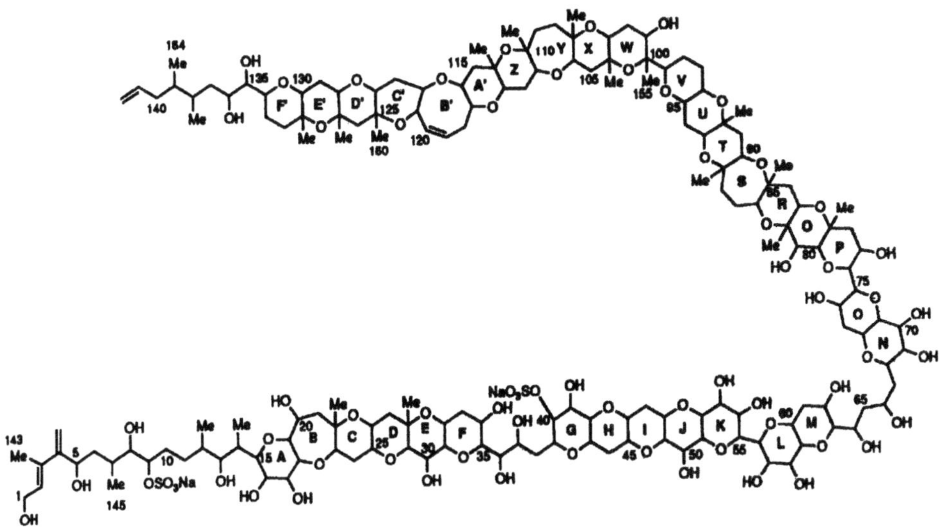

Figure 1 Maitotoxin-1.

Three maitotoxins, MTX-1, -2, and -3, have been extracted from cultures of three strains (FP, NQ1, and WC1/1) of *G. toxicus*. MTX-2 has a molecular weight of 3298 for the sodium salt, whereas MTX-3 is smaller (MW: 1060 for the disodium salt). The chemical structures of MTX-2 and -3 are unknown (Holmes and Lewis, 1994).

Extraction of maitotoxins: Cultures of *G. toxicus* are extracted with aqueous methanol (80%). The extract is then partitioned successively against hexane, diethylether and 1-butanol. MTXs are present in the butanol fraction and then purified by HPLC (see in Murata *et al.*, 1993).

Detection of maitotoxin-1: *In vivo* testing on animals, mice in general, assesses the global toxicity. Rapid and more sensitive assays such as *in vitro* cell bioassays (Diogène *et al.* 1994; 1995) and immunoassays (Bignami *et al.*, 1996) have been developed. Recently, a quantitative method for determination of MTX-1 by capillary electrophoresis has been performed (Bouaïcha *et al.*, 1997a).

2.2 Okadaic acid and derivatives

Okadaic acid (OA) was initially isolated from the Pacific and Caribbean sponges, *Halichondria okadai* and *H. melanodocia,* respectively. Four natural derivatives of okadaic acid, dinophysistoxins 1–4 (DTXs-1,2,3,4), are known at this date. DTX-1 was first isolated from the mussel *Mytilus edulis*. Later, OA and DTXs were shown to be produced by the dinoflagellates *Prorocentrum lima* and *Dinophysis fortii* and accumulated in variable small amounts by filter feeders such as shellfish.

OA and DTXs associated with DSP have been determined to be large polyether compounds as shown in Figure 2 (Murata *et al.*, 1982). The activity is attached to the acidic function (Holmes *et al.*, 1990).

Extraction of okadaic acid and derivatives: Several methodologies have been proposed to extract DSP toxins from cultures of *P. lima* or shellfish tissue:

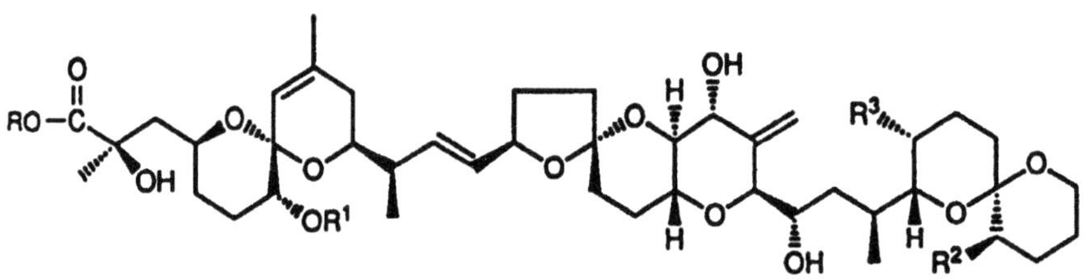

Okadaic acid (OA) : R = R^1 = R^2 = H, R^3 = Me

Dinophysistoxin-1 (DTX-1) R = R^1 = H, R^2 = R^3 = Me

DInophysistoxin-2 (DTX-2) : R = R^1 = R^3 = H, R^2 = Me

DInophysistoxin-3 (DTX-3) : R = H, R^1 = Acyl, R^2 = H or Me, R^3 = H or Me

Dinophysistoxin-4 (DTX-4) : R^1 = R^2 = H, R^3 = Me, R =

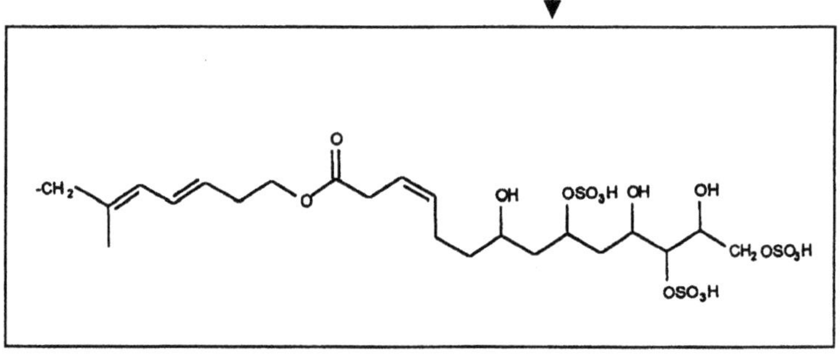

Figure 2 Okadaic acid and dinophysistoxins.

these are extracted with acetone or with aqueous methanol (80%), and this extract is partitioned successively against hexane and chloroform or hexane, diethylether, and 1-butanol. OA, DTX-1, DTX-2, and DTX-3 have been identified in the ether fraction, and traces of these lipid-soluble toxins can also be present in the butanol fraction. DTX-4 has been detected in the butanol fraction only (see in Quilliam, 1995).

Detection of OA and derivatives: As for other toxins, there are several methods for the detection of OA involving *in vivo* and *in vitro* bioassays as well as antibody (ELISA) and enzymatic (phosphatase activity inhibition) assays (see in Bouaïcha *et al.*, 1997).

Numerous methods for the HPLC determination of DSP toxins have been evaluated. They mostly involve derivatization of the carboxylic acid moiety of the compounds to form highly fluorescent derivatives, which are then separated by reverse-phase chromatography. The method developed by Lee *et al.* (1987) for the determination of OA and DTX-1 has received the most attention up to now. A new method based on capillary electrophoresis has recently been proposed (Bouaïcha *et al.*, 1997b).

2.3 Microcystins

Cyanobacterial heptapeptide hepatotoxins named microcystins occur worldwide in freshwater. They have been found and characterized from *Microcystis aeruginosa* (Kützing) Lemmermann, *Microcystis viridis* Lemmermann, *Anabaena flos-aquae* Brebisson, *Anabaena* spp., *Oscillatoria agardhii* Gomont and *Nostoc* sp. (Brown) Lemmermann. Approximately 50 microcystin variants have been identified (Carmichael, 1992). Accumulation in mussels in estuaries has been described.

The general structure of microcystins (Fig. 3) corresponds to cyclo (D-Ala^1-X^2-D-MeAsp^3/D-Asp^3-Z^4-Adda^5-D-Glu^6-Mdha^7/Dha^7). X and Z are variable L-amino acids [in MCYST-LR, X = L-Leucine (L) and Z = L-arginine (R)]; R^1 and R^2 are H (demethylmicrocystins) or CH_3; D-MeAsp is D-erythro-β-methylaspartic acid; Adda is (2S, 3S, 8S, 9S)-3-amino-9-methoxy-2,6,8-trimethyl-10-phenyldeca-4,6-dienoic acid and Mdha is N-methyldehydroalanine, Dha is dehydroalanine.

The configuration of C-4 and C-6 in the Adda group and the glutamyl carboxyl are essential to toxicity. Contrary to OA, covalent binding to phosphatase 1 has been demonstrated (Mac Kintosh *et al.*, 1995).

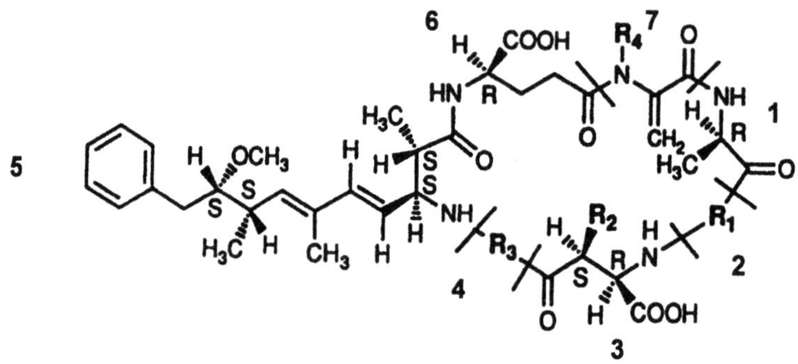

MCYST-YR; R_1 = Tyr, R_2 = CH_3; R_3 = Arg; R_4 = CH_3 (MW = 1044)
MCYST-LR; R_1 = Leu, R_2 = CH_3; R_3 = Arg; R_4 = CH_3 (MW = 994)
MCYST-RR; R_1 = Arg, R_2 = CH_3; R_3 = Arg; R_4 = CH_3 (MW = 1037)

Figure 3 Microcystins.

Protocol 1 Extraction of microcystins (see in Bouaïcha *et al.* 1996)

1. Microcystins are extracted from wet or frozen cyanobacteria cells using aqueous methanol (80%).
2. The supernatant (crude extract) is then evaporated to dryness under vacuum and redissolved in ethanol for analysis.
3. Solid phase extraction (SPE) is generally used to concentrate microcystins from waterbloom samples. Reversed-phase (C18) silica cartridges are very useful.
4. Microcystins are eluted with 0.1% TFA (trifluoroacetic acid) in methanol from C18 silica cartridges.

Detection of microcystins: The most commonly used method is the one based on mouse bioassay. Other detection methods include *in vitro* bioassays, immunological based assays such as ELISA, a protein phosphatase inhibition assay and quantitative instrumental methods [HPLC coupled with (UV) or mass spectral detection]. Micellar Electrokinetic Chromatography has been proposed (see in Bouaïcha *et al.*, 1996).

2.4 Recommendations

Maitotoxin, OA, microcystin-LR, and microcystin-RR standards can be obtained from Sigma (Saint-Quentin Fallavier, France) and microcystin-YR from Calbiochem Novabiochem (La Jolla, CA, USA). All these toxins are soluble in methanol and ethanol where they are very stable at –20°C. For immediate experimental use maitotoxin and microcystins can also be dissolved in water.

Commercially available microcystins are usually very pure. On the other hand, maitotoxin and okadaic acid are of questionable purity.

3 Toxin Mode of Action

3.1 Maitotoxin 1

Maitotoxin 1 toxicity

MTX-1 has a LD_{50} in mice of 50 ng/kg at 24 h (i.p.) and induces necrosis. In all investigated cell systems, MTX-1 showed the same pattern of toxicity, which appears at very low concentration depending on cell type (Neutral Red Uptake$_{50}$ after 24 h: NRU_{50} = 3 pM with fibroblasts FR3T3) and increases very steeply. Lactico dehydrogenase release became detectable after 1 h at 45 pM with the same cells and the same toxin sample, which indicates that the toxin produces holes in the plasma membrane at low levels. MTX-1 provokes necrosis as *in vivo* (Ammar *et al.* unpublished data).

No tissue specificity has been described as expected with an unknown ubiquitous membrane target, since the hydrophilic MTX-1 is claimed not to enter cells.

Its toxicity can vary very strongly and must be evaluated for each sample always under the same conditions.

Maitotoxin 1, its effects on calcium, and phosphoinositide breakdown

Calcium-dependent cell responses have been intensively investigated with MTX-1, but MTX-1 samples may be questionable. Since the hydrophilic molecule is believed not to enter cells, a possible interference of the toxin with signal transduction pathways has been considered. While a direct effect on cAMP synthesis has been discarded, many positive responses have been described related to Ca^{2+} influx and phosphoinositide breakdown. Ca^{2+} outside cells is necessary for MTX-1 effect (see in Gusovski and Daly, 1990, and Durand-Clément *et al.*, 1991).

Electrophysiological investigations and studies with ^{45}Ca or fluorochromes have shown a dose-dependent increase of Ca^{2+} influx or $Ca^{2+}{}_i$. Other ions could possibly be involved (particularly Na^+), as indicated by electrophysiology. In almost every experiment, inorganic Ca^{2+} channel blockers such as Cd^{2+}, Mn^{2+}, and La^{2+} antagonized the effect of MTX-1 on Ca homeostasis. In contrast, organic channel blockers such as verapamil or dihydropyridins were not always effective.

The kinetics of Ca^{2+} intracellular increase induced by MTX-1 is considered as different from that provoked by Ca^{2+} ionophores. While the entrance of Ca^{2+} mediated by an ionophoretic substance is abrupt, the increase of $Ca^{2+}{}_i$ concentration caused by MTX-1 is slow. But such a result has been obtained on cell suspensions. When the increase in $Ca^{2+}{}_i$ has been analyzed by scanning laser cytometry cell by cell, the response of synchronized FR3T3 fibroblasts was shown to be very homogeneous in samples and to depend on the stage of the cell division cycle (Berreur-Bonnenfant *et al.*, 1994). The data suggest that the slow $Ca^{2+}{}_i$ increase observed with cell suspensions corresponds with a cell population response and is due to the asynchrony of the experimented cellular system.

Phosphoinositide (PIP) breakdown is activated by MTX-1 in all the cell models studied. This breakdown is dose-dependent. In general, channel blockers do not interfere with MTX action except Cd^{2+} or high levels of Mn^{2+} in PC-12 cells.

3.2 Okadaic acid

Okadaic acid as a tool to investigate serine-threonine phosphorylations
As soon as OA was demonstrated to inhibit serine-threonine phosphatases 2A and 1 (Bialojan and Takai, 1988), the toxin was intensively used to discover the potential role of these phosphatases in various cellular processes. PP3 is also inhibited (Honkanen et al., 1994)

Cell responses to OA, once observed, led to analysis of the phosphorylated state of specific proteins determined by genetic or other kinds of studies to play a key role in the physiological process studied, the analysis performed both in control and OA-treated cells or in *in vitro* acellular systems.

One of the most successful examples concerns the cell cycle and the transition from G2 to mitosis (see in Dunphy, 1994). Serine-threonine phosphorylations of protooncogenic proteins such as c-fos and c-jun (Schönthal *et al.*, 1991; Adunyah *et al.*, 1992) have been investigated. The signalling pathways have been studied with OA, also demonstrating that serine-threonine phosphorylations concern Marks proteins for example (Clarke *et al.*, 1993).

Okadaic acid as a tool to investigate mechanisms of carcinogenesis
OA and DTX-1 led to a new class of tumor promoters: In 1988 (Suganuma *et al,* 1988) demonstrated in a two-stage carcinogenesis experiment with rodents that OA can be a tumor promoter different from phorbol esters such as TPA (12-O-tetradecanoylphorbol-13-acetate), which activates protein kinase C.

From studies involving OA and its derivative DTX1 and other marine toxins, Japanese scientists proposed a new class of tumor promoters that inhibit serine-threonine phosphatases 1 and 2A (see in Fujiki and Suganuma, 1993).

In fact, OA as well as TPA provoke induction of the Tumor Necrosis Factor α. OA would be able to act as a tumor promoter via this endogenous promoter (Fujiki and Suganuma, 1994).

OA induces genotoxic effects: No mutagenecity of OA could be demonstrated with the Ames test on *Salmonella typhimurium* (Aonuma *et al.*, 1991). No initiation potential for OA was shown when experimenting with BALB/3T3 cells. However, an increase in mutation frequencies in some loci has been shown; in addition, an induction of sister chromatid exchange in the presence of 5-bromodesoxyuridine has been observed (see in Fessard *et al.*, 1996; in Nakagama *et al.*, 1997). Thus, the toxin could induce mutagenic effects in eukaryotic cells such as point mutations and clastogenesis.

Recently OA was demonstrated to induce DNA adduct formation at low non toxic concentrations (nM or even less) with the ^{32}P-post labelling technique in different cell types but also in treated zebrafish embryos (Fessard *et al.*, 1996; Huynh *et al.*, 1998). Comet assay seems to confirm clastogenesis potential (Fessard *et al.*, unpublished data). In addition OA was shown to produce DNA minisatellite sequence rearrangement in the progeny of treated NIH3T3 cells (Nakagama *et al.*, 1997).

OA as a complete carcinogen: In animals OA alone was not shown to induce tumors at a very significant level (Suganuma *et al.*, 1988). However some clones from NIH3T3 cells treated with OA only produced tumors when implanted into nude mice (Nakagama *et al.*, 1997). In addition, the formation of DNA adducts suggests that OA behaves as a complete carcinogen.

All the results indicate interrelations between serine-threonine phosphorylations, DNA damage, and genomic instability. Thus, the induction of DNA adducts, DNA breaks, and genomic instability by OA deserves further investigations which could result in modification of present sanitary regulations.

3.3 Microcystins

The heptapeptide microcystins, which inhibit phosphatases 1 and 2A (and also PP3, Honkanen et al., 1994), are potent hepatotoxins (Carmichaël and Falconer, 1993). The microcystin LR has been demonstrated to be a liver promoter (Nishiwaki-Matsushima *et al.*, 1992) and is considered to belong to the okadaic class of tumor promoters (Fujiki and Suganuma, 1993). However, DNA damage in mouse liver *in vivo* has recently been described (Rao and Bhattacharya, 1997).

The strong hepatotoxic specificity of the microcystins relates to their entry into cells which takes place *via* a specific transport system, the bile acid carrier found in freshly isolated hepatocytes (see in Falconer, 1993; Falconer and Humpage, 1996). Freshly prepared enterocytes of the small intestine, which are also damaged in the presence of microcystin, probably possess the same type of transport for bile acids (Falconer *et al.*, 1992).

The microcystins, which have a lipophilicity depending on their amino acid composition, should be interesting tools to investigate bile acid transport system, which is still not well understood.

These toxins also constitute promising tools for analyzing mechanisms of carcinogenesis in parallel with OA, but this type of study requires sensitive cell lines in addition to fresh hepatocytes or enterocytes. Comparative experiments with various rat or human hepatocyte cell lines seem to confirm that biliary canalicules are necessary for a toxic effect of the microcystin-LR (Huynh *et al.*, unpublished data).

4 Conclusion

The three types of toxins – maitotoxin, OA, and microcystins – are useful experimental tools. OA is the easiest to handle but its carcinogenic potency requires caution. Because of its high steep toxicity, MTX-1 must be used in a narrow range of low concentrations; in addition, its stability in aqueous solutions has to be controlled. The peptidic microcystins are promising molecules, but sensitive cell lines are necessary for further investigation.

These toxins are considered to be potential carcinogens, but as an issue in public health, the risk to humans through seafood consumption is higher with OA and its derivatives since MTX-1 is claimed to accumulate with the producing microalgae in the intestine of herbivorous fish only and would not accumulate along the food web. Microcystins are hazardous to humans and cattle *via* drinking water. However, their presence in shellfish in addition to OA has to be considered.

References

Adunyah S, Unlap T, Franklin C, Kraft A (1992) Induction of differentiation and *c-jun* expression in human leukemic cells by okadaic acid, an inhibitor of protein phosphatases. J. Cell. Physiol. 151: 415–26

Anouma S, Ushijima T, Nakayasu M *et al* (1991) Mutation induction by okadaic acid, a protein phosphatase inhibitor, in CHL cells, but not in *S. typhimurium.* Mutat. Res. 250: 375–81

Berreur-Bonnenfant J, Ammar M, Dubreuil A *et al* (1994) Modulation of fibroblast response to maitotoxin along the cell division cycle. Cell Biol Toxicol 10: 423–27

Bialojan C, Takai A (1988) Inhibitory effect of a marine sponge toxin, okadaic acid, on protein phosphatases: specificity and kinetics. Biochem J. 256: 283–90

Bignami G, Waller D, Yasumoto T (1996) Antibodies to maitotoxin elicited by immunization with toxin-fragment conjugates. Toxicon 34: 1393–97

Bouaïcha N, Ammar M, Hennion M-C, Sandra P (1997a) A new method for determination of maitotoxin by capillary zone electrophoresis with ultraviolet detection. Toxicon 35: 955–62

Bouaïcha N, Hennion M-C, Sandra P (1997b) Determination of okadaic acid by micellar electrokinetic chromatography with ultraviolet detection. Toxicon 35: 273–81
Bouaïcha N, Rivasseau C, Hennion M-C, Sandra P (1996) Detection of cyanobacterial toxins (microcystins) in cell extracts by micellar electrokinetic chromatography. J. Chromatogr, B, 685: 53–57
Carmichael WW (1992) A status report on planktonic cyanobacteria (blue-green algae) and their toxins. United States Environnemental Protection Agency, EPA/600/R-92/079. Office of Research and Development, Washington, DC 600: 32–33
Carmichael WW, Falconer IR (1993) Disease related to freshwater algal blooms. Algal Toxins in Seafood and Drinking Water, (Falconer IR, ed) pp 187–209, London: Academic Press
Clarke P, Siddhanti S, Cohen P, Blackshear P (1993) Okadaic acid-sensitive protein phophatases dephosphorylate MARCKS, a major protein kinase C substrate. FEBS Lett 336: 37–42
Diogène J, Dubreuil A, Breittmayer P, Puiseux-Dao S (1994) Cytotoxic quantification of maitotoxin-like activity from the dinoflagellate *Gambierdiscus toxicus*. Toxicol *in Vitro* 8: 37–45
Diogène J, Fessard V, Dubreuil A, Puiseux-Dao S (1995) Comparative studies of the actin cytoskeleton response to maitotoxin and okadaic acid. Toxicol *in Vitro* 9: 1–10
Dunphy WG (1994) The decision to enter mitosis. Trends Cell Biol 4: 202–07
Durand-Clément M, Diogène J, Puiseux-Dao S (1991) Recent advances on the mode of action of maitotoxin. Proceeding of Symposium on Marine Biotoxins, (Frémy JM, ed) pp 75–82, Paris: CNEVA
Falconer IR (1993) Mechanism of toxicity of cyclic peptide toxins from blue-green algae. Algal Toxins in Seafood and Drinking Water, (Falconer IR ed) pp 177–86, London: Academic Press
Falconer I, Dornbusch M, Moran G, Yeung S (1992) Effect of the cyanobacterial (blue-green algal) toxins from *Microcystis aeruginosa* on isolated enterocytes from the chicken small intestine. Toxicon 30: 790–93
Falconer IR, Humpage AR (1996) Tumor promotion by cyanobacterial toxins. Phycologia 35 (6 Supplement): 74–79
Fessard V, Grosse Y, Pfohl-Leszkowicz A, Puiseux-Dao S (1996) Okadaic acid treatment induces DNA adduct formation in BHK21 C13 fibroblasts and HESV keratinocytes. Mutation Res 361: 133–44
Fujiki H, Suganuma M (1993) Tumor promotion by inhibitors of protein phosphatases 1 and 2A: the okadaic class of compounds. Adv Cancer Res 61: 143–94
Fujiki H, Suganuma M (1994) Tumor Necrosis Factor α, a new tumor promoter, engendered by biochemical studies of okadaic acid. J. Biochem. 115: 1–5
Gusovsky F, Daly J (1990) Maitotoxin: a unique pharmacological tool for research on calcium-dependent mechanisms. Biochem. Pharmacol. 39: 1633–39
Honkanen RE, Codispoti BA, Tse K, Boynton A (1994) Characterization of natural toxins with inhibitory activity against serine/threonine protein phosphatases. Toxicon 32: 339–350
Holmes CFB, Luu HA, Carrier FJ, Schmitz FJ (1990) Inhibition of protein phosphatase-1 and 2A with acanthifolicin: comparison with diarrhetic shellfish toxins and identification of a region on okadaic acid important for phosphatase inhibition. FEBS Lett. 270: 216–18
Holmes MJ, Lewis RJ (1994) Purification and characterisation of large and small maitotoxins from cultured *Gambierdiscus toxicus*. Nat. Toxins 2: 64–72
Huynh C, Pinelli E, Puiseux-Dao S *et al* (1998) Okadaic acid and DNA adduct formation. Harmful Algae (Reguera B, Blanco J, Fernandez ML,Wyatt T, eds) Xunta de Galicia and Intergovernmental Oceanographic Commission of Unesco, pp 581–83.
Lee JS, Yanagi T, Kenma R, Yasumoto T (1987) Fluorometric determination of diarrhetic shellfish toxins by high-performance liquid chromatography. Agr. Biol. Chem. 5: 1877–81
Mac Kintosh RW, Dalby KN, Campbell DG *et al* (1995) The cyanobacterial toxin microcystin binds covalently to cystein-273 on protein phosphatase-1. FEBS Lett. 371: 236–40
Murata M, Gusovsky F, Sasaki M *et al* (1991) Effect of maitotoxin analogues on calcium influx and phosphoinositide breakdown in cultured cells. Toxicon 29: 1085–96

Murata M, Naoki H, Iwashita T *et al* (1993) Structure of maitotoxin. J. Am. Chem. Soc. 115: 2060–62

Murata M, Shimatini M, Sugitani H *et al* (1982) Isolation and structural elucidation of the causative toxin of the diarrhetic shellfish poisoning. Bull. Jap. Soc. Sci. Fish. 48: 549–52

Nakagama H, Kaneko S, Shima H *et al* (1997) Induction of minisatellite mutation in NIH 3T3 cells by treatment with the tumor promoter okadaic acid. Proc. Natl. Acad. Sci. USA 94: 10813–16

Nishiwaki-Matsushima R, Ohta T, Nishiwaki S *et al* (1992) Liver cancer promotion by the cyanobacterial cyclic peptide toxin microcystin-LR. J. Cancer Res. Clin. Oncol. 118: 420–424

Rao PVL, Bhattacharya R (1997) The cyanobacterial toxin microcystin LR induced DNA damage in mouse liver *in vivo*. Toxicology 114: 29–36

Quilliam M (1995) Analysis of diarrhetic shellfish poisoning toxins in shellfish tissue by liquid chromatography with fluorometric and mass spectrometric detection. J. AOAC Int. 78: 555–70

Schönthal A, Tsukitani Y, Feramisco J (1991) Transcriptional and post-transcriptional regulation of c-*fos* expression by the tumor promoter okadaic acid. Oncogene 6: 423–30

Sugamuna M, Fujiki H, Suguri H *et al* (1988) Okadaic acid: an additional non-phorbol 12-tetradecanoate-13-acetate type tumor promoter. Proc. Natl. Acad. Sci. USA. 85: 1768–71

3 Sea Anemone Toxins, Acting on Na^+ Channels and K^+ Channels: Isolation and Characterization

Lászlo Béress, Thomas Bruhn, Judith Sánchez-Rodríguez, Elmar Wachter and Hugues Schweitz

Contents

Methods and Tools in Biosciences and Medicine
Animal Toxins, ed. by H. Rochat and M.-F. Martin-Eauclaire

1 Introduction

Sea anemones are nettling animals; they belong to the phylum Cnidaria, class Anthozoa, family Actiniidae. From an evolutionary point of view they represent the simplest level of organization within the metazoa. All sea anemones possess stinging organelles called nematocysts which play an important role in paralysing prey and also in defence. They contain potent toxins with an immediate paralysing action on crustaceans and fishes after injection.

Since the early work of the French physiologist Charles Richet (Richet, 1903) on the extraction of two toxic substances, thalassin and congestin, from the tentacles of the sea anemone *Anemonia sulcata* (Fig. 1), there has been a steady increase in work on the isolation of biologically active compounds from the sea anemones and also from other coelenterates (for review see Halstead 1965; Béress, 1982).

The partial purification of the first crab-paralysing polypeptide toxin isolated from the homogenate of tentacles of a sea anemone (*Condylactis gigantea*) was reported by Shapiro (1968). A systematic study devoted to the isolation and characterization of sea anemone toxins especially from *Anemonia sulcata* was initiated 30 years ago by L. Béress and R. Béress in 1966–67 at the Zoological Station in Naples. This work led to the partial purification of two toxins, toxin I (ATX I) and toxin II (ATX II) already in 10–50 mg quantities (Béress and Béress, 1971). Both toxins are polypeptides of middle molecular weights (MW 5000) paralysing crustaceans (*Carcinus maenas*) and mammals (mice) after i.m. and i. v. application.

Figure 1 *Anemonia sulcata.*

Improved methods (Béress *et al.*, 1975a,b) resulted in large-scale purification of the toxins mentioned above up to electrophoretical homogeneity together with a smaller sea anemone toxin called toxin III (ATX III). Using the same isolation technique four toxins from the sea anemone *Condylactis aurantiaca* (CTX I, II, III, and IV) and two from the sea anemone *Bolocera tuediae* (BTTX I and II) have also been isolated (Béress *et al.*, 1976; Béress and Zwick, 1980). Inspired by the studies mentioned above, two further sea anemone toxins (APA, APB) have been purified from the Pacific sea anemone *Anthopleura xanthogrammica*, one toxin (APC) from the sea anemone *Anthopleura elegantissima* (Norton *et al.*, 1976; 1978) and four from *Heteractis magnifica*, previously named *Radianthus paumotensis* (Schweitz et al., 1985). The sequences of most of the named toxins have been elucidated. With the exception of ATX III, which is a short toxin, the sequences of the sea anemone toxins mentioned are homologous (see Tab. 1). All toxins are interconnected by three disulfide bridges stabilizing the structure of the molecules. Their molecular weights are ranging from 2933 Dalton (ATX III) up to 5630 Dalton (CTX III).

Shapiro and Lilleheil (1969) and Lilleheil and Shapiro (1969) were the first to demonstrate that the polypeptide toxin isolated of the sea anemone *Condylactis gigantea* in concentrations from 6–200 µg/ml prolongs the action potential duration up to several hundred milliseconds in lobster giant axons and crayfish stretch receptor cells.

The mechanism of this action has been elucidated by Narahashi *et al.* (1969) and Murayama *et al.* (1972). Using voltage clamp techniques at the crayfish giant axon they have shown that the *Condylactis* toxin selectively affects the Na^+-channel of the membranes by prolonging the time constant of the Na^+-channel inactivation.

As attempts to reproduce the original isolation procedure for the *Condylactis gigantea* toxin were unsuccessful (Yost and ÓBrian, 1978), and as ATX II (*Anemonia sulcata* Toxin II) became available in pure form in gram amounts only after publication of Béress and Béress (1974) and Béress *et al.* (1975b), detailed investigations of the mode of action of a sea anemone toxin on neuromuscular transmission (Rathmayer *et al.*, 1975; Rathmayer and Béress, 1976; Tazieff-Depierre *et al.*, 1976), on Na^+ ion channels (Romey and Lazdunski, 1975; Romey *et al.*, 1976; Bergmann *et al.*, 1976; Conti *et al.*, 1976) and on the mammalian heart muscle (Alsen, 1975; Alsen *et al.*, 1976; Ravens, 1976) could start only after 1975 with toxins originated in our laboratory. All these results gave rise to a great number of further investigations on the mode of action of ATX II in different kind of excitable membranes especially on voltage-dependent Na^+-channels. Toxin II (ATX II) from *Anemonia sulcata* and other homologous sea anemone toxins act on the h-gate of the voltage-dependent Na^+-channel by slowing down the inactivation, a process which gives rise to the prolongation of action potential duration (Romey et al., 1976) and to the enormous increase of transmitter release from the nerve terminal (Abita et al., 1977). ATX II and the other Na^+ channel active sea anemone toxins act also on mammalian heart muscle, evoking a dose-dependent positive inotrope effect

Table 1 Amino acid sequences of the first purified sea anemone toxins (ATX I: Wunderer and Eulitz, 1978; ATX II: Wunderer *et al.*, 1976; APA: Tanaka *et al.*, 1977; APB: Reimer *et al.*, 1985; APC: Norton *et al.*, 1978; ATX III: Béress *et al.*, 1977, Martinez *et al.*, 1977; AS V: Scheffler *et al.*, 1982)

	1	2	3	4	5	6	7	8	9	10	11	12	13	14	15	16	17	18	19	20	21	22	23	24	25
ATX I / As I	Gly	Ala	Ala/Pro	Cys	Leu	Cys	Lys	Ser	Asp	Gly	Pro	Asn	Thr	Arg	Gly	Asn	Ser	Met	Ser	Gly	Thr	Ile	Trp	Val	Phe
ATX II / As II	Gly	Ile/Val	Pro	Cys	Leu	Cys	Asp	Ser	Asp	Gly	Pro	Ser	Val	Arg	Gly	Asn	Thr	Leu	Ser	Gly	Ile	Ile	Trp	Leu	Ala
As V	Gly	Val	Pro	Cys	Leu	Cys	Asp	Ser	Asp	Gly	Pro	Ser	Val	Arg	Gly	Asn	Thr	Leu	Ser	Gly	Ile	Leu	Trp	Leu	Tyr
AP-A / Ax I	Gly	Val	Ser	Cys	Leu	Cys	Asp	Ser	Asp	Gly	Pro	Ser	Val	Arg	Gly	Asn	Thr	Leu	Ser	Gly	Thr	Leu	Trp	Leu	Tyr
AP-B / Ax II	Gly	Val	Pro	Cys	Leu	Cys	Asp	Ser	Asp	Gly	Pro	Arg	Pro	Arg	Gly	Asn	Thr	Leu	Ser	Gly	Ile	Leu	Trp	Phe	Tyr
AP-C	Gly	Val	Pro	Cys	Leu	Cys	Asp	Ser	Asp	Gly	Pro	Ser	Val	Arg	Gly	Asn	Thr	Leu	Ser	Gly	Ile	Leu	Trp	Leu	Ala
ATX III / As III	Arg	Ser	Cys	Cys	Pro	Cys	Tyr	Trp	Gly	Gly	Cys	Pro	Trp	Gly	Gln	Asn	Cys	Tyr	Pro	Glu	Gly	Cys	Ser	Gly	Pro
ATX IV / As IV	Arg	Ser	Cys	Cys	Pro	Cys	Tyr	Trp	Gly	Gly	Cys	Pro	Trp	Gly	Gln	Asn	Cys	Tyr	Pro	Glu	Gly	Cys	Ser	Gly	Pro

	26	27	28	29	30	31	32	33	34	35	36	37	38	39	40	41	42	43	44	45	46	47	48	49
ATX I / As I			Gly	Cys	Pro	Ser	Gly	Trp	Asn	Asn	Cys	Glu	Gly	Arg	Ala		Ile	Ile	Gly	Tyr	Cys	Cys	Lys	Gln
ATX II / As II			Gly	Cys	Pro	Ser	Gly	Trp	His	Asn	Cys	Lys	Lys	His	Gly	Pro	Thr	Ile	Gly	Trp	Cys	Cys	Lys	Gln
As V			Gly	Cys	Pro	Ser	Gly	Trp	His	Asn	Cys	Lys	Lys	His	Lys	Pro/Gly	Thr	Ile	Gly	Tyr	Cys	Cys	Lys	
AP-A / Ax I	Pro	Ser	Gly	Cys	Pro	Ser	Gly	Trp	His	Asn	Cys	Lys	Ala	His	Gly	Pro	Thr	Ile	Gly	Trp	Cys	Cys	Lys	Gln
AP-B / Ax II	Pro	Ser	Gly	Cys	Pro	Ser	Gly	Trp	His	Asn	Cys	Lys	Ala	His	Gly	Pro	Asn	Ile	Gly	Trp	Cys	Cys	Lys	Lys
AP-C			Gly	Cys	Pro	Ser	Gly	Trp	His	Asn	Cys	Lys	Ala	His	Gly	Pro	Thr	Ile	Gly	Trp	Cys	Cys	Lys	Gln
ATX III / As III			Lys	Val																				
ATX IV / As IV																								

even in nanomolar concentrations (Renaud et al., 1986). For review see Rathmayer, 1979; Lazdunski et al, 1980; Lazdunski and Renaud, 1982; Alsen, 1983; Béress, 1988.

We should like to emphasize that contemporary independent studies on the isolation and structure elucidation of the toxins APA and APB of the Pacific sea anemone *Anthopleura xanthogrammica* and that of APC from *Anthopleura elegantissima* by Norton et al. (1976, 1978) and also the investigation of the mode of action of the named toxins on the mammalian heart muscle by Shibata et al. (1976), ending in similar results as those with ATX II (Renaud et al., 1986; for review see Norton, 1981; Béress, 1988; Norton 1991), helped to motivate us were of great value for our work.

The basic feature of the sea anemone toxin ATX II is that it acts similarly to purified α-scorpion toxins by sharing a common receptor site associated with the action potential sodium ionophore (Romey and Lazdunski, 1975; Catterall and Béress, 1978; Couraud et al., 1978; Warashina et al., 1988; Loret et al., 1994). The particular feature of ATX II and of type I sea anemone toxins is that they are able to discriminate between Na^+-channels of different animal species cells, whereas α-scorpion toxins and some of the type II sea anemone toxins do not have this discriminating power (Frelin et al., 1984; Schweitz et al., 1985). Moreover, α-scorpion toxins bind only to the polarized form of Na^+-channel, whereas ATX II and type I sea anemone toxins also bind to the depolarized form of the channel. Their binding sites are also 10 times more numerous (Vincent *et al.*, 1980; Strichartz and Wang, 1986; Catterall, 1988. For review see Lazdunski *et al.*, 1980; Norton, 1991).

Anemonia sulcata toxins can only bind to the open form of the Na^+-channel (Romey et al., 1976), which they maintain open. This particular mode of action reveals "silent channels" in embryonic cells. Silent channels are channels that do not generate action potentials under physiological conditions (Romey *et al.*, 1979, 1980).

It is far beyond the scope of the present work to refer all the extensive efforts which have been achieved in isolation and structure elucidation of toxins, and meanwhile also on other biologically active proteins and polypeptides, e.g., haemolytic polypeptides (Devlin, 1974; Hessinger and Lenhoff, 1974; Bernheimer and Avigard, 1976; Kem, 1988a,b; Meinardi *et al.*, 1995), proteinase inhibitors (Fritz *et al.*, 1972; for review see Béress, 1982; Tschesche and Kolkenbrock, 1984; Kolkenbrock and Tschesche, 1987), blood pressure-depressing antiviral polypeptides (Béress *et al.*, 1985; Driscoll *et al.*, 1989a,b), ichthyotoxic antitumor proteins (Cline *et al.*, 1995), and even palytoxin (Mahnir *et al.*, 1992), a polyketide, isolated in the last 2 decades from sea anemones.

Special attention, however, is due to the recent discovery of K^+-channel peptides in sea anemones. They were discovered nearly at the same time in different laboratories and in different sea anemones species. They belong to two different structural classes. Small homologous 35–36-amino acid peptides are BgK from *Bunodosoma granulifera* (Aneiros *et al.*, 1993), ShK from *Stichodactyla helianthus* (Castañeda et al., 1995; Pennington *et al.*, 1995), and AsKS, called kaliseptin, from *Anemonia sulcata* (Schweitz *et al.*, 1995). Longer homologous 58–59-amino acid peptides (AsKC1, AsKC2, AsKC3), called kalicludines, were also found in *Anemonia sulcata* (Schweitz et al., 1995). Kalicludines are structurally homologous both to dendrotoxins, the strong K^+-channel blockers from Mamba venoms, and to the basic pancreatic trypsin inhibitor. They have both also properties: they block voltage-dependent K^+-channels and they inhibit trypsin activity. All these peptides have been found to be active on the cloned K^+-channels Kv1, 2 and Kv1, 3, but the complete spectrum of action is still to be determined.

The structure and structure-function relationship of sea anemone peptides that interact with Na^+-channels was initiated by Barhanin *et al.* (1981) who showed that the site responsible for the fixation of ATX II to the Na^+-channel is distinct from the site responsible for the toxicity of this peptide. The next studies were reviewed by Warashina and Fujita (1983); Lazdunski *et al.* (1987); Kem (1988a,b); Kem *et al.* (1990); Norton (1991); Cohen and Barchi (1993); Kem *et al.* (1996). The structure and structure-function relationship of sea anemone peptides that interact with K^+-channels are much more recent (Schweitz *et al.*, 1995; Harvey *et al.*, 1996; Tudor *et al.*, 1996).

The aim of the present work is to summarize our most significiant efforts in the isolation of pure toxins and other biologically active polypeptides from sea anemones. In contrast to recent publications reporting the purification of sea anemone toxins from electrically stimulated nematocytes (Malpezzi *et al.*, 1993), we always failed to obtain sea anemone toxins in relative purity and measurable quantity by this "toxin-milking" method. Our isolation always started with the homogenization of the entire sea anemone specimens because the tentacles and surprisingly also the body filaments contained the toxin in the same concentration (see also Béress and Béress, 1971).

The complexity of the problem of the toxin purification makes a comparative presentation of the standardized isolation methods necessary.

2 Materials and Chemicals

Chromatographic buffers, solvents

Acetonitril gradient grade	Merck	100030
Acetonitril isocratic grade	Merck	114291
Methanol gradient grade	Merck	106007
2-Propanol gradient grade	Merck	101040
Trifluoracetic acid for spectroscopy Uvasol	Merck	108262

- Ammonium acetate buffer pH 6.0 (0.01 mol/l)
- Ammonium acetate buffer pH 6.0 (0.05 mol/l)
- Ammonium acetate buffer pH 6.0 (0.1 mol/l)
- Ammonium acetate buffer pH 6.0 (0.3 mol/l)
- Ammonium acetate buffer pH 6.0 (0.5 mol/l)
- Ammonium acetate buffer pH 6.0 (1.0 mol/l)
- Ammonium acetate buffer pH 6.0 (0.3 mol/l)
- Ammonium acetate buffer pH 8.0 (0.05 mol/l)
- Ammonium acetate buffer pH 8.0 (0.3 mol/l)
- Ammonium acetate buffer pH 8.0 (1.0 mol/l)
- Ammonium acetate buffer pH 9.0 (1.0 mol/l)
- Ammonium hydrogencarbonat-buffer pH 8.6 (0.1 mol/l)
- Phosphate buffer pH 5.0 (0.02 mol/l)
- Phosphate buffer pH 6.0 (0.04 mol/l)
- Phosphate buffer pH 5.0 (0.05) mol/l
- Tris/hydrochloric acid pH 8.0 (0.05mol/l)

Gels

Sephadex G10	Pharmacia Biotech	17-0010
Sephadex G25 superfine	Pharmacia Biotech	17-0031
Sephadex G25 fine	Pharmacia Biotech	17-0032
Sephadex G50 fine	Pharmacia Biotech	17-0042
Sephadex G50 medium	Pharmacia Biotech	17-0043
Bio-Gel P2 200-400 mesh (wet)	BIO-RAD	1500155
Bio-Gel P4 200-400 mesh (wet)	BIO-RAD	1500455
Bio-Gel P6 200-400 mesh (wet)	BIO-RAD	1500750

Ion exchangers

SP Sephadex C-25	Pharmacia Biotech	17-0240
QAE Sephadex A-25	Pharmacia Biotech	17-0190
Fractogel EMD SO_3^-- 650 M	Merck	116882
Whatman CM C-52	Whatman	

HIC material

Serdolit AD-2 research grade	Serva	42829

HPLC columns

Nucleosil ET 250 × 4 mm 100-5 C18	Machery and Nagel

Other equipment

- homogenizers
- centrifuges
- chromatographic columns
- UV-detectors (l = 226, 254, 280 nm)
- fraction collectors
- peristaltic pumps
- xy-plotter or data system
- HPLC system
- distilled or deionized water system
- rotary evaporators
- lyophilizers

3 Toxin Isolation

3.1 Isolation of toxins from the sea anemone *Anemonia sulcata*

Protocol 1 Isolation of toxins from the sea anemone *A. sulcata* by alcoholic extraction, batchwise separation, gel filtration, and ion-exchange chromatography (Béress et al., 1975b)

Alcoholic extraction

1. Freshly collected sea anemones (2000 specimens taken from the bay of Naples) were centrifuged for 10 min at 1500 x g. The supernatant seawater containing approximately 40% of the total weight of the sea anemones could be discarded because of its relatively low toxicity.
2. From the residue 5000 g was suspended in 5 l 96% ethanol and homogenized in several lots in a 4-l steel Cenco Waring blender.
3. The homogenate was adjusted to pH 6 and heated to 60°C for 2 min and cooled to room temperature. The denatured proteins along with the cell fragments were centrifuged (1320 x g), and the brown supernatant was collected.
4. The residue was homogenized a second time with 3 l of 50% ethanol, heated to 50°C and centrifuged.
5. The combined supernatant was divided and dialysed for 6 h against deionized water in visking dyalysis tubes (MWCO 12–14,000), centrifuged (1320 x g), and filtrated on a SEITZ-T1000 filter.
6. The extract was adjusted to pH 6.5 and diluted with deionized water to a conductivity of 4 ms/cm.

Batchwise separation and ion-exchange chromatography

7. 150 g dry carboxymethyl cellulose (Serva) was added while stirring for 30 min thereby adsorbing the strongly basic batch I polypeptides (proteinase inhibitors and K^+ channel toxins).
8. After filtration on a SEITZ-T1000 filter and washing the retained ion exchanger with distilled water, the filtrate was adjusted with acetic acid to pH 5 and further diluted with deionized water to a conductivity of 2 ms/cm.
9. A second batch of 150 g dry CM-cellulose was added, onto which the batch II polypeptides were bound at the same conditions.
10. The CM-cellulose was filtered, and from the filtrate at pH 3 (adjusted with HCl) and 2 ms/cm conductivity the less basic batch III polypeptides were adsorbed on 500 g (wet weight) SP Sephadex C-25.

    Notes:
    After this treatment all toxicity was adsorbed quantitatively.

11. The protein fractions of batches I, II, and III were each eluted from the cation exchangers with 2l 1 mol/l sodium chloride solution (pH 8) from the filter funnel, concentrated, and partially desalted by ultrafiltration on an Amicon µM05 filter, or by dialysis (MWCO 3500) and concentration at reduced pressure, to final volumes of 70 ml each.

Gel filtration

12. The three toxic fractions were gel-filtered on Sephadex G-50 each in 1 mol/l sodium chloride (pH 8) as eluting solvent (Fig. 2A).

The toxicity was always concentrated in the second void volume (in the third peak) of the gel filtration. In case of batch III proteins a considerable crab-paralysing activity was also found in the last (the sixth) chromatographic faction (Fig. 2A). Fraction 3 of the Sephadex G-50 filtrates of batches I, II and III were desalted by ultrafiltration and further separated on SP Sephadex C-25 (Fig. 2B) by elution of the proteins at linear increasing buffer concentration. The sea anemone toxins I, II, and III can be obtained after desalting the toxic fractions by gel filtration on Sephadex G-25 in distilled water as eluting solvent. The purification of toxin III required two additional gel filtration steps on Sephadex G-10 and Biogel P2. Fraction 3 of the Sephadex G-50 filtrate of batches I, II, and III were further separated on SP Sephadex C-25 (Fig. 2B), by which toxin II was obtained in electrophoretically pure form. Toxin I, appearing only in batch III, required additional chromatography on QAE Sephadex A-25 (Fig. 2C), on Sephadex G-25 superfine (Fig. 2D) and SP Sephadex C-25 (Fig. 2E) for its purification. Both toxins I and II were rechromatographed on SP Sephadex C-25 under equilibrium conditions (Fig. 2E, F). The pure fractions were desalted on Sephadex G-25 and lyophilized. As the batch I and batch II proteins contained

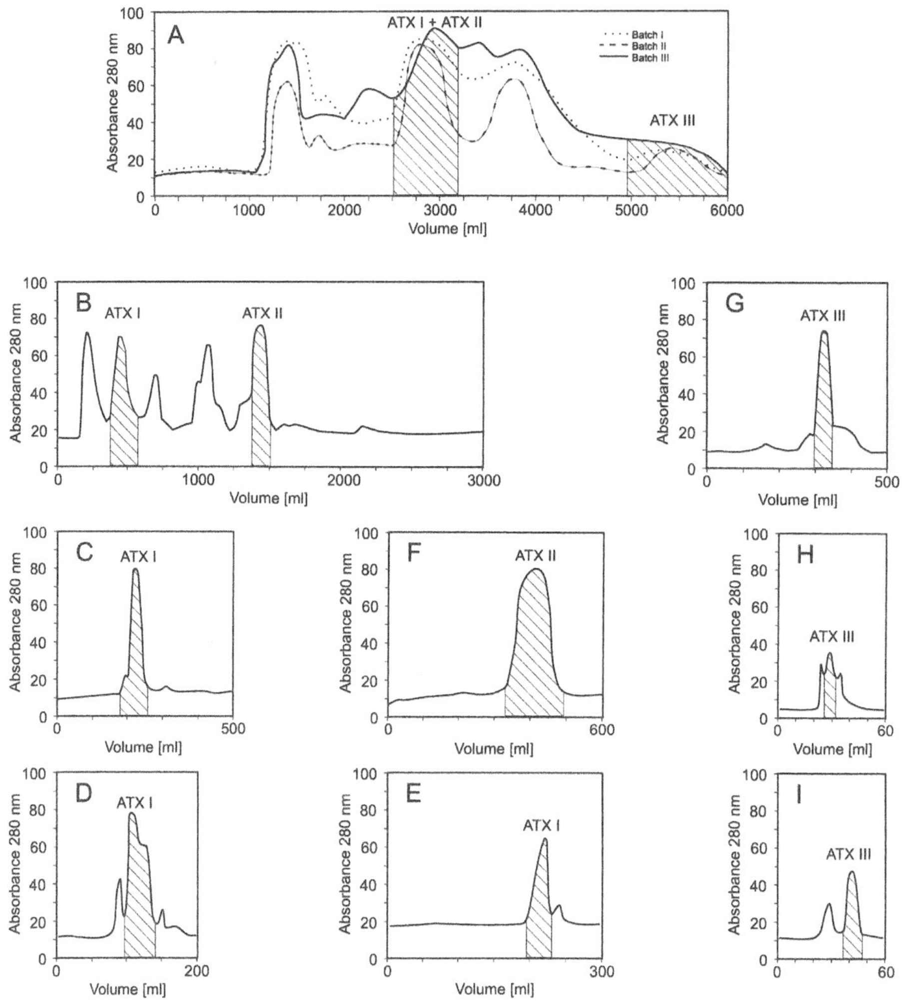

Figure 2 Purification of the toxins ATX I, ATX II, and ATX III from *Anemonia sulcata.*

A) Gel-filtration of the proteins of the batches I – III on Sephadex G-50 (7 x 140 cm) in 1 mol/l sodium chloride, 0.05 mol/l Tris /HCl buffer, pH 8.0, flow rate 5 ml/min. The toxic fractions (shaded) were desalted and lyophilized. B) Cation-exchange chromatography of fraction A3 on SP Sephadex C-25 (3 x 100 cm) eluting solvent was 2.4 l 40 mmol/l phosphate buffer (pH 6.0), linear sodium chloride gradient (0.08 – 0.38 mol/l), flow rate 3 ml/min. Fraction 2 is ATX I and fraction 7 is ATX II; both were desalted and lyophilized. C) Anionexchange chromatography on QAE Sephadex A-25 of fraction B2 (ATX I). Column 2 x 100 cm; eluting solvent was 1 l 50 mmol/l Tris/HCl buffer (pH 8.0); linear gradient of 0.05 – 0.35 mol/l sodium chloride; flow rate 1 ml/min. The toxic fraction was desalted and lyophilized. D) Gel-filtration on Sephadex G-25 superfine of the toxic fraction from C. Column: 2 x 60 cm, eluting solvent was distilled water; flow rate: 0.3 ml/min. The toxic fraction 2 was lyophilized. E) Cation-exchange chromatography of Fraction D2 on Sephadex C-25. Column 1 x 100 cm; eluting solvent was 300 ml 20 mmol/l phosphate buffer (pH 5.0); linear gradient of 0.04 – 0.1 mol/l sodium chloride; flow rate 0.5 ml/min. The toxic fraction was desalted and lyophilized. F) Cation-exchange chromatography under isocratic conditions of ATX II (fraction B7) on SP only minor amounts of toxins, for demonstration only the isolation of toxins from the batch III proteins is presented (see Fig. 5). All three toxins were purified up to electrophoretic homogeneity. The yield of toxins from the three batches was batch I: 62 mg ATX II; batch II: 270 mg ATX II; batch III: 300 mg ATX I, 412 mg ATX II, 7 mg ATX III.

The strongly basic proteinase inhibitors and the K^+-channel-blocking toxins can be obtained from the strongly basic batch I proteins (Fritz et al., 1972, Béress et al., 1975b, Schweitz et al., 1995).

3.2 Isolation of isotoxins from the sea anemone *A. sulcata*

Previous sequence studies of ATX I (Wunderer and Eulitz, 1978), ATX II (Wunderer et al., 1976) and also on a later isolated new toxin V from *A. sulcata* ASV (Schweitz et al., 1981; Scheffler et al., 1982) indicated that the named toxins were not completely pure proteins, each having distinct microheterogeneities in their own sequence. ATX I contained Ala-Pro in position 3, ATX II Ile-Val in position 2 and ASV Pro-Gly in position 39 (Tab. 1), indicating the presence of at least two very related, but principally different toxin molecules in each fraction. These so-called isotoxins could not be separated from each other, even by use of the finest ion-exchange chromatographic and gel filtration techniques. Only the final use of the HPLC on an RP C-18 material enabled the complete purification of the sea anemone (*A. sulcata*) toxins into different isotoxins, a fact which gave rise to a reinvestigation of our former, already standarized sea anemone toxin purification procedure (see Protocol 1).

The isolation of ATX I and ATX II has been achieved with a slight modification of our Protocol 1 described previously.

Sephadex C-25. Column: 2 x 100 cm; eluting solvent was 0.25 mol/l sodium chloride in 0.25 mmol/l phosphate buffer (pH 6), flow rate, 1 ml/min. The toxic fraction was desalted and lyophilized. G) Gel filtration on Sephadex G-25 superfine of ATX III (fraction A6) . Column: 3.5 x 50 cm; eluting solvent was distilled water; flow rate, 6 ml/min. The toxic fraction G3 was lyophilized. H) Gel filtration of fraction G3 on Sephadex G-10. Column 1 x 50 cm; eluting solvent was distilled water; flow rate, 0.4 ml/min. Toxic fraction 2 was lyophilized. I) Gel-filtration of fraction 1.3 mg H2 on Biogel P-2 400 mesh. Column: 0.6 x 150 cm; eluting solvent was 0.1 mol/l ammonium hydrogen carbonate buffer pH 8.6; 1flow rate 1.2 ml/h.

Protocol 2 Isolation of isotoxins from the sea anemone *A. sulcata* by a modified toxin purification method and by final use of HPLC (Wachter *et al.*, 1987)

1. Five kilograms wet *A. sulcata* were homogenized and extracted with 5 l ethanol at pH 4.5 and 20°C.
2. After centrifugation (2000 × g , 10 min) the supernatant was collected and the residue reextracted with 5 l 50% ethanol and centrifuged again.

   Notes:
   The 1.5-kg residue can be reextracted with 3 l of 50% ethanol containing 350 g sodium chloride solved at pH 9 for the isolation of all nontoxic, very strong basic sea anemone polypeptides, proteinase inhibitors, potassium channel-blocking polypeptides (not shown here).
3. The first two supernatants were combined and concentrated, at reduced pressure, to one liter.

   Notes:
   The foaming can be avoided if the evaporation starts by cooling with ice.
4. After the addition of 10 l acetone to the 1 l toxin concentrate at 20°C, the toxin with the sea anemone proteins became precipitated, and the lipids solved.
5. After the precipitation was completed at –20°C standing for 12 h, the acetone layer was discarded and the toxin precipitate redissolved in 2 l distilled water and dialysed two times against 20 l distilled water for 8 h (MWCO 12–14,000).
6. The non-dialysable tube content was centrifuged (4000 × g, 5 min) and the toxic supernatant adjusted to pH 3.5.
7. After dilution with water to a conductivity of 2 ms/cm, the supernatant was passed through an SP Sephadex C-25 column (7 × 50 cm, pH 3.5) equilibrated in distilled water.

   Notes:
   During this treatment the toxins together with all the other basic sea anemone proteins were adsorbed.
8. The column was washed with distilled water, and the adsorbed proteins were eluted in one step with 2 l of a 1 mol/l ammonium acetate buffer (pH 9).
9. The eluate was dialysed for 5 h against 3 × 10 l distilled water, and the toxic tube content was concentrated at reduced pressure to 140 ml.
10. The concentrated eluate was divided into two parts and gel-filtered on a Sephadex G-50 column (size 7 × 140 cm) equilibrated with 1 mol/l acetic acid (Fig. 3A).
11. The total toxicity was concentrated again in the second void volume in the third chromatographic fraction (Fig. 3A) resulting after concentration at reduced pressure and lyophilization of 9 g of grey-brown crude sea anemone toxin.

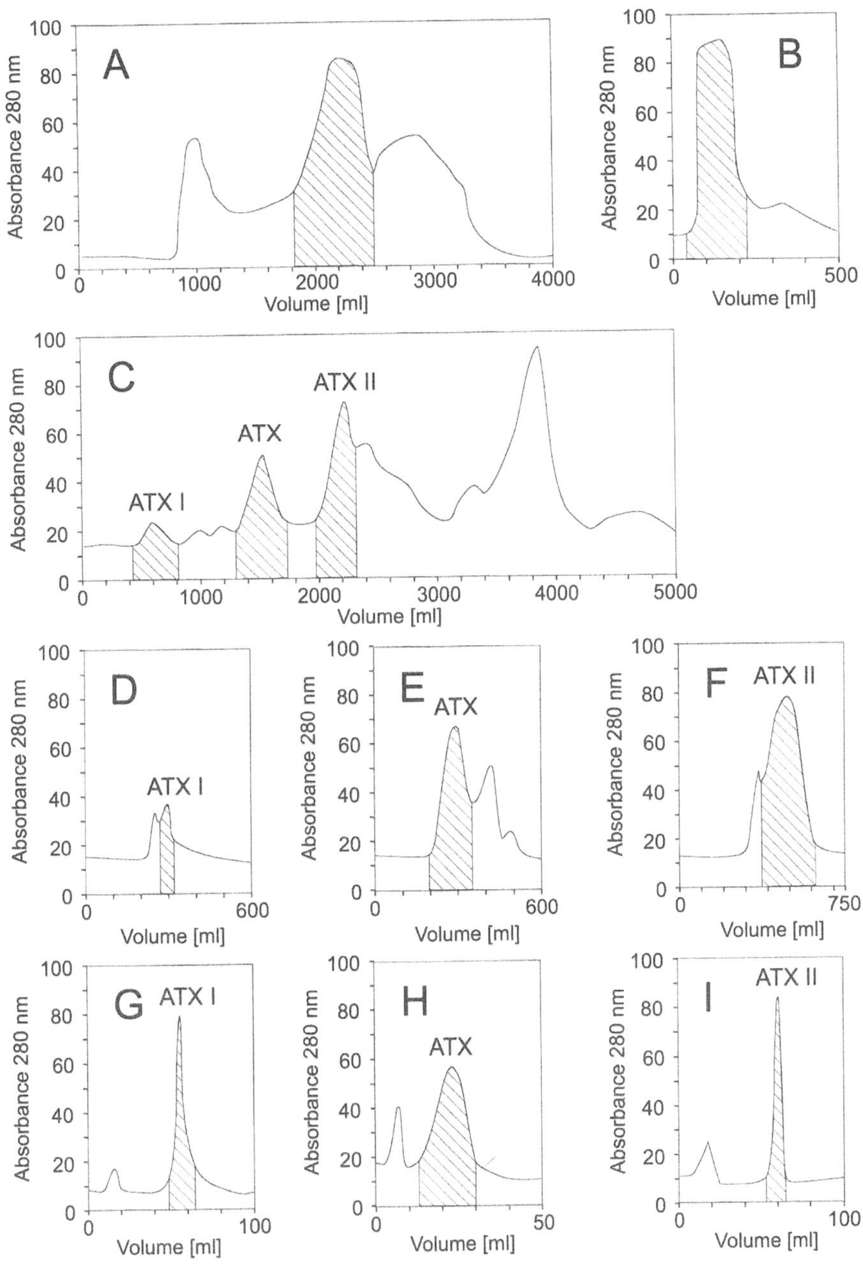

Figure 3 Isolation of the Anemonia sulcata toxins ATX I , ATX, and ATX II.

A) Gel-filtration on Sephadex G-50. Column 7 x 140 cm. Solvent: 0.5 mol/l acetic acid. B) Anion exchange chromatography of A3 on QAE Sephadex A-25. Column: 3 x 40 cm. Solvent 0.05 mol/l ammonium acetate pH 8.0. C) Cation-exchange chromatography of B1 on SP Sephadex C-25. Column: 3 x 40 cm. Solvent: gradient elution 0.05 – 0.3 ammonium acetate. D, E, F) Gel filtration of C1, C3, C5 on Sephadex G-25 Column: 3 x 40 cm. Solvent: 0.3 mol/l acetic acid.

G, H, I) Cation-exchange chromatography of D2, E1, and F2 on Whatman C52 CM cellulose. Column (for 5 mg toxin): 0.5 x 10 cm. Solvent: gradient elution ammonium acetate (pH 6.0) 0.01 – 0.1 for G, 0.05 – 0.3 for H and I.

12 This material was dissolved in 500 ml distilled water, and the pH was adjusted to 8, whereby precipitation occured.

13 The precipitated impurities were eliminated after 2 h standing at 0°C by centrifugation (5000 × g, 5 min), and the toxin solution was chromatographed on a QAE Sephadex A-25 column (size 3 × 40 cm) equilibrated with 0.01 mol/l ammonium acetate buffer (pH 8) (Fig. 3B).

Notes:
With this treatment all basic protein passed through, and the coloured nontoxic contaminating acidic polypeptides were adsorbed on the column.

14 The solution with the basic proteins was adjusted to pH 6 with acetic acid and then freeze-dried.

Notes:
This crude decolorated sea anemone toxin, a mixture of polypetides of a molecular weight of about 5000 Da, is the starting material of further toxin isolations (see Protocol 4).

15 The toxins were separated from each other and also from accompanying strong basic nontoxic anemone polypeptides by ion-exchange chromatography on an SP Sephadex C-25 column (size 3 × 40 cm) using an ammonium acetate buffer with a linear concentration gradient of 0.01–0.1–0.3 mol/l.

Notes:
With this gradient the sea anemone toxins were eluted successively (Fig. 3C).

16 The strong basic nontoxic sea anemone polypeptides were thereafter eluted by use of a buffer concentration gradient from 0.3 to 1.0 mol/l.

17 All fractions were tested on toxicity on the shore crab *Carcinus maenas*. The toxic fractions were collected, concentrated at reduced pressure to 10 ml, and desalted on a Sephadex G-25 column (3 × 50 cm) by use of 0.5 mol/l acetic acid as running solvent (Fig. 3D–F).

18 After concentration and lyophilization of the three toxin samples, each was solved in distilled water and rechromatographed on the cation exchanger CM-cellulose (Whatman CMC52). For ATX I a linear ammonium acetate gradient of pH 6 from 0.01 to 0.1 mol/l was used (Fig. 3G).

19 ATX and ATX II were repurified with an ammonium acetate gradient from 0.05 to 0.3 mol/l at pH 6 (Fig. 3H, I).

20 The toxin fractions were desalted again by gel filtration on Sephadex G-25, lyophilized, and proved for homogeneity by paa gel electrophoresis (pH 8.6).

21 Even if they proved to be homogeneous under this condition, HPLC on a Vydac RP C-18 column by use of a linear gradient from 10 to 50% acetonitrile in water revealed the inhomogeneity of the three toxin samples (Fig. 4E).

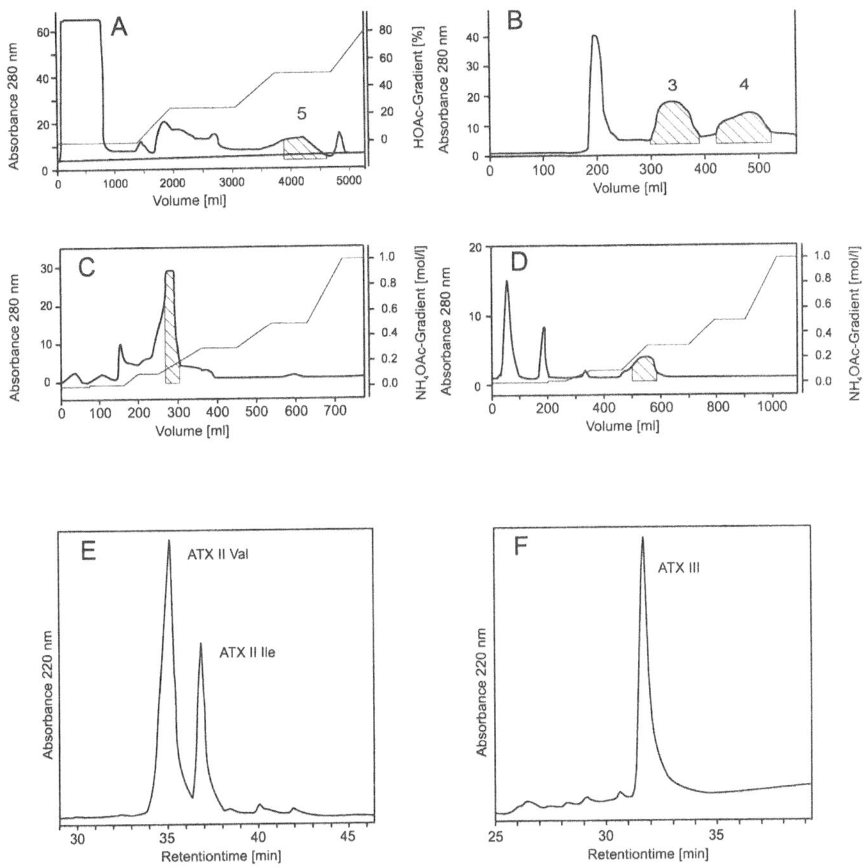

Figure 4 Purification of the crude sea anemone extract from *Anemonia sulcata* on Serdolit AD-2.

A) The column (2.5 x 30 cm) was equilibrated with water, and the crude extract was loaded and eluted in this solvent, followed by an acetic acid gradient from 0 to 80%. Elution was monitored by measuring the adsorbance at 278 nm, flow rate, 9 ml/min. Fraction size, 28 ml. The collected toxic fraction 5 is shaded. B) Gel-filtration of the main toxic fraction A5 on a Sephadex G50 column. Serdolit AD-2 fraction A5 was gel-filtered in 20% acetic acid on Sephadex G50 (3 × 120 cm). 21 ml per tube was collected, and the flow rate was 3 ml/min. Adsorbance at 278 nm. The shaded fractions B3 and B4 are toxic when injected into shore crabs. C) Sephadex fraction B3 was rechromatographed on the cation-exchanger Fractogel EMD SO_3^-650 M (0.7 x 10 cm) equilibrated with 0.01 M ammonium acetate pH 5.5. After elution of the unbound material, a gradient from 0.01 to 1.0 M ammonium acetate (pH 5.5) was used at a flow rate of 1.4 ml/min; fractions of 7 ml were collected. The absorbance was monitored at 278 nm. The collected toxic fraction is shaded. D) Sephadex fraction B4 was rechromatographed on the cation-exchanger Fractogel EMD SO_3^-650 M (0.7 x 10 cm) equilibrated with 0.01 M ammonium acetate pH 5.5. After elution of the unbound material, a gradient from 0.01 to 1.0 M ammonium acetate (pH 5.5) was used at a flow rate of 1.4 ml/min; fractions of 7 ml were collected. The absorbance was monitored at 278 nm. The collected toxic fraction is shaded. E) HPLC of the toxic fraction of C on a Macherey-Nagel Nucleosil 100-5 C18 4 x 250 mm column; a linear gradient from 10 to 70% A in B in 70 min was used. (A: 5% acetonitrile in 0.1% trifluoroacetic acid; B: 95% acetonitrile in 0.1% trifluoroacetic acid.) The absorbance was monitored at 220 nm. F) HPLC of the toxic fraction of D on a Macherey-Nagel Nucleosil 100-5 C18 4 x 250 mm column; a linear gradient from 10 to 70% A in B in 70 min was used. (A: 5% acetonitrile in 0.1% trifluoroacetic acid; B: 95% acetonitrile in 0.1% trifluoroacetic acid.) The absorbance was monitored at 220 nm.

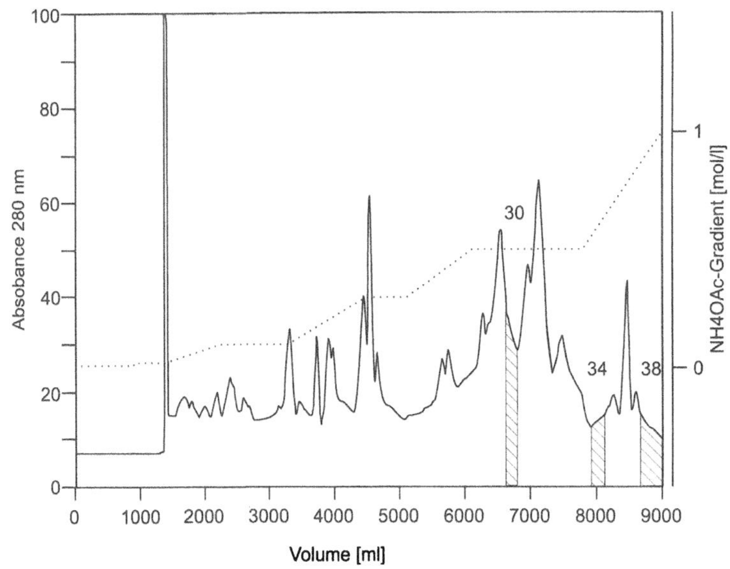

Figure 5 Ion-exchange chromatography on SP Sephadex C-25 of 6 g crude toxic material obtained from *Anemonia sulcata*.

The 2.5 x 140 cm column was equilibrated with 0.01 mol/l ammonium acetate buffer, pH 4.5. The crude toxic fraction was loaded and eluted with this buffer followed by a stepwise gradient as indicated in the figure. The final elution was carried out with 1 mol/l sodium choride (not shown). Elution was monitored by measuring the absorbance at 280 nm, flow rate 225 ml/h, fraction size, 8 ml. The shaded fractions were further purified by HPLC, resulting the kalicludines and the kaliseptine.

Final purification of the isotoxins of ATX I, ATX and ATX II have been achieved by the preparative HPLC on a reversed-phase Vydac RP-C18 column with a linear gradient of acetonitrile in water from 10 to 50% (for details see Wachter *et al.*, 1987). The small sea anemone toxin ATX III was not isolated by this procedure.

3.3 Isolation of ATX II and ATX III from the sea anemone *A. sulcata*

From a crude watery sea anemone extract the polystyrene resin Serdolit AD-2 is able to bind quantitatively all the toxicity after passing the extract through a column filled with an adequate quantity of the adsorbent. The toxins are eluted by concentrations from 25 to 80% of an organic with water miscible solvent (acetic acid, acetonitrile, ethanol). This method allows a high degree of specific purification of the toxins already in the first chromatographic step.

Protocol 3 Isolation of ATX II and ATX III from the sea anemone *A. sulcata* extract by adsorption on the polystyrene resin Serdolit AD-2 (Serva) and following elution with acetic acid followed by ion-exchange chromatography, gel filtration, and HPLC on C18 reversed-phase material (Sanchez *et al.*, 1996).

1. 700 g wet sea anemones (*A. sulcata*) were homogenized and extracted with 1500 ml 50% acetic acid.
2. After centrifugation and concentration of the extract in vacuum to 1/10 of the original volume, the proteins were precipitated with 10 parts of acetone.
3. This precipitate was dissolved in 2.0 mol/l sodium chloride, centrifuged (25,000 × g) and the supernatant was applied on a 2.5 × 30 cm column of Serdolit AD-2, whereby the crude sea anemone extract could be easily fractionated.
4. The toxins which were adsorbed on the Serdolit could be eluted by a water/acetic acid gradient with an increasing acetic acid concentration from 0–80%.

   Notes:
   They eluted predominantly at a concentration of 50% (Fig. 4A).
5. Fractions of 28 ml were collected at a flow rate of 9 ml/min.
6. The resulted fractions were concentrated and gel-filtrated on a Sephadex G-50 column (3.0 × 120 cm) equilibrated with 20% acetic acid (Fig. 4B).
7. The elution was carried out in the same solvent; fractions of 21 ml/min were collected at a flow rate of 3 ml/min.
8. After lyophilization the toxic fractions were rechromatographed on the cation exchanger Fractogel EMD SO_3^-- 650 M (0.7 × 10 cm) equilibrated with 0.01 mol/l ammonium acetate, pH 5.5.
9. After the elution of the unbound nontoxic material, a linear gradient from 0.01 to 1.0 mol/l ammonium acetate was used at pH 5.5 (Fig. 4C, D). At a flow rate of 1.4 ml/min, fractions of 7 ml were collected.
10. The collected fractions were desalted on Sephadex G-25 (1.7 × 57 cm) in 1.0 M acetic acid as eluting solvent at a flow rate of 1.3 ml/min; fractions of 4 ml were collected and lyophilized.
11. The purity of the fractions was determined by HPLC on ET 250/4 Nucleosil 100-5 C18 (4 × 250 mm), using a linear gradient of 10–70% acetonitrile in 0.1% trifluoroacetic acid for 70 min at a flow rate of 1 ml/min (Fig. 4E, F).

3.4 Isolation of two different classes of sea anemone toxins which block voltage-sensitive K^+ channels

The first steps of the purification were performed exactly in the same way described for the isolation of the HPLC-purified Na^+-channel toxins of *A. sulcata* (see Protocol 2; Wachter *et al.*, 1987).

Protocol 4 Isolation of two different classes of sea anemone toxins which block voltage-sensitive K^+ channels (Schweitz *et al.*, 1995).

1. 12 g of the decolorated crude sea anemone toxin (see Protocol 2) was dissolved in 120 ml of 1 mol/l sodium chloride and gel-filtrated in two parts on a Sephadex G-50 medium column (7 × 140 cm) equilibrated in 1 mol/l sodium chloride.
2. The main crab-paralysing fractions of the two gel filtrations were combined, dialysed (MWCO 12–14,000, 5 h), concentrated at reduced pressure and finally desalted on Sephadex G-25 (column size 7 × 70 cm) in 0.3 mol/l acetic acid as eluting solvent, concentrated at reduced pressure and lyophilized.
3. 6 g of this crude toxin was separated into its 38 main polypeptide components by ion-exchange chromatography on SP Sephadex C-25 (column size 2.5 × 140 cm) (see Fig. 5).
4. Fractions which were potentially active on K^+ channels are shaded. As eluting solvent ammonium acetate buffers (pH 4.5, 0.01–1 mol/l) were used alternatively in a linear gradient followed by equilibrium conditions at each step (0.01–0.1-0.1-0.3-0.3-0.5-0.5-1 mol/l) with each 2:2l+2l of the eluting solvent.
5. Each fraction was concentrated at reduced pressure, desalted on Sephadex G-25 fine in 0.3 mol/l acetic acid as eluting solvent, and lyophilized.
6. 100 mg of fraction 30 (AsKS: *A. sulcata* Kaliseptine) from the ion-exchange chromatography was chromatographed under equilibrium conditions on Fractogel EMD SO_3^--650 M (1 × 40 cm) with ammonium acetate (pH 4.5) as eluting solvent.

   Notes:
   The eluted material appeared in three peaks; after desalting on Sephadex G-25 and lyophilization, only the first peak shows activity.
7. 11.3 mg of this fraction was loaded on an Ultrasphere ODS 5 μm column (10 × 250 mm) and eluted over 60 min at 3 ml/min with a linear gradient from 15 to 40%; 0,5% trifluoroacetic acid plus 0.85% triethylamine, plus 10μl/l b-mercaptoethanol in acetonitril.

   Notes:
   Activity was found in the first peak eluting at 17% D between 5.0 and 6.0 min.

8. This peak was lyophilized and loaded on the same column. A linear gradient of 5 up to 20% in 60 min was applied at the same flow rate of 3 ml/min. The toxin AsKS (*A. sulcata* Kaliseptine) eluted at 14% elution buffer between 34 and 36 min (recovery 1.45 mg).

The further two potassium channel blocking-polypeptide fractions 34 and 38 (AsKC1 and AsKC2, *A. sulcata* Kalicludine 1,2) have been purified under analogous conditions. For details see Schweitz *et al.*, (1995).

The kaliseptine (AsKS) and the three kalicludines have been sequenced (Schweitz *et al.*, 1995). AsKS contains 36 amino acids, and its sequence is homologous to the sequence of Bgk toxin, a potassium channel blocker, isolated from previously the sea anemone *Bunodosoma granulifera* (Castañieda *et al.*, 1995). The anemone kalicludines (AsKC1, AsKC2) each contain 58 amino acids and AsKC3 59 amino acids; their sequence is homologous to the sequence of DTX 1 (dendrotoxin 1) and to BPTI (bovine pencreatic trypsin inhibitor) (Schweitz *et al.*, 1995).

4 Remarks

Because of their highly selective action on the Na^+-channel inactivation process, their positive inotrope effect on mammalian heart muscle and also because of their blocking action on K^+-channels, the different types of polypeptide toxins, especially ATX II, isolated from *A. sulcata* and from other sea anemones have become extremely useful tools in the neurobiological, electrophysiological, and pharmacological research in the last 30 years. The additional presence of further polypeptides, as polyvalent proteinase inhibitors of the Kunitz type, an elastase inhibitor, and blood pressure-depressing antiviral active polypeptides in the named speciment, underline its scientific importance. This fact and the great experimental problems concerning the isolation of the sea anemone toxins gave rise to this compendium summary of the most significant publications on the field and also of the isolation procedures.

Because sea anemone toxins have to be purified generally from extracts of the whole homogeneous sea anemones, the isolation of pure sea anemone toxins is consequently really a very difficult task, compared with the isolation of pure polypeptide toxins from easily available crude snake venoms and crude scorpion venoms.

Highly selective methods had to be developed in order to isolate the toxins from the sea anemone extract, a mixture of a large number of nontoxic water-soluble organic and inorganic compounds, including proteins, polysaccharides, nucleic acids, amino acids, and salts.

The extraction of the sea anemone cell homogenates with water-miscible organic solvents (ethanol, acetic acid) of 50% concentration proved to be the optimal extraction condition, always resulting in a quantitative yield of the toxin in the clear red-brown supernatant after centrifugation. The warming up to 60°C below pH 6 resulted in denaturation of the labile inactive sea anemone proteins without any loss of the toxicity. After elimination of the denaturated proteins, the sea anemone toxins can be separated from the accompanying lipids after concentration at reduced pressure by following precipitation with 10 parts of acetone (see Protocol 2) or by dialysis and centrifugation/filtration (Protocol 1).

Elimination of the salts and the low molecular weight nontoxic water-soluble organic impurities by dialysis rendered the nondialysable toxin-containing sea anemone protein mixture suitable for treating with cation exchangers for binding the basic toxic proteins. These bind either stepwise at different pH values (pH 6.5, 5, and 3) (see Protocol 1), or in one step at pH 3 (Protocol 2). The elution of the toxins from the adsorbent either by 1 mol/l sodium chloride (Protocol 1) or by 1 mol/l ammonium acetate (Protocol 2) is quantitative, and the Na^+-channel and also the K^+-channel active sea anemone toxins always appear together with the proteinase inhibitors and the blood pressure-depressing polypeptides in the second void volume (fraction 3) of the gel filtration on Sephadex G-50. The use of 1 mol/l acetic acid as eluting solvent by the former gel filtration facilitates the isolation of the crude toxin, which elutes before the salts; thus it can be isolated by direct concentration and lyophilization. On the other hand, protein separation by gel filtration on Sephadex G-50 in 1 mol/l sodium chloride is about 10% better, because of the higher ionic strength of this solvent which counteracts molecular aggregation and protein adsorption on the gel matrix. For this reason a rechromatography of the crude toxin obtained by Protocol 2 is recommended (see Protocol 4) before its further purification by ion-exchange chromatography.

For elution of the different basic sea anemone polypeptides from the cation exchangers SP Sephadex C-25 and Fractogel EMD SO_3^--650 M sodium phosphate (pH 6), sodium acetate (pH 5–6) and ammonium acetate buffers between pH 4.5 and pH 6 have been used. The most frequently used buffer was ammonium acetate at pH 5.2. It neither crystallizes nor precipitates during the concentration, and so the fractions can be desalted on Sephadex G-25 immediately. The HPLC should be always the last step of sea anemone toxin purification. In no case was it possible to isolate sea anemone toxins in a highly pure form without HPLC because of the presence of the different "isotoxins" within each "classically purified" toxin fraction. In this respect the sea anemone toxins behave like the scorpion toxins whose final purification is also very difficult because of the same phenomenon (Görög *et al.*, 1994)

It is also possible to concentrate sea anemone toxins from the crude sea anemone extract already by one chromatographic step in one fraction, by using polystyrene adsorbent Serdolit AD-2 with a water-acetic acid gradient. For total purification of the toxin, the use of the obligatory chromatographic techniques, such as anion-exchange chromatography on QAE Sephadex A-25, gel filtration on Sephadex G-50, cation-exchange chromatography on SP Sephadex

C-25 or on Fractogel EMD SO_3^--650 M, desalting on Sephadex G-25, and finally HPLC on an RP-C18 material is nessesary.

The three different toxin isolation methods were elaborated for the optimization of the yield and purity of toxins. Even if Protocol 2 was preferably used for large-scale purification of toxins ATX I and ATX II , toxin III of *A. sulcata* was never isolated by us in this procedure. On the other hand, the use of Protocol 3 was not advantageous for the isolation of ATX I, which we could not obtain by this procedure.

The great advantage of pure sea anemone toxins in comparison with pure α-scorpion toxins is, that the former can be prepared in large quantities according to our method and thus can be developed as genuine tools.

The use of each of the three procedures mentioned can be applied with individual modifications to isolate all kind of biologically active basic polypeptides of middle molecular weight (MW 4000–10,000 Da) from sea anemones, scorpion venoms, snake venoms, all kind of cell homogenates (L. Béress, personnel observation), and even to isolate palytoxin, a nonproteinic water-soluble highly toxic marine polyketide (Béress *et al.*, 1983).

Though the scientific value of toxin isolation often gets lost between the convictive electrophysiological and pharmacological studies dealing with the investigation of the mode of action of natural toxins on ion channels and on all kind of excitable membranes, it has to be kept in mind that no clearly defined results can be obtained without the use of pure and well-characterized toxins, whose isolation and characterization are a complicated, difficult, and time-consuming task even now.

5 Troubleshooting

Column blocked: In all forms of chromatography good resolution and long column life depend on the sample, which must be free from particulate matter. "Dirty" samples have to be cleaned by filtration or centrifugation before being applied on the column. This is especially important when chromatography is performed on small particle matrices such as HPLC materials. For particle size of 45–10 μm samples should be filtrated through a 0.45-μm filter. For smaller particle sizes, sterile filtration or for extra-clean samples a 0.22-μm filter is required.

Centrifugation at 10,000 × g for 15 min can also be used to prepare clean samples. After preparation (filtration or centrifugation), the sample should be clear and free of particles and of visible contamination of lipids. Later on they can be can be adsorbed on charecoal and centrifuged or filtrated.

Gel filtration: The ratio of the protein solution volume to the gel volume should be 1:50 or even less for gel filtration on Sephadex G-50 and 1:20 for desalting on Sephadex G-25. The running solvent should be of well-defined pH and ionic composition. Frequently 1 mol/l acetic acid was used, but also solvents with 0.15 mol/l or greater ionic strength up to 1 mol/l sodium chloride or 1 mol/l ammonium acetate can be used to counteract molecular association or binding of protein to the gel matrix.

Be sure to pack the column in the right way, that no gas bubbles remain inside, and that the distribution of the gel in the column becomes regular.

Sample volume: Do not use sample volumes larger than 1–5% of the total bed volume for good resolution. For gel filtration of molecules with little deviation in the molecular weight, use sample volumes of 1–2% of the gel bed volume; for desalting, use up to 5% of the bed volume as sample volume.

Sample viscosity: The viscosity may limit the quantity of sample that can be applied on a column. A high sample viscosity causes instability of the zone and irregular flow pattern. The limit of usable viscosity us reached with protein concentrations of approximately 5%. Concentration of ammonium acetate solutions generates solutions with high viscosity; in these cases it is preferable to delute and to divide the sample volume in parts and desalt in different runs if no larger columns are available.

Ion exchanger: For good chromatographic resolution the sample mass should be no more than 10% of the capacity of the exchanger.

If gradient elution techniques are used the sample mass applied is of far more importance than the sample volume. Large volumes of dilute solution, such as pooled fractions from a gel-filtration step or after dialysis, can be applied directly to the ion exchanger without prior concentration. The substances can be concentrated by binding on the top of the column. To enable this, the ionic conductivity of the solution applied should be lower than the conductivity of your starting buffer, and the pH should be adequately adjusted.

Acknowledgement

Many thanks to Prof. Dr. O. Wassermann for his permanent interest and generous support almost from the beginning of our work. L. Béress is very grateful to H. Fritz for initial help and encouragement and to R. Béress and J. Zwick for excellent technical assistance during the first 10 years of work. Thanks are also due to E. Wachter, E. Habermann, W. Rathmayer, C. Alsen, M. Ravens, J. Schmidtmayer, F. Hucho, G.E. de Motta, R.S. Norton and C.H. Wu for their cooperation during the long years. The idea to produce sea anemone toxins and distribute them worldwide emerged from B. Hille; many thanks to him.

One of us (H. Schweitz) is grateful to Prof. M. Lazdunski for his generous help and his cooperation for such a long time.

Supported by the Deutsche Forschungsgemeinschaft, Fonds der Chemischen Industrie, and by CNRS. The scholarship support for Judith Sánchez-Rodriguez was provided by CONACyT.

References

Abita J-P, Chicheportiche R, Schweitz H, Lazdunski M (1977) Effects of neurotoxins (veratridine, sea anemone toxine, tetrodotoxin) on transmitter accumulation and release by nerve terminals in vitro. *Biochemistry* 16: 1838–1844

Alsen C (1975) Cardiotoxic effect of two toxins isolated from the sea anemone. Naunyn-Schmiedebergś Arch. Pharmacol. 278: 105

Alsen C (1983) Biological significance of peptides from Anemonia sulcata. Fed. Proc. 42: 101–109

Alsen C, Béress L, Fischer K et al (1976) The action of a toxin from the sea anemone (Anemonia sulcata) upon mammalian heart muscles. Naunyn. Schmiedebergś Arch. Pharmacol. 295: 55–62

Aneiros A, Garcia I, Martinez JR et al (1993) A potassium Channel Toxin from the secretion of the sea anemone Bunodosoma granulifera. Isolation, amino acid sequence, and biological activity. Biochem. Biophys. Acta 1157: 86–92

Barhanin J, Hugues M, Schweitz H et al (1981) Structure-function relationship of sea anemone toxin II from Anemonia sulcata. J. Biol. Chem. 256: 5764–5769

Béress L (1982) Biologically active compounds from coelenterates. Pure Appl Chem. 54: 1981–1994

Béress L (1988) Sea anemone toxins as tools for physiological, pharmacological, and biophysical research. Poisonous and Venomous Marine Animals of the World, 2nd ed (Halstead BW, ed) pp 150–161: Princeton: Darwin Press

Béress L, Béress R (1971) Reinigung zweier krabbenlähmender Toxine aus der Seeanemone Anemonia sulcata. Kiel. Meeresforsch. 27: 117–127

Béress L, Béress R (1974) Purification of biological active polypeptides from the sea anemone Anemonia sulcata. 9th FEBS Meeting, abstract no. F127

Béress L, Zwick J (1980) Purification of two crab-paralysing polypeptides from the sea anemone Bolocera tuediae. Marine Chem. 8: 333–338

Béress L, Béress R, Wunderer G (1975a) Isolation and characterization of three polypeptides with neurotoxic activity from Anemonia sulcata. FEBS Lett. 50: 311–314

Béress L, Béress R, Wunderer G (1975b) Purification of three polypeptides with neuro- and cardio-toxic activity from the sea anemone Anemonia sulcata. Toxicon 13: 359–367

Béress R, Béress L, Wunderer G (1976) Purification and characterization of four polypeptides with neurotoxic activity from Condylactis aurantiaca. Hoppe-Seylers Z. Physiol. Chem. 357: 409–414

Béress L, Wunderer G, Wachter E (1977) Amino acid sequence of toxin III from Anemonia sulcata. Hoppe Seylerś Z. Physiol Chem. 358: 985–988

Béress L, Zwick J, Kolkenbrock HJ et al (1983) A method for the isolation of the caribbean Palytoxin (C-PTX) from the coelenterate (zoanthid) Palythoa caribaeorum. *Toxicon* 21: 285–290

Béress L, Doppelfeld IS, Etschenberg E et al (1985) Polypeptide, Verfahren zu ihrer Herstellung und ihre Verwendung als blutdrucksenkende Wirkstoffe, patent DE 33 24 689 A1

Bergmann C, Dubois JM, Rojas E, Rathmayer W (1976) Decreased rate of sodium conductance inactivation in the node of Ranvier induced by polypeptide toxin from sea anemone. Biochem. Biophys. Acta 455: 173–184

Bernheimer AW, Avigard LS (1976) Properties of a toxin from the sea anemone Stoichatis helianthus including specific binding to sphingomyelin. Proc. Natl Acad. Sci. USA 73: 467–471

Castañeda O, Soltalongo V, Amon AM et al (1995) Characterization of a potassium channel toxin from the Caribbean sea anemone Stichodactyla helianthus. Toxicon 33: 603–613

Catterall WA, Béress L (1978) Sea anemone toxin and scorpion toxin share a common receptor site associated with the action potential sodium ionophore. JBiol. Chem. 253: 7393–7396

Catterall WA (1988) Structure and function of voltage sensitive ion channels. Science 242: 50–61

Cline EL, Wiebe LI, Young JD, Samuel J (1995) Toxic effect of a novel protein UpI from the sea anemone Urtzicina pisivora. Pharm. Res. 32: 309–314

Cohen SA, Barchi RL (1993) Voltage dependent sodium channels. Int Rev Cytol 137: 55–103

Conti F, Hille B, Neumcke B et al (1976) Conductance of the sodium channel in myelinated nerve fibres with modified sodium activation. J. Physiol. 262: 729–742

Couraud F, Rochat H, Lissitzky S (1978) Binding of scorpion and sea anemone neurotoxins to a common site related to the action potential Na+ ionophore in neuroblastoma cells. Biophys. Res. Commun. 83: 1525–1530

Devlin JP (1974) Isolation and partial purification of hemolytic toxin from sea anemones Stoichactis helianthus. J. Pharm. Sci. 63: 1478–1480

Driscoll PC, Clore GM, Béress L, Gronenborn AM (1989a) A proton nuclear magnetic resonance study of the antihypertensive and antiviral protein BdS I from the sea anemone Anemonia sulcata: sequential and stereospecific resonance assignment and secondary structure. Biochemistry 28: 2178–2187

Driscoll PC, Gronenborn AM, Béress L, Clore GM (1989b) Determination of the three-dimensional solution structure of the antihypertensive and antiviral protein BdS I from the sea anemone Anemonia sulcata: a study using nuclear resonance and hybrid distance geometry-dynamical simulation anneling. Biochemistry 28: 2188–2198

Frelin C, Vigne P, Schweitz H, Lazdunski M (1984) The interaction of sea anemone and scorpion neurotoxins with tetrodotoxin-resistant Na+ channels in rat myoblasts. A comparison with Na+ channels in other excitable and nonexcitable cells. Mol. Pharmacol. 26: 70–74

Fritz H, Brey B, Béress L (1972) Polyvalente Isoinhibitoren für Trypsin, Chymotrypsin, Plasmin und Kallikrein aus Seeanemonen (Aktinaria), Isolierung, Hemmverhalten und Aminosäurezusammensetzung. Hoppe-Seyler' s Z. physiol. Chem. 353: 19–30

Görög T, Bruhn T, Draus E, Béress L (1995) Isolation of crab paralysing polypeptides from the venom of the scorpion Leiurus quinquestriatus hebraeus. Toxicon 33: 295

Halstead BW (1965) Phylum coelenterata. Poisonous and venomous marine animals of the world, 1st edn (Halstead BW, ed) vol 1: pp 297–371: Washington D.C.: United States Government Printing Office

Harvey AL, Rowan EG, Vatanpour H et al (1996) Potassium channel neurotoxins from sea anemones. Biochemical Aspects of Marine Pharmacology, 1st edn (Lazarovici P, Spira ME, Zlotkin E, eds) pp 121–131: Fort Collins, Colorado: Alaken

Hessinger D, Lenhoff HM (1974) Degradation of red cell membrane phospholipids by sea anemone nematocyst venoms. Toxicon 12: 379–383

Kem WR (1988a) Peptide chain toxins of marine animals. Biomedical importance of marine organisms (Fautin D, ed) Mem. Calif. Acad. Sci. 13: 69–83

Kem WR (1988b) Sea anemone toxins: structure and action. The Biology of Nematocyts, (Hessinger D, Lenhoff H, eds) pp 375–405: New York: Academic Press

Kem WR, Pennington MW, Dunn BM (1990) Sea anemone polypeptide toxins affecting sodium channels: initial structure-activity investigations. Marine Toxins: Origins, Structure, and Pharmacology, (Hall S, Strichartz G, eds) pp 279–289: Washington D.C.: Am. Chem. Soc.

Kem WR, Pennington MW, Krafte DS, Hill RJ (1996) Sea anemone toxins affecting sodium channels: are the similarities greater than the differences? Biochemical Aspects of Marine Pharmacology, 1st edn (Lazarovici P, Spira ME, Zlotkin E, eds) pp 98–120: Fort Collins, Colorado: Alaken

Kolkenbrock H, Tschesche H (1987) A new inhibitor of elastase from the sea anemone Anemonia sulcata. Hoppe-Seyler' s Z. physiol. Chem. 368: 93–99

Lazdunski M, Renaud JF (1982) The action of cardiotoxins on cardiac plasma membranes. Ann. Rev. Physiol. 44: 463–473

Lazdunski M, Balerna M, Chicheportiche R et al (1980) Marine neurotoxins to study the voltage dependent sodium channel in excitable membranes. Neurotoxins: fundamental and clinical advances (Chubb IW, Geffen LB) pp 111–119: Adelaide: Adelaide Univ. Union Press

Lazdunski M, Frelin C, Barhanin J et al (1987) Polypetide toxins as tools to study voltage sensitive Na+ channels. Ann N. Y. Acad. Sci. 479: 204–220

Lilleheil G, Shapiro BI (1969) The mechanism of plateau formation by anemone toxin. Comp. Biochem. Physiol. 30: 281–293

Loret EP, Del Valle RMS, Manbuelle P et al (1994) Positively charged amino acid residues located similarly in sea anemone and scorpion toxins. J. Biol. Chem. 269: 16785–16788

Mahnir VM, Kozlovskaya EP, Kalinnovsky AI (1992) Sea anemone Radianthus macrodactylus: a new source of Palytoxin. Toxicon 30: 1449–1456

Malpezzi EL, De Freitas JC, Muramato K, Kamiya H (1993) Characterization of polypetides in sea anemone venom collected by a novel procedure. Toxicon 31: 853–864

Martinez G, Kopeyan C, Schweitz H, Lazdunski M (1977) Toxin III from Anemonia sulcata: Primary structure. FEBS Lett 83: 247–252

Meinardi E, Florin-Christensen M, Paratcha G et al (1995) The molecular basis of the self/non self selectivity of a coelenterate toxin. Biochem. Biophys. Res. Comm. 216: 348–354

Murayama K, Abbot NJ, Narahashi T, Shapiro BI (1972) Effects of allethrin and Condylactis toxin on the kinetics of sodium conductance of crayfish axon membranes. Comp. Gen. Pharmacol. 3: 391–400

Narahashi T, Moore JW, Shapiro BI (1969) Condylactis toxin: interaction with nerve membrane ionic conductances. Science 163: 680–681

Norton RS (1991) Structure and structure-function relationship of sea anemone proteins that interact with the sodium channel. Toxicon 29: 1051–1084

Norton TR (1981) Cardiotonic polypeptides from Anthopleura xanthogrammica (Brandt) and A. elegantissima (Brandt). Fed. Proc. 40: 21–25

Norton TR, Shibata S, Kashiwagi M, Bentley I (1976) Isolation and characterization of the cardiotonic polypeptide anthopleurin-A from the tentacles of the sea anemone Anthopleura xanthogrammica. J. Pharm. Sci. 65: 1368–1374

Norton TR, Kashiwagi M, Shibata S (1978) Anthopleurin A, B and C cardiotonic polypeptides from the sea anemones Anthopleura xanthogrammica (Brandt) and Anthopleura elegantissima (Brandt). "Drugs and foods from the sea, myth or reality?" (Kaul PN, Sindermann CJ, eds) pp 37–50: University of Oklahoma Press

Pennington MW, Byrnes ME, Zaydenberg I et al (1995) Chemical synthesis and characterization of ShK toxin a potent potassium channel inhibitor from a sea anemone. Int. J. Peptide Protein Res. 46: 354–358

Rathmayer W (1979) Sea anemone toxins: Tools in study of excitable membranes. Advances in Cytopharmacology, vol. 3 (Ceccarelli B, Clementi F) pp 335–344: New York: Raven Press

Rathmayer W, Béress L (1976) The effect of toxins from Anemonia sulcata (coelenterata) on neuromuscular transmission and nerve action potentials in crayfish (Astacus leptodactylus). J. Comp. Physiol. 109: 373–382

Rathmayer W, Jessen B, Béress L (1975) Effects of toxins of sea anemones on neuro muscular transmission. Naturwissenschaften 62: 538–539

Ravens U (1976) Electromechanical studies of an Anemonia sulcata toxin in mammalian cardiac muscle. Naunyn-Schmiedebergś Arch. Pharmacol. 296: 73–78

Reimer NS, Yasunobu CL, Yasunobu KT, Norton TR (1985) Aminoacid sequence of the Anthopleura xanthogrammica heart stimulant, anthopleurin-B. J. Biol. Chem. 260: 8690–8693

Renaud J-F, Fosset M, Schweitz H, Lazdunski M (1986) The interaction of polypeptide neurotoxins with tetrodotoxin-resistant Na+ channels in mammalian cardiac cells. Correlation with inotropic and arrhytmic effects. Eur. J. Pharmacol. 120: 161–170

Richet C (1903) Des poisons contenue dans les tentacles des actinies (Congestine et Thalassine). Compt. Rend. Soc. Biol. 55: 701–707

Romey G, Lazdunski M (1975) Scorpion and sea anemone neurotoxins action on axonal membranes. Proceedings of the fifth International Congress on Biophysics, p 503: Copenhagen

Romey G, Abita JP, Schweitz H et al (1976) Sea anemone toxin A tool to study molecular mechanisms of nerve conduction and exhibition secretion coupling. Proc. Natl. Acad. Sci. USA 73: 4055–4059

Romey G, Jacques Y, Schweitz H et al (1979) The sodium channel in non-impulsive cells. Interaction with specific neurotoxins. Biochim. Biophys. Acta 556: 344–353

Romey G, Renaud JF, Fosset M, Lazdunski M (1980) Pharmacological properties of the interaction of a sea anemone polypeptide toxin with cardiac cells in culture. J. Pharmacol. Exp. Ther. 213: 607–615

Sánchez J, Bruhn T, Aneiros A et al (1996) A simple method in the search for bioactive polypeptides in a sea anemone (Anemonia sulcata). Toxicon 34: 1361–1366

Scheffler JJ, Tsugita A, Linder G et al (1982) Amino acid sequence of toxin V from Anemonia sulcata. Biochem. Biophys. Res. Comm. 107: 272–278

Schweitz H, Bidard J-N, Frelin C et al (1985) Purification, sequence, and pharmacological properties of sea anemone toxins from Radianthus paumotensis. A new class of sea anemone toxins acting on the sodium channel. Biochemistry 24: 3554–3561

Schweitz H, Vincent JP, Barhanin J et al (1981) Purification and pharmacological proberties of eight sea anemone toxins from Anemonia sulcata, Anthopleura xanthogrammica, Stoichactis giganteus, and Actinodendron plumosum. Biochemistry 20: 5245–5252

Schweitz H, Bruhn T, Guillemare E et al (1995) Kalicludins and kaliseptine: two different classes of sea anemone toxins for voltage-sensitive K+ channels. J. Biol. Chem. 270: 25121–25126

Shapiro BI (1968) Purification of a toxin from the tentacles of the sea anemone Condylactis gigantheа. Toxicon 5: 253–258

Shapiro BI, Lilleheil G (1969) The action of sea anemone toxine on crustacean neurons. Comp. Biochem. Physiol. 28: 1225–1241

Shibata S, Norton TR, Izumi T (1976) A polypeptide (AP-A) from sea anemone (Anthopleura xanthogrammica) with potent inotropic action. J. Pharmacol. Exp. Ther. 199: 298–309

Strichartz GR, Wang GK (1986) Rapid voltage-dependent dissotiation of scorpion α-toxins coupled to Na channel inactivation in amphibia myelinated neurons. J. Gen. Physiol. 88: 413–435

Tanaka M, Honin M, Yasunobu KT, Norton TR (1977) Amino acid sequence of the Anthopleura xanthogrammica heart stimulant, Anthopleurin A. Biochemistry 16: 204–208

Tazieff-Depierre F, Choncavy M, Métezeau P (1976) Properties pharmacologiques des toxins isolées de lánemone de mer. C.R. Acad. Sci. Paris 283: 699–702

Tschesche H, Kolkenbrock HJ (1984) An inhibitor of elastase from Anemonia sulcata. Chemistry of peptides and proteins, vol 2 (Voelter W, Bayer E, Ovchinnikov YA, Wünsch E, eds) pp 349–355: Berlin: Walter de Gruyter

Tudor JE, Pallaghy PK, Pennigton MW, Norton RS (1996) Solution structure of ShK toxin, a novel potassium channel inhibitor from a sea anemone. Nature Struct Biol 3: 317–320

Vincent JP, Balerna M, Barhanin J et al (1980) Binding of sea anemone toxin to receptor sites associated with gating system of sodium channels in synaptic nerve endings in vitro. Proc. Natl. Acad. Sci. USA 77: 1646–1650

Wachter E, Klostermann G, Binder A, Béress L (1987) Isolation of isotoxins from the sea anemone Anemonia sulcata by HPLC chromatography. Receptors and Ion Channels (Ovchinnikov YA, Hucho F, eds) pp 313–322: Berlin: Walter de Gruyter

Warashina A, Fujita S (1983) Effects of sea anemone toxins on the sodium inactivation process in crayfish axons. J. Gen. Physiol. 81: 305–323

Warashina A, Ogura T, Fujita S (1988) Binding properties of sea anemone toxins to sodium channels in the crayfish giant axon. Comp Biochem Physiol 90: 351–359

Wunderer G, Eulitz M (1978) Amino acid sequence of toxin I from Anemonia sulcata. Eur. J. Biochem. 89: 11–17

Wunderer G, Machleidt W, Wachter E (1976) Toxin II from Anemonia sulcata: the first sequence of a coelenterate toxin. Hoppe Seylers Z. Physiol. Chem. 357: 239–240

Yost GA, ÓBrian RD (1978) Isolation of the two components of Condylactis toxin. Arch. Biochem. Biophys. 185: 483–487

4 Nemertine Toxins

William R. Kem

Contents

1 Introduction

1.1 Nemertine biology

Nemertines are a generally toxic phylum of carnivorous worms which are rarely observed by marine biologists as they are generally of small size (0.5–3 cm is a common length) and usually hide under rocks or burrow in sand and mud. Over 800 species have been described by taxonomists (Gibson, 1972). Most inhabit the intertidal zone, although others occur in a variety of other habitats including land, freshwater and the open sea. Being soft-bodied and unable to rapidly escape from predators, they produce integumentary toxins which serve as a chemical defense against predators (Kem, 1971, 1985, 1988a). The phylum is roughly divided into two large groups, the hoplonemertines which possess a proboscis stylet (mineralized armature), and the anoplans, which lack a stylet. The hoplonemertine stylet is used to puncture the skin of the prey (usually annelids or crustaceans) at multiple sites, thus facilitating entry of toxins into the circulatory system of the organism. The relative scarcity of nemertines in most marine habitats makes them relatively difficult to collect. Another problem is species identification, as the external morphologies of some species are so similar that the preparation of tissue sections for histological examination may be necessary. Nonetheless, this phylum represents an unusually rich source of toxins, including alkaloids, peptides, and proteins, most still awaiting investigation.

Methods and Tools in Biosciences and Medicine
Animal Toxins, ed. by H. Rochat and M.-F. Martin-Eauclaire

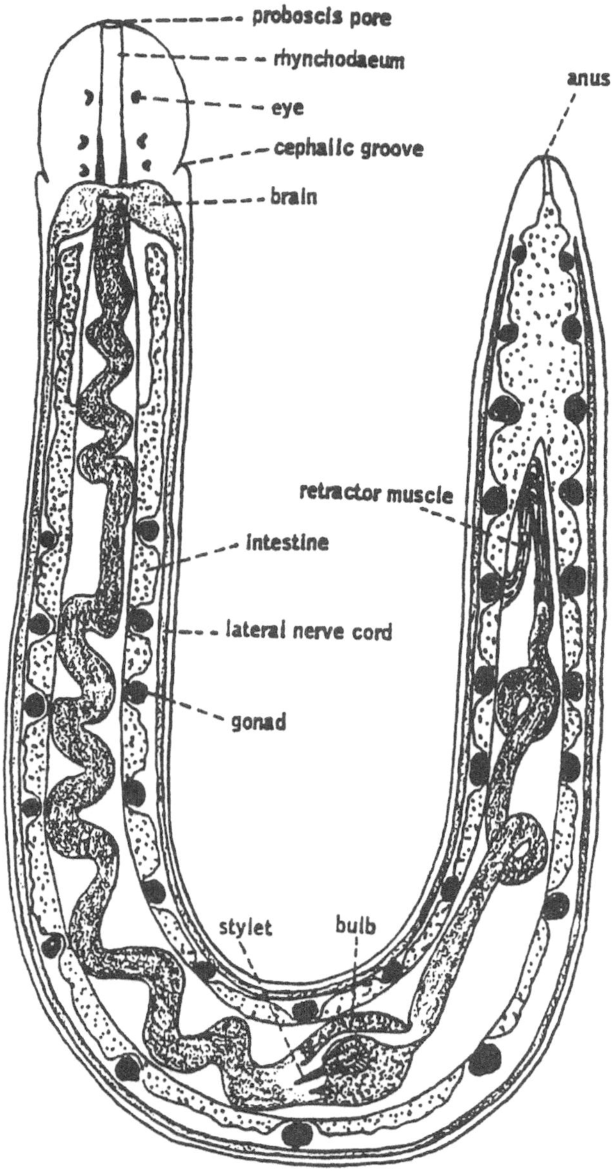

Figure 1 Anatomy of an Enoplan nemertine showing the three regions of its proboscis.

The median proboscis contains a mineralized stylet which is used to puncture the skin of the prey organism, while the anterior proboscis epithelial glands synthesize and secrete alkaloid toxins. The integument covering the rest of the worm is continuous with the secretory epithelium of the proboscis and also produces toxins used as a chemical defence against predators.

1.2 Nemertine toxins

The Belgian pharmacologist Bacq discovered the existence of toxins in nemertines during the mid-1930s while searching for invertebrate neurotransmitters. An aqueous homogenate of a small littoral nemertine (*Amphiporus lactifloreus*), like acetylcholine (ACh), contracted isolated frog skeletal muscle and stimulated the cat cervical autonomic ganglion. However, unlike ACh, this activity was stable in highly alkaline solution and was soluble in organic solvents under basic conditions. On the basis of this limited profile, Bacq (1937) correctly guessed that "amphiporine" was an alkaloid similar to nicotine! In the same study, he also observed neurotoxic activity in several anoplans which differed from "amphiporine" activity in lacking nicotinic effects and in only slowly passing across a dialysis membrane.

Over 30 years passed before another study was made on nemertine toxins (Kem, 1969). During this period of time many new chromatographic, amino acid analysis, peptide sequencing and spectroscopic methods for structure elucidation were introduced. This made it possible to work with smaller quantities of natural products. Even in the 1960s, it was necessary to focus upon abundant and relatively toxic nemertine species so that sufficiently large toxin samples could be obtained for structural and pharmacological characterization. HPLC systems were not yet available. Proton NMR spectroscopy required several milligrams of a small compound, whereas today only several micrograms would be necessary. Mass spectrometers were very sensitive, but the types of spectra which could be recorded were still limited, and MS peptide sequencing was impractical. Nevertheless, it was possible to demonstrate that the hoplonemertine half of the phylum possesses a variety of pyridine alkaloids and to identify several compounds (Kem, 1971; Kem *et al.*, 1976; Kem, 1988c). In contrast, the anoplan nemertines were found to contain peptide and protein toxins (Kem, 1971, 1985, 1988a).

This chapter will describe the known nemertine toxins and suggest some potential uses for them in neurobiological research and medicine.

2 Enoplan (Hoplonemertine) Toxins

2.1 Anabaseine

Anabaseine was first isolated from a Pacific nemertine, *Paranemertes peregrina* (Kem et al., 1971). The "peregrine" designation aptly describes this moderately large (length >15 cm) species, since it openly glides along the surface of the mud flat at low tide searching for its annelid prey, in full view of potential predators such as sea gulls, raccoons, and other large predators. Later, this compound was also found in certain ant venoms (Wheeler et al., 1981). While anabaseine is chemically similar to anabasine, an alkaloid occurring in tobacco plants, it differs by possessing an imine double bond in the otherwise saturated ring (Fig. 2). An imine-enamine tautomerism makes the beta-carbon of the tetrahydropyridyl ring lie within the same plane as the alpha-carbon and the imine N. This system, in turn, is conjugated with the pi electrons of the pyridyl ring. Therefore, the two rings of anabaseine are approximately co-planar. This

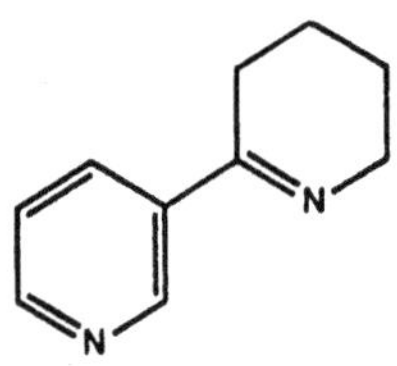

Anabaseine

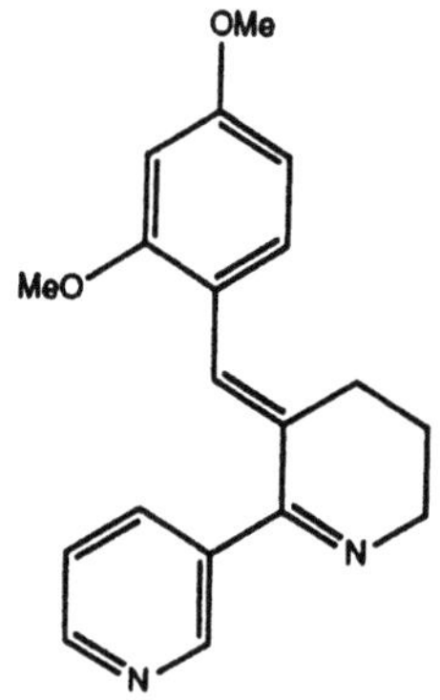

DMXB-Anabaseine (GTS-21)

Figure 2 Covalent structures of anabaseine (*Paranemertes* toxin) and its 2,4-dimethoxybenzylidene derivative, GTS-21.

While anabaseine stimulates a wide variety of vertebrate nicotinic receptors, the latter compound selectively stimulates alpha7-type nicotinic receptors occurring in the mammalian brain. It is currently undergoing clinical tests for possible use in treating neurodegenerative diseases.

contrasts with nicotine and anabasine, whose two rings are approximately at right angles with respect to each other in aqueous solution.

First we consider the synthesis of the compound and its salient chemical properties. Then we shall summarize its most important pharmacological properties. Two other nemertine alkaloids which have received much less attention will also be presented. Finally, a drug candidate (the synthetic anabaseine analog, DMXB-anabaseine) for treatment of neurological disorders will be described.

Synthesis

Anabaseine was first obtained as a intermediate in the synthesis of anabasine by two Austrian tobacco chemists (Spath and Mamoli, 1936). A mixed aldol-like condensation reaction between nicotinic acid ethyl ester and N-benzoyl piperidone yielded the expected diketone, which rearranged in the presence of concentrated HCl at high temperature to anabaseine hydrochloride. Conversion of the salt to the free base, extraction of the free base with organic solvent, and purification by distillation or chromatography generally provided anabaseine in a very low yield (Kem et al., 1971; Kem, 1973). Subsequently, several modifications have been made to provide a more efficient synthesis and isolation (Bloom, 1990; Zoltewicz and Cruskie, 1995); these papers should be consulted by those wishing to prepare the compound, which is not yet commercially available. The protection of the piperidone nitrogen can be accomplished with a variety of chemical groups. Direct crystallization of the ammonium-ketone open-chain form as a dihydrochloride salt from the final reaction was found to be much more efficient than the older workup methods. Synthetic anabaseine dihydrochloride (MW 251) obtained in this manner exists as the ammonium-ketone form and possesses one molecule of water. While stable as the dried salt, to avoid any decomposition aqueous solutions of the toxin should be refrigerated when not in use and replaced after 1–2 weeks. The cationic forms of anabaseine are quite soluble in protic solvents such as water, methanol, and ethanol, but the less hydrophilic free base is best dissolved in nonaqueous solvents such as alcohols, acetone, or ethyl acetate.

Chemical properties of anabaseine

The chemical simplicity of anabaseine is more apparent than real, as this compound occurs in several different forms in the presence of water (Kem, 1969). At neutral pH there are three forms in roughly equal concentrations (Zoltewicz et al., 1989). These are the free base (cyclic imine), the monocationic cyclic iminium, and the monocationic ammonium-ketone. This multiplicity complicated our initial attempts at determining which of these forms interact with nicotinic receptors based upon the pH dependence of anabaseine potency, so stable analogs of each form were prepared so that their pharmacological properties could be examined. The 2,3'-bipyridyl, which is predicted to possess chemical conformation similar to the cyclic imine form of anabaseine, was selected as an analog of the unionized form, while 2-(3,4,5,6-tetrahydropyrimidinyl)-3-pyridine, called PTHP, was selected as a stable, permanently ionized form similar to the cyclic iminium form. To obtain stable open-chain forms of anabaseine, the open-chain nitrogen of anabasine was di- or trimethylated. Of the various stable analogs, only PTHP displayed an ability to contract skeletal muscle and to bind to brain nicotinic receptors (Kem et al., 1994). While the unionized and open-chain forms of anabaseine lacked nicotinic phamacological activity, it is possible that they possess activity upon some other receptors.

Pharmacological properties of anabaseine

A survey of the actions of anabaseine upon a variety of vertebrate nicotinic receptors was recently reported (Kem et al., 1997). Like nicotine, anabaseine stimulates all nicotinic receptors to some degree and thus must be classified as a nonselective nicotinic agonist. However, it preferentially stimulates the same nicotinic receptors (skeletal muscle and brain alpha7 subtypes) which display high affinities for the snake toxin alpha-bungarotoxin. In contrast, nicotine preferentially stimulates central neuronal receptors containing beta2 and beta4 subunits. The most interesting difference between these two alkaloids relates to their efficacy of action upon the two major neuronal alpha7 and alpha4-beta2 receptors occurring in the brain. Anabaseine is a full agonist at the alpha7 receptor but a very weak (low efficacy or maximal response) agonist at the alpha4-beta2 subtype of receptor. Nicotine is a partial agonist, relative to ACh, at both of these receptors, but its maximum effect on the alpha4-beta2 receptor is much greater than upon the alpha7 receptor. Since nicotine also binds to alpha4-beta2 receptors at much lower (about 100-fold) concentrations than to the alpha7 receptor, its *in vivo* effects at smoking concentrations are probably mediated through the beta2 subunit-containing receptors.

The whole animal (mouse) toxicity of anabaseine is very similar to that of nicotine (Kem et al., 1976). Because of its lack of receptor selectivity, few *in vivo* studies have been carried out with anabaseine, other than a comparison with nicotine of its ability to cause prostration when injected into the lateral ventricle of the rat brain (Kem et al., 1997). The significantly higher potency of nicotine in causing prostration is consistent with the notion that beta2 subunit-containing nicotinic receptors primarily mediate this behavior. In the rat frontoparietal cortex, anabaseine elevated both norepinephrine and acetylcholine levels but did not affect serotonin and dopamine levels (Summers et al., 1997). Again, these alterations are most likely mediated through nicotinic receptors containing beta2 subunits.

Anabaseine actions on invertebrate nicotinic receptors

Anabaseine also affects a variety of invertebrate nicotinic receptors. Marine annelids which are the usual prey for *Paranemertes* are paralyzed, as are crustaceans and insects. Nicotinic cholinergic receptors primarily reside on central neurons in these organisms. 2,3'-Bipyridyl, a largely unionized analog of anabaseine, is even more active than anabaseine in paralyzing crustaceans (Kem et al., 1976). While it does not cause paralysis, nemertelline (a tetrapyridyl found in *Amphiporus angulatus*), in common with anabaseine and 2,3'-bipyridyl, stimulates an unusual receptor in the stomatogastric muscle of the crayfish, which is apparently a chloride channel (Kem et al., unpublished results). At present this is the only known action of this complex alkaloid, which is the most abundant pyridine in this species of *Amphiporus*. Nemertine body wall muscles also contain nicotinic receptors, but they only respond to anabaseine at exceptionally high concentrations. A resistance to its own toxin is probably advantageous to the worm (Kem, 1997).

Effects on crustacean chemoreceptors

A variety of pyridine compounds including anabaseine and 2,3'-bipyridyl stimulate chemoreceptor neurons in crayfish walking legs which influence feeding behavior (Hatt and Schmiedel-Jacob, 1984). Marine aquarium observations indicated that spiny lobsters attack but subsequently reject living *Amphiporus angulatus*. Anabaseine and 2,3'-bipyridyl were found to be two of the most active compounds in stimulating similar pyridine receptors on spiny lobster leg nerves (Kem, Hatt and Ache, unpublished results). The nemertine alkaloids, by acting upon these chemoreceptors, may act as repellants against certain predators.

2.2 DMXB-Anabaseine (GTS-21): A synthetic anabaseine drug candidate

While anabaseine is a broad spectrum nicotinic agonist, a variety of 3-substituted anabaseines have been found to possess higher nicotinic receptor selectivity (Kem et al., in preparation). Here we shall only consider 3-(2,4-dimethoxybenzylidene)-anabaseine, whose code name is GTS-21 (de Fiebre et al., 1995; Meyer et al., 1997; Kem, 1998). This compound is of special interest because it has been shown to be a "cognition enhancer" as well as neuroprotective, and is being tested in humans.

Chemistry

GTS-21 (Fig. 1) is readily prepared by reaction of 2,4-dimethoxybenzaldehyde with anabaseine in acidic alcohol at elevated temperature, in a manner similar to the preparation of 3-(4-dimethylaminobenzylidene)-anabaseine (Kem, 1971, 1976). The resulting product can be precipitated and recrystallized using less polar solvents. Unlike anabaseine, the 3-arylidene-anabaseines do not hydrolyze to open-chain forms at physiological pH like anabaseine, but they are moderately photolabile and must thus be protected from strong light. In principle, they can adopt two possible conformations with respect to the vinyl portion of the arylidene ring, namely E or Z. By NMR we have shown that the E form is preferred in aqueous solution (Zoltewicz et al., 1993). Only in the presence of intense light does the E to Z conversion become significant. The Z form does not display significant affinity for the alpha4-beta2 receptor (Kem et al., unpublished results).

Pharmacology

In contrast to anabaseine, DMXB-anabaseine is only agonistic upon one known nicotinic receptor subtype, the neuronal homo-oligomeric alpha7 receptor, which is primarily found in the brain. It also acts as an antagonist at the brain alpha4-beta2 nicotinic receptor. Only at much higher concentrations does it act as a weak antagonist at other nicotinic receptors. What makes this compound of considerable interest is its selective stimulation of a CNS receptor whose physiological function has been very difficult to investigate in the past. Initially this receptor was only recognized by its ability to bind alpha-bungarotoxin; much later, after cloning and expression in cultured cells, it was found to be physiologically active as a ligand-gated ion channel with high permeability for calcium ions, even at normal membrane resting potentials. The fact that DMXB-anabaseine is capable of enhancing performance in cognitive tasks indicates that alpha7 receptors can modulate cognition. Additionally, it has also been shown that this anabaseine derivative, like nicotine, enhances auditory gating in mice. Since the effects of both compounds are prevented by prior administration of BTX, alpha7 receptors probabaly mediate this action. Besides affecting cognitive functions, DMXB-anabaseine also displays neuroprotective properties (Martin et al., 1994). Finally, it also is a moderately potent antagonist at mouse 5-HT_3 receptors (Machu et al., 1996).

Behavioral actions

It has been over a decade since a drastic decrease in nicotinic cholinergic receptors was reported in Alzheimer's patients (Whitehouse et al., 1986). This stimulated considerable academic and pharmaceutical interest in the development of nicotinic agonists which could selectively stimulate the remaining brain nicotinic receptors involved in cognitive and other critical mental functions. GTS-21 is a relatively unique drug candidate which posssesses mixed agonist (alpha7)/antagonist (beta2-containing receptors) activity. Its effects upon cognitive behavior have been investigated by four different laboratories using a variety of mammals. Initially it was observed that the compound enhanced passive avoidance performance in rats (Meyer et al., 1994) and active avoidance in mice (Arendash et al., 1995) and acquisition of conditioned eyeblink reflex in aging rabbits (Woodruff-Pak et al., 1994). Memory of more intricate learning tasks such as water and radial maze performance by rats (Bjugstad et al., 1996) and delayed matching by monkeys (Briggs et al., 1997) was also enhanced, suggesting that the compound may be able to enhance cognition in aging humans, particularly Alzheimer's patients. The compound is currently in clinical trials.

3 Toxins of Anoplan Nemertines

3.1 Neurotoxins

There are two taxonomic classes of anoplan nemertines. Paleonemertines, considered as the most primitive representatives, thus far have been found to contain only protein cytolysins of fairly large molecular size; none have been purified. Some heteronemertines, besides containing protein cytolysins, also possess peptide neurotoxins (Kem, 1971). The only neurotoxins which have been isolated belong to a large (>1 m, 20 g) Atlantic coast species, *Cerebratulus lacteus*.

Biochemistry

These so-called *Cerebratulus* B toxins have molecular sizes of ~ 6000, similar to the scorpion neurotoxins, and are cross-linked by four disulfide bonds (Kem, 1976; Blumenthal and Kem, 1977). The sequences (Fig. 3) of the two most

```
                                             10
B-II  Ala-Ser-Ser-Thr-Trp-Gly-Gly-Ser-Tyr-Hyp-Ala-Cys-Glu-Asn-Asn-Cys-Arg-Lys-
B-IV  Ala-Ser-Ala-Thr-Trp-Gly-Ala-Ala-Tyr-Hyp-Ala-Cys-Glu-Asn-Asn-Cys-Arg-Lys-

         20                             30
B-II  Gln-Tyr-Asp-Asp-Cys-Ile-Lys-Cys-Gln-Gly-Lys-Trp-Ala-Gly-Lys-Arg-Gly-Lys-
B-IV  Lys-Tyr-Asp-Leu-Cys-Ile-Arg-Cys-Gln-Gly-Lys-Trp-Ala-Gly-Lys-Arg-Gly-Lys-

                 40                               50
B-II  Cys-Ala-Ala-His-Cys-Ala-Val-Gln-Thr-Thr-Ser-Cys-Asn-Asp-Lys-Cys-Lys-Lys-His
B-IV  Cys-Ala-Ala-His-Cys-Ile-Ile-Gln-Lys-Asn-Asn-Cys-Lys-Gly-Lys-Cys-Lys-Lys-Glu
```

Figure 3 Amino acid sequences of the two *Cerebratulus lacteus* neurotoxins, B-II and B-IV. Identical sequence regions are enclosed within boxes (Blumenthal and Kem, 1976; Blumenthal et al., 1981).

abundant and active isotoxins, B-II and B-IV, were reported some time ago (Blumenthal and Kem, 1976; Blumenthal et al., 1981). Both are very basic peptides and contain a single residue of hydroxyproline at position 10. Unlike the scorpion and sea anemone peptide toxins whose secondary structures are largely composed of antiparallel beta-strands, the B toxins are devoid of beta-sheet structure but rich in alpha-helix (Kem et al., 1990). Recently, the secondary and tertiary structures of B-IV were determined by NMR (Hansen et al., 1992; Barnham et al., 1997). It can be observed that there are two long stretches of helix, represented by positions 11–23 and 34–49. The two helices are connected by a loop consisting of two inverse gamma-turns and a beta-turn. The entire sequence, 11–49, thus constitutes a rather unique helical hairpin structure Fig. 4).

Pharmacology

The two B toxins are potent toxins when injected into crustaceans, especially crayfish. Initially the animal displays tremors (including flipping of the tail), but then convulses in a massive contracture of the limbs and tail. In a few minutes the contractural paralysis is replaced by flaccid paralysis and eventually death. This toxicity of the *Cerebratulus* B toxins (and unpurified *Lineus* toxins) seems confined to the crustacean nervous system. The limb contracture can also be observed in the perfused crayfish cheliped. The nerve terminals seem to be the site of action, since tetrodotoxin effectively blocks the action of toxin B-IV, even when the peptide is applied directly on the muscle. At relatively high (0.1–1.0 μM) concentrations toxin B-IV also affects the evoked compound

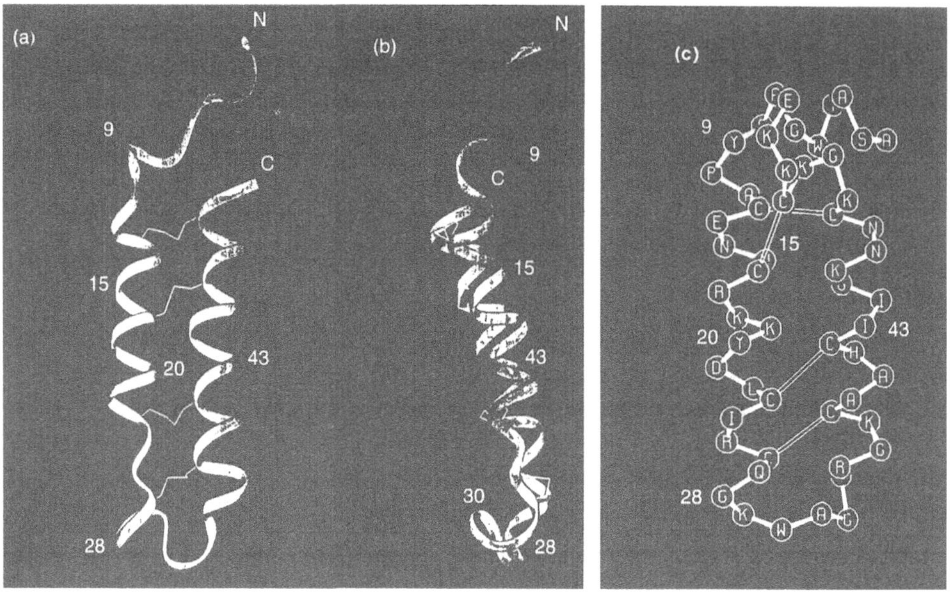

Figure 4 Two different views of the NMR-derived solution backbone structure of *Cerebratulus* neurotoxin B-IV (from Barnham et al., 1997).

The NMR-derived helical hairpin structure shown in parts a and b is very similar to a similar structure (part c) initially predicted from the sequence of B-IV by Dr. Robert Williams, Uniformed Services University, Bethesda, MD. The local structure at the top of part c was incorrect because of an error in the orginal disulfide bond identification, which was corrected by the NMR analysis (see Barnham et al., 1997).

action potential recorded from isolated crab leg nerves, causing some repetitive spiking which also probably contributes to the prolongation of the compound action potential (Kem, unpublished results). The current conception of how the toxins act is that they cause the opening of at least a small population of sodium channels, and this causes neuronal repetititive spiking which causes a massive release of excitatory (and perhaps inhibitory) neurotransmitters at neuromuscular synapes. In arthropods the skeletal muscles are generally electrically inexcitable, but are innervated along their entire length by excitatory glutamatergic and inhibitory GABAergic synapes. A more intensive analysis with intracellular recording techniques would certainly be desirable. B-IV was inactive when tested on a variety of isolated axons and neurons from various noncrustacean groups, including vertebrates and molluscs (Kem, 1976).

Lieberman and Blumenthal (1986) measured the binding of iodinated toxin B-IV to membranes prepared from Maine lobster (*Homarus americanus*) muscle. High affinity binding was observed. The specific binding could not be displaced by scorpion alpha-toxin. Further experiments are required to identify the membrane receptor involved. While they suceeded in chemically cross-linking the toxin to the membranes, the solubilized membrane proteins labelled by the iodinated toxin possessed a smaller molecular size (about 40,000) than would be predicted for the sodium channel alpha-subunit, which contains the binding site for scorpion and sea anemone peptide toxins. This result could be explained in several ways. Membrane proteins can be degraded by endogenous proteases unless a cocktail of inhibitors was added. This seems unlikely to be the reason here, since several inhibitors were added to the incubation saline. Other alternative explanations are that the toxin either binds predominantly to a B subunit of the sodium channel or that it interacts with some other membrane protein, perhaps another ion channel.

Structure-activity studies

Considerable data are available implicating some amino acid side chains in the toxic action of toxin B-I. The initial studies utilized a chemical modification approach and focused upon the few aromatic residues (two tyrosyls, two tryptophanyls). By manipulating the conditions of the reactions, it was possible to differentially label the two Tyr and two Trp residues, the former by nitration and the latter by alkylation. By bioassaying the toxin samples at different degrees of modification, it was deduced that Tyr9 and Trp30 are probably involved in receptor binding, since their modification did not affect secondary structure as measured by CD spectroscopy (Blumenthal and Kem, 1980a; Blumenthal, 1980).

During the past decade Blumenthal's lab has utilized molecular biological methods to express B-IV in *Escherichia coli* and to obtain mutants for structure-activity relationship (SAR) studies (Howell and Blumenthal, 1989, 1991). Initial experiments showed that replacement of the hydroxyproline at position 10 with Pro did not affect crayfish paralytic activity, nor did replacement of Ala residues at either position 3 or 8 with serine. In fact, the toxicity of B-IV was enhanced by simultaneous substitution of serine at positions 3 and 8. Arg17 has been implicated in receptor binding as toxicity was undetectable but the CD spectrum unchanged when glutamine, Ala, or Lys was substituted. Replacement of Arg25 with Lys reduced crayfish toxicity 400-fold (Wen and Blumenthal, 1996). Some of the side chains implicated in toxicity of B-IV include the guanidinyl side chains of Args 17, 25, and 34, and the aromatic side chains of Trp 30, and Tyr 9. It was rather surprising that the implicated residues are found along the entire length of one surface of the toxin. This could imply that the toxin binds to an extensive portion of receptor surface, in contrast with scorpion and sea anemone toxins, which bind to sodium channels but at a different site. Alternately, modification of certain residues such as Tyr9 and Trp30 may have altered the folded structure sufficiently to deleteriously affect activity without affecting the CD spectrum. Experiments with other toxin mutants, coupled with more intensive NMR structural analyses of tertiary structure, should provide further insights regarding the binding surface of this toxin.

3.2 Cytotoxins

Lytic toxins are practically ubiquitous among all living organisms, including bacteria, plants, and animals. Almost all animal venoms include some lytic substances which enhance the penetration of the other toxins into the circulation of the affected organism. So far, cytolytic proteins have only been found in anoplan nemertines.

Biochemistry

The heteronemertine *Cerebratulus* contains at least four homologous protein lysins called A toxins (Kem and Blumenthal, 1978). Only one of these toxins, A-III, has been fully sequenced (Fig. 5) and shown to contain three disulfide bonds (Blumenthal and Kem, 1980b; Blumenthal, 1980b). CD and Raman spectroscopy have revealed the presence of approximately 60% alpha-helix and 10% B-sheet in toxin A-III (Kem et al., in preparation). The C-terminal portion which is not cross-linked by a disulfide bond is thought to exist as a helical hairpin structure. Because of its amphipathic nature this region of toxin sequence may interact with membrane lipids and possibly participate in the formation of a pore.

Pharmacology

At sublytic concentrations *Cerebratulus* toxin A-III blocks the squid axon sodium channel and affects the kinetics of opening and closing of the potassium channel. Then it begins to increase the so-called leakage conductance, which is a measure of membrane integrity (Kem, in preparation). These differential effects on sodium versus potassium channels are interesting and deserve further study.

Perhaps due to its ability to disrupt lipid bilayers, toxin A-III also inhibits brain phospholipid sensitive $Ca2^{+}$-dependent protein kinase *in vitro*, as do other peptide toxins which possess a detergent-like action (Kuo et al., 1983).

The lytic effects of this toxin have been investigated using isolated erythrocyte suspensions and liposomes (Kem and Blumenthal, 1978; Blumenthal, 1982, 1984, 1985; Kem, 1994). Human erythrocytes are much more sensitive than rat cells. However, the latter may be made more sensitive by replacing external ions with uncharged solutes (Liu and Blumenthal, 1988a). The lytic potency of the toxin is also enhanced by binding of oleic acid, a fatty acid liberated by coincubation with phospholipase A2 (Liu and Blumenthal, 1988b). Apparently there are two portions of the toxin sequence which insert into the lipid aspect of the cell membrane. Radiolabelled oleic acid was covalently attached to the toxin in the presence of a carbodiimide (Liu and Blumenthal, 1991). Blumenthal (1982, 1984) also obtained further evidence for this by proteolytic nicking experiments with membrane-inserted toxin. It is thought that

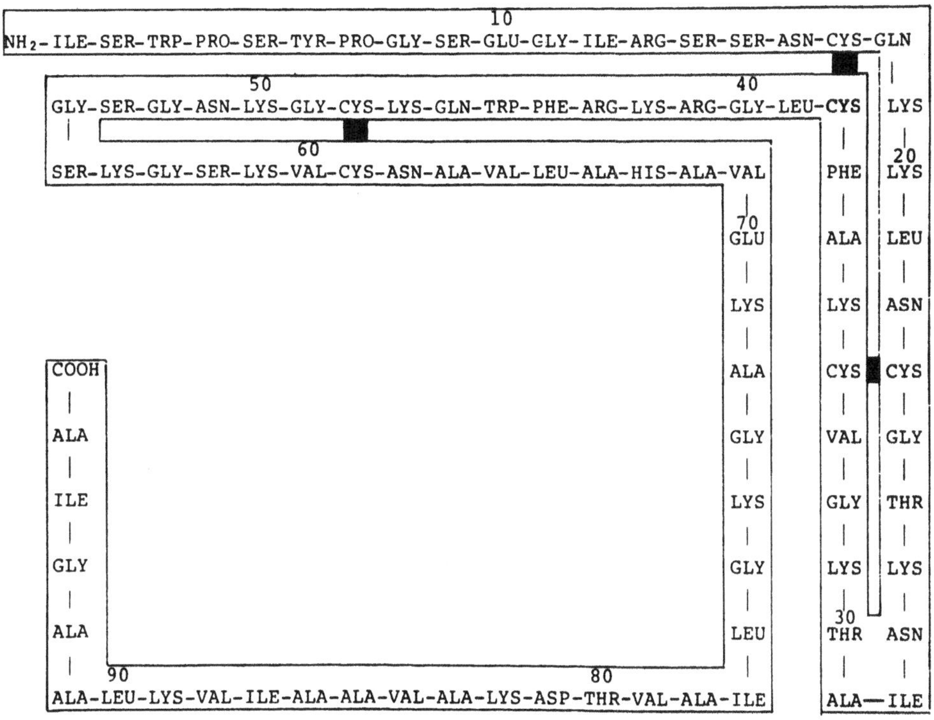

Figure 5 Amino acid sequence of the major *Cerebratulus lacteus* cytolysin, toxin A-III (Blumenthal and Kem, 1980b).

The three disulfide bridges are also highlighted (Blumenthal, 1980b). The C-terminal one-third of this sequence is predicted to exist as an amphipathic helical hairpin which inserts within the lipid bilayer aspect of the cell membrane and in some as yet unknown manner permeabilizes the membrane, possibly through formation of a pore.

the C-terminal third of the molecule forms a helical hairpin structure capable of spanning the entire thickness of the membrane. Further experiments are desirable to test this hypothesis. Considering the relatively small size of this protein, molecular level insights on its mechanism of membrane insertion may be more forthcoming than for endogenous membrane proteins and for the larger bacterial protein cytolysins.

4 Concluding Comments

Clearly one can expect that this invertebrate phylum will not only provide additional variants of the above-mentioned toxins, but also other types of toxins not yet encountered. Besides providing some unique tools for biomedical research, some of the toxins (particularly alkaloids like anabaseine) may become useful models for drug design. Some toxins are also providing chemical and biochem-

Table 1 Comparison of the relative activities of anabaseine, nicotine, and GTS-21 on several vertebrate nicotinic receptors (summarized from results presented in de Fiebre et al., 1995, and Kem et al., 1997).

Receptor type	Anabaseine	Nicotine	GTS-21
Central:			
Alpha7 (Rat)	Full Agonist	Weak Partial Agonist	Partial Agonist
Alpha4-Beta2 (Rat)	Weak Partial Agonist	Strong Partial Agonist	Antagonist
Peripheral:			
Sympathetic (Rat PC12)	Full Agonist	Full Agonist	Weak Antagonist
Skeletal Muscle (Frog)	Full Agonist	Full Agonist	Weak Antagonist

ical structural insights. For instance, the NMR investigation of anabaseine has provided an improved understanding of the stability of tetrahydropyridine compounds under aqueous conditions. Norton's NMR solution structural analysis of *Cerebratulus* toxin B-IV has revealed a paired helical structure which is unique among the known neurotoxins. Although the narrow phylogenetic activity of the *Cerebratulus* B neurotoxins may limit their utility as molecular probes, it seems likely that other homologous variants of these toxins will eventually be found which act upon homologous ion channels of vertebrates and other animals.

Nemertines are a phylum of carnivorous marine worms which produce a variety of alkaloidal, peptidic, and proteinaceous toxins. All of the toxins likely serve as chemical defenses against potential predators; hoplonemertine toxins are also used offensively to capture prey. Only a few species have been investigated to date, so toxins in the remaining 800+ species await investigation. The "amphiporine" activity discovered by Bacq corresponds to pyridine alkaloids with nicotinic properties. One of these, anabaseine, was subsequently found in the venom glands of certain ants. It stimulates a wide variety of mammalian nicotinic receptors, but particularly those blocked by alpha-bungarotoxin. A synthetic derivative, DMXB-anabaseine (also called GTS-21), improves memory in experimental animals and is currently in clinical trials for Alzheimer's disease. Two other pyridines, 2,3'-bipyridyl and the tetrapyridyl nemertelline, are alkaloids which stimulate anion-selective nicotinic receptors in the crustacean foregut musculature and thus presumably interfer with food ingestion. Anoplan species, which lack an armed proboscis, secrete peptide and protein toxins which either target neuronal ion channels or permeabilize and eventually lyse a variety of cells. The *Cerebratulus* B toxins are basic peptides of MW

~6,000 which cause convulsions and paralysis at very low doses, apparently by activating sodium channels at normal resting membrane potentials. These toxins possess predominantly helical secondary structures which are different from other known sodium channel toxins. Several side chains important for toxicity have been identified by site-directed mutagenesis of a synthetic gene for toxin B-IV. The cytolytic A toxins are approximately twice the size of the B toxins and are predicted to possess a helical hairpin structure important for membrane disruption and possibly channel formation. These three classes of toxins, each presenting a unique chemical architecture, can serve as molecular probes or as models for drug design. DMXB-anabaseine (GTS-21) is an example of the latter type compound.

Acknowledgements

The investigations which provided the data referred to in this article were supported by NIH, NSF and Taiho Pharmaceutical Company grants. The author is grateful to the many students, colleagues, and technical personnel who have contributed immeasurably to this endeavor. To stimulate wider use of these toxins as chemical tools, small (milligram) quantities of anabaseine dihydrochloride and other pure nemertine toxins may be obtained from the author, based on the costs of their preparation (it should be noted that these toxins, in milligram amounts, do not pose any threat to human health).

References

Arendash GW, Sengstock GJ, Sanberg R, Kem WR (1995) Improved learning and memory in aged rats with chronic administration of the nicotinic receptor agonist GTS-21. Brain Res 674: 252–259

Bacq ZM (1937) L'"amphiporine" et la "nemertine," poisons des vers nemertiens. Arch. Int. Physiol. 44: 190–204

Barnham KJ, Dyke TR, Kem WR, Norton RS (1997) Structure of neurotoxin B-IV from the marine worm Cerebratulus lacteus: a helical hairpin cross-linked by disulphide bonding. J. Mol. Biol. 268: 886–902

Bjugstad KB, Mahnir VM, Kem WR, Arendash GW (1996) Long-term treatment with GTS-21 or nicotine enhances water maze performance in aged rats without affecting the density of nicotinic receptor subtypes in neocortex. Drug Devel. Res. 39: 19–28

Bloom LB (1990) Influence of solvent on the ring-chain hydrolysis equilibrium of anabaseine and synthesis of anabaseine and nicotine analogues. Ph.D. Dissertation, University of Florida

Blumenthal KM (1980a) Inactivation of Cerebratulus lacteus toxin B-IV concomitant with tryptophan alkylation. Arch. Biochem. Biophys. 203: 822–826

Blumenthal KM (1980b) Disulfide bonds of Cerebratulus lacteus toxin A-III. J. Biol Chem. 255: 8273–8274

Blumenthal KM (1982) Structure and action of heteronemertine toxins: membrane penetration of Cerebratulus lacteus: toxin A-III. Biochem. 21: 4229–4233

Blumenthal KM (1984) Release of liposomal markers by Cerebratulus lacteus toxin A-III. Biochem. Biophys. Res. Commun. 121: 14–18

Blumenthal KM, Kem WR. (1976) Primary structure of Cerebratulus lacteus toxin B-IV. J. Biol. Chem. 251: 6025–6029

Blumenthal KM, Kem WR (1977) Disulfide bonds of Cerebratulus lacteus toxin B-IV. J. Biol. Chem. 252: 3328–3331

Blumenthal KM, Kem WR (1980a) Inactivation of Cerebratulus lacteus toxin B-IV by tyrosine nitration. Arch. Biochem. Biophys. 203: 816–821

Blumenthal KM, Kem WR (1980b) Primary structure of Cerebratulus lacteus toxin A-III. J. Biol. Chem. 255: 8266–8272

Blumenthal KM, Keim PS, Heinrikson RL, Kem WR (1981) Structure and action of heteronemertine polypeptide toxins. Amino acid sequence of Cerebratulus lacteus toxin B-II and revised structure of toxin B-IV. J. Biol. Chem. 256: 9063–9067

Briggs CA, Anderson DJ, Brioni JD et al (1997) Functional characterization of the novel neuronal nicotinic acetylcholine reeceptor ligand GTS-21 Vitro and Vivo. Pharmacol. Biochem. Behav. 57: 231–241

De Fiebre CM, Meyer EM, Henry JC et al (1995) Characterization of a series of anabaseine-derived compounds reveals that the 3-(4)-Dimethylaminocinnamylidine derivative (DMAC) is a selective agonist at neuronal nicotinic alpha 7/[^{125}K] alpha-Bungarotoxin receptor subtypes. Mol. Pharmacol. 47: 164–171

Gibson, R (1972) Nemerteans. London: Hutchinson University Library

Hansen PE, Kem WR, Bieber AL, Norton RS (1992) ^{1}H-NMR study of neurotoxin B-IV from the marine worm Cerebratulus lacteus. Solution properties, sequence-specific resonance assignments, secondary structure and global fold. Eur. J. Biochem. 210: 231–240

Hatt H, Schmiedel-Jacob I (1984) Electrophysiological studies of pyridine-sensitive units on the crayfish walking leg. I. Characteristics of stimulatory molecules. J. Comp. Physiol. 154A: 855–863

Howell ML, Blumenthal KM (1989) Cloning and expression of a synthetic gene for Cerebratulus lacteus neurotoxin B-IV. J. Biol. Chem. 264: 15268–15273

Howell ML, Blumenthal KM (1991) Mutagenesis of Cerebratulus lacteus neurotoxin B-IV identifies NH_2-terminal sequences important for biological activity. J. Biol. Chem. 266: 12884–12888

Kem WR (1969) A Chemical Investigation of Nemertine Toxins. Ph.D. Dissertation, University of Illinois (Urbana)

Kem WR (1971) A study of the occurrence of anabaseine in Paranemertes and other nemertines. Toxicon 9: 23–32

Kem WR (1973) Biochemistry of Nemertine Toxins, Marine Pharmacognosy: Marine Biotoxins as Probes of Cellular Function (Martin, D.F, Padilla, G. M, Eds.). Monographs on Cell Biology Series, pp 37–84: New York Academic Press

Kem WR (1976) Purification and characterization of a new family of polypeptide neurotoxins from the heteronemertine Cerebratulus lacteus (Leidy). J. Biol. Chem. 251: 4184–4192

Kem WR (1985) Structure and action of nemertine toxins. Am Zoologist 2: 99–111

Kem WR (1988a) Worm toxins. Handbook of Natural Toxins, vol 4: Marine Toxins and Venoms (Tu, ed). pp 253–378: Marcel Dekker

Kem WR (1988b) Peptide chain toxins of marine animals. Biomedical Importance of Marine Organisms (Fautin, D, ed) Mem. Calif. Acad. Sci. 13: 69–83

Kem WR (1988c) Pyridine alkaloid distribution in the hoplonemertines. Hydrobiol. 156: 145–151

Kem WR (1994) Structure and membrane actions of a marine worm cytolysin, Cerebratulus toxin A-III. Toxicology 87: 189–203

Kem WR (1998) Alzheimer's drug design based upon an invertebrate toxin (anabaseine) which is a potent nicotinic receptor antagonist. Invertebrate Neurosci (in press)

Kem WR (1997) Nemertine body wall and proboscis longitudinal muscles possess unique nicotinic receptors. Fifth Intern. Conf. Invertebrate Neurochem. Neurophysiol. Mtg, Eilat, Israel (Abstr.)

Kem WR, Abbott BC, Coates RM (1971) Isolation and structure of a hoplonemertine toxin. Toxicon 9: 15–22

Kem WR, Blumenthal KM (1978) Purification and characterization of the cytotoxic Cerebratulus A toxins. J. Biol. Chem. 253: 5725–5757

Kem WR, Scott KN, Duncan JH (1976) Hoplonemertine worms: a new source of pyridine neurotoxins. Experientia 32: 684–686

Kem WR, Tu C-K, Williams RW et al (1990) Circular dichroism and laser Raman spectroscopic analysis of the secondary structure of Cerebratulus lacteus toxin B-IV. J. Prot. Chem. 9: 433–443

Kem WR, Mahnir VM, Bloom LB, Gabrielson BJ (1994) The active form of the nicotinic receptor agonist anabaseine is the cyclic iminium cation. 11th World Congress on Animal, Plant, and Microbial Toxins, Tel Aviv

Kem WR, Mahnir VM, Papke R, Lingle C (1997) Anabaseine is a potent agonist upon muscle and neuronal alpha-bungarotoxin sensitive nicotinic receptors. J. Pharmacol. Exper. Therap. 283: 979–992

Kuo JF, Raynor RL, Mazzei GJ et al (1983) Cobra polypeptide cytotoxin I and marine worm polypeptide cytotoxin A-IV are potent and selective inhibitors of phospholipid sensitive Ca^{2+}-dependent protein kinase. FEBS Lett. 153: 183–186

Lieberman DL, Blumenthal KM (1986) Structure and action of heteronemertine polypeptide toxins. Specific cross-linking of Cerebratulus lacteus toxin B-IV to lobster axon memebrane vesicles. Biochim. Biophys. Acta 855: 1–48

Liu J, Blumenthal KM (1988a) Membrane damage by Cerebratulus lacteus cytolysin A-III: effects of monovalent and divalent cations on A-III hemolytic activity. Biochim. Biophys. Acta 937: 153–160

Liu J, Blumenthal KM (1988b) Functional interaction between Cerebratulus lacteus cytolysin A-III and phospholipase A2: implications for the mechanism of cytolysis. J. Biol. Chem. 263: 6619–6624

Liu J, Blumenthal KM (1991) Identification of oleic acid binding sites in cytolysin A-III from the heteronemertine Cerebratulus lacteus. Toxcon 29: 13–20

Machu TK, Strahlendorf J, Kem WR (1996) Nicotinic receptor ligands antagonize 5-HT_3 receptors expressed in Xenopus oocytes. J. Neurosci. 22: 1780

Mahnir VM, Lin B, Prokai-Tatrai K, Kem WR (1998) Pharmacokinetics and urinary excretion of DMXBA (GTS-21), a compound enhancing cognition. Biopharm. Drug Dispos (in press)

Martin EJ, Panickar KS, King MA et al (1994) Cytoprotective actions of 2: 4-dimethoxybenzylidene anabaseine in differentiated PC12 cells and septal cholinergic neurons. Drug Dev. Res. 31: 135–141

Meyer EM, de Fiebre CM, Hunter BE et al (1994) Effects of anabaseine related analogs on rat brain nicotinic receptor binding and on avoidance behavior. Drug Dev. Res. 31: 135–141

Meyer EM, Tay ET, Papke RL et al (1997) 3-[2: 4-Dimethoxybenzylidene]anabaseine (DMXB) selectively activates rat alpha7 receptors and improves memory-related behaviors in a mecamylamine-sensitive manner. Brain Res. 768: 49–56

Spath E, Mamoli L (1936) Eine Neue Synthese Des D,L-Anabasins. Chem. Ber. 69: 1082–1085

Summers K, Kem WR, Giacobini E (1997) Nicotinic agonist modulation of neurotransmitter levels in the rat frontoparietal cortex. Jap. J. Pharmacol. 74: 139–146

Wen PH, Blumenthal KM (1996) Role of electrostatic interactions in defining the potency of neurotoxin B-IV from Cerebratulus lacteus. J. Biol. Chem. 271: 29752–29758

Wheeler JW, Olubajo O, Storm CB, Duffield RM (1981) Anabaseine: venom alkaloid of Aphaenogaster ants. Science 211: 1051–1052

Whitehouse RJ, Price DL, Clark AW et al (1986) Nicotinic acetylcholine binding in Alzheimer's disease. Brain Res. 371: 146–151

Woodruff-Pak DS, Li Y-T, Kem WR (1994) A nicotinic receptor agonist (GTS-21), eyeblink classical conditioning, and nicotinic receptor binding in rabbit brain. Brain Res. 645: 309–317

Zoltewicz JA, Cruskie MP Jr (1995) A superior synthesis of cholinergic anabaseine. OPPI Briefs 27: 510–513

Zoltewicz JA, Bloom LB, Kem WR (1989) Quantitative determination of the ring-chain hydrolysis equilibrium constant for anabaseine and related tobacco alkaloids. J. Org. Chem. 54: 4462–4468

Zoltewicz JA, Prokai-Tatrai K, Bloom LB, Kem WR (1993) Long range transmission of polar effects of cholinergic 3-arylideneanabaseines. Conformations calculated by molecular modelling. Heterocycles 35: 171–179

5 Conus Peptides and Their Iodinated Derivatives as Probes for Ion Channels and Receptors

Lourdes J. Cruz, J. Michael McIntosh, Julita S. Imperial and William R. Gray

Contents

1 Introduction

The genus of venomous marine snails known as *Conus* has been a very rich resource for the development of biochemical tools in cell biology. Since the isolation of α-conotoxin GI from the venom of the fish-hunting species *Conus geographus* (Cruz et al., 1978), many other peptides have been purified from its venom. Among these are several peptides that inhibit acetylcholine receptors (α-conotoxins), peptides that block voltage-sensitive sodium channels (μ-conotoxins), peptides which block voltage-sensitive calcium channels (ω-conotoxins), an inhibitor of the NMDA receptor (conantokin-G), and an agonist of the vasopressin receptor (conopressin) (Gray et al., 1988; Olivera et al., 1990; Myers et al., 1993). Examination of the venom components of other species has revealed new classes of peptides with different modes of action.

Conus peptides are widely used as ligands in pharmacology and neuroscience because of their high selectivity and strong affinity for ion channels and receptors. Noteworthy in this respect are the ω-conotoxins, which selectively inhibit different Ca channel subtypes at the presynaptic termini, thereby inhibiting neurotransmitter release (Olivera et al., 1994). Likewise, μ-conotox-

Methods and Tools in Biosciences and Medicine
Animal Toxins, ed. by H. Rochat and M.-F. Martin-Eauclaire

in GIIIA has been reported as the most specific ligand for site 1 of the skeletal subtype of voltage-sensitive sodium channels (Cruz et al., 1985; Gonoi et al., 1987). It discriminates between the TTX-sensitive and TTX-insensitive skeletal sodium channels 50 times better than tetrodotoxin itself.

Each *Conus* venom has a profile of about 50–200 different neuroactive peptides. The majority of the peptides are relatively small (consisting of about 12–30 amino acid residues) and rich in disulfide bonds. In addition, *Conus* venoms contain a heterogeneous group of peptides that are not disulfide-rich (e.g., the conantokins), large polypeptides (>10 kDa), and small molecules such as biologically active amines. The word "conotoxin" has been used as a generic term for the multiply disulfide-bonded *Conus* neuropeptides. They are folded in a compact structure where the cysteine-rich backbones serve as scaffolds for the orientation of amino acid side groups that interact with the target site at the ion channel or receptor. It should be noted that many "conotoxins" are not "toxic" in the sense of being able to produce death or severe disability. Indeed, a number of "conotoxins" are being investigated for therapeutic uses in humans; one such peptide, ω-conotoxin MVIIA, will soon be available for treatment.

The relatively small size of conotoxins provides a big advantage with respect to chemical synthesis. The synthetic ability is particularly important if the peptide is present only in very limited amounts in the venom or if its primary structure is obtained by complementary DNA (cDNA) sequencing. Once the synthetic material is available, the conotoxin can be derivatized for specific experimental purposes. Radioiodinated conotoxins are used for binding experiments (e.g., Cruz et al., 1986; Cruz et al., 1987), in autoradiography to identify the anatomical location of the target (e.g., McIntosh et al., 1992; Albensi et al., 1993), and for cross-linking studies (e.g., Myers et al., 1991) to identify the macromolecular target. Likewise, fluorescently labeled conopeptides have been used to localize the distribution of macromolecules specifically targeted by the peptide (e.g., Mills et al., 1994; Haugland, 1996).

In this chapter, we present a list of macromolecular targets that can be studied using *Conus* peptides. A list of known conotoxin sequences and commercially available peptides and derivatives is given. We expect the list to expand continuously as more and more of the 500 *Conus* species are studied by various research groups. Finally, some general considerations in handling the peptides and procedures for preparing radioiodinated derivatives of the conopeptides are presented.

2 *Conus* peptides and their macromolecular targets

So far, the list of known macromolecular targets of over 40 *Conus* peptides includes six types of ion channels and receptors (see Tab. 1). Tables 2 and 3 give

Table 1 Classes of *Conus* Peptides and Their Macromolecular Targets

Target	Class	Mode of Action
Acetylcholine receptor	α-Conotoxins	Competitive inhibitor
	αA-Conotoxins	Competitive inhibitor
	ψ-Conotoxins	Noncompetitive inhibitor
Sodium channels	μ-Conotoxins	Blocker; competes with STX and TTX for site 1
	μO-Conotoxins	Blocks current; does not compete with STX and TTX
	δ-Conotoxins	Delays inactivation of Na channels by binding to site VI
Calcium channels	ω-Conotoxins	Block current of specific subtypes
Potassium channels	κ-Conotoxins	Blocker
NMDA receptor	Conantokins	Inhibitor
Vasopressin receptor	Conopressins	Agonist
Neurotensin receptor	Contulateins	Agonist (Craig et al., 1999)

the primary structures of conotoxins (Myers et al., 1993) and their subtype specificities and/or site preferences. ω-Conotoxin GVIA of *Conus geographus* remains the most widely used tool for specifically blocking N-type calcium currents (Olivera et al., 1994). It has the same specificity as ω-conotoxin MVIIA of *C. magus* in the $\alpha 1_B$ subunit of the N-type channel. On the other hand, ω-conotoxins SVIB of *C. striatus* and MVIIC and MVIID of *C. magus* block P/Q and N-type cal-cium channels (Hillyard et al., 1992; Monje et al., 1993). Among the μ-conotoxins, GIIIA, GIIIB and GIIIC interact with site 1 of skeletal muscle Na channel (μ1 subtype), whereas μ-conotoxin PIIIA of *C. purpurascens* acts on rat brain type II Na channel (Cruz et al., 1985; Shon et al., 1998). Several conotoxins discovered more recently (μO and δ-types) act on sodium channels of molluscan neurons as well as vertebrates (Fainzilber et al., 1995; McIntosh et al., 1995; Terlau et al., 1996; Shon et al., 1994; Woodward et al., 1990; Fainzilber et al., 1994). Of the *Conus* peptides that act on acetylcholine receptors, several α and αA-conotoxins inhibit the skeletal muscle nACh receptor (Gray et al., 1981; McIntosh et al., 1982; Zafaralla et al., 1988; Myers et al., 1991; Ramilo et al., 1992; Martinez et al., 1995; Hopkins et al., 1995; Jacobsen et al., 1997), and a few act on neuronal subtype of ACh receptors (Cartier et al., 1996; Fainzilber et al., 1994).

Preparation of radiolabeled or fluorescent-labeled derivatives greatly expands the utility of conotoxins as tools in cell biology. As peptides, conotoxins have a number of groups that can be modified. Tyrosine and histidine residues can be readily radioiodinated. The amino group at the N-terminus and lysine residues can be modified with the Bolton-Hunter reagent (1973; Thompson et al., 1987) to acquire an iodinatable phenolic group. Likewise, amino groups can be modified with fluorescent probes attached to acylating groups such as succinimidyl esters, carboxylic acids, sulfonyl halides, etc. Currently available fluorescent labels provide a selection of absorption spectra in the range ~300–

Table 2 Structure and Specificity of Conotoxins That Act on Ion Channels

Peptide[‡]	Primary Structure	Channel Specificity
Conotoxins that block voltage-sensitive calcium channels (Olivera et al., 1994)		
ω-GVIA	CKSOGSSCSOTSYNCCRSCNOYTKRCY*	N-type Ca channel ($\alpha 1_B$ subunit)
ω-GVIIA	CKSOGTOCSRGMRDCCTSCLLYSNKCRRY*	N-type Ca channel ($\alpha 1_B$ subunit)
ω-MVIIA	CKGKGAKCSRLMYDCCTGSCRSGKC*	N-type Ca channel ($\alpha 1_B$ subunit)
ω-MVIIB	CKGKGASCHRTSYDCCTGSCNRGKC*	N-type Ca channel ($\alpha 1_B$ subunit)
ω-MVIIC	CKGKGAPCRKTMYDCCSGSCGRRGKC*	P/Q- ($\alpha 1_A$) and N-type Ca channel
ω-MVIID	CQGRGASCRKTMYNCCSGSCNRGRC*	P/Q- ($\alpha 1_A$) and N-type Ca channel
ω-SVIA	CRSSGSPCGVTSICCGRCYRGKCT*	P/Q- ($\alpha 1_A$) and N-type Ca channel
ω-SVIB	CKLKGQSCRKTSYDCCSGSCGRSGKC*	P/Q- ($\alpha 1_A$) and N-type Ca channel
Conotoxins that block voltage-sensitive sodium channels (Cruz et al., 1985; Fainzilber et al., 1995; McIntosh et al., 1995)		
μ-GIIIA	RDCCTOOKKCKDRQCKOQRCCA*	μ1 subtype skeletal muscle Na channel, (site 1)
μ-GIIIB	RDCCTPPRKCKDRRCKPMKCCA*	μ1 subtype skeletal muscle Na channel, (site 1)
μ-GIIIC	RDCCTPPRKCKDRRCKPMKCCA*	μ1 subtype skeletal muscle Na channel, (site 1)
μO-MrVIA	ACRKKWEYCIVPIIGFIYCCPGLICGPFVCV^	Na channels in molluscan neurons (~100 nM) and block of rapidly inactivating Ca currents at >1 μM Type II Na channels and channels in rat hippocampal cells
μO-MrVIB	ACSKKWEYCIVPILGFVYCCPGLICGPFVCV^	Same
Ligands that delay inactivation of voltage-sensitive sodium channels (site VI) (Shon et al., 1994; Woodward, et al., 1990; Fainzilber et al., 1994)		
δ-GmVIA	VKPCRKEGQLCDPIFQNCCRGWNCVLFCV^	Na channels in molluscan neurons; shifts voltage-dependent activation curve to more (-) potentials and inactivation curve to more (+) potentials
δ-TxVIA	WCKQSGEMCNLLDQNCCDGYCIVLVCT^	Na channels in molluscan neurons; binds with no apparent physiological effects on mammalian Na^+ channels but protects against toxic effects of other toxins binding to same site
δ-PVIA	EACYAOGTFCGIKOGLCCSEFCLPGVCFG	Rat brain type II Na channel; rat hippocampal neurons; vertebrate neuromuscular junction
Potassium channel blocker (Shon et al., 1998; Terlau et al., 1997)		
κ-PVIIA	CRIONQKCFQHLDDCCSRKCNRFNKCV^	Shaker potassium channels

‡ The capital letter or capital letter followed by a small letter after the name of the toxin class indicates the *Conus* species from which the toxin was isolated: G, *C. geographus*; M, *C. magus*; S, *C. striatus*; Mr, *C. marmoreus*; Gm, *C. gloriamaris;* Tx, *C. textile*; P, *C. purpurascens*.

* Peptides amidated at the C-terminus; ^peptides with a free carboxyl end. Standard one-letter symbols are used to represent amino acid residues except O, which represents hydroproline.

Table 3 Structure and Specificity of Conopeptides Acting on Ligand-Gated Ion Channels

Peptide‡	Primary Structure	Specificity and Target Site
Conotoxins that inhibit skeletal muscle ACh receptors		
Cys pattern: CC—C—C (Gray et al., 1981; McIntosh et al., 1982; Zafaralla et al., 1988; Myers et al., 1991; Ramilo et al., 1992; Martinez et al., 1995)		
α-GI	ECCNPACGRHYSC*	Mouse α/δ subunit interface; *Torpedo* α/γ? subunit interface
α-GIA	ECCNPACGRHYSCGK*	Mouse muscle, δ vs. γ specificity not determined
α-GIB	ECCNPACGRHYSCKG*	Mouse muscle, δ vs. γ specificity not determined
α-GII	ECCHPACGKHFSC*	Mouse muscle, δ vs. γ specificity not determined
α-MI	GRCCHPACGKNYSC*	Mouse α/δ AChR interface
α-SI	ICCNPACGPKYSC*	
α-SIA	YCCHPACGKNFDC*	*Torpedo* α/δ AChR interface
α-SII	GCCCNPACGPNYGCGTSCS^	
α-EI	RDOCCYHPTCNMSNPQIC*	*Torpedo* α/δ subunit interface
Cys pattern: CC - - - C - - C - C - - - C (Jacobsen et al., 1997; Hopkins et al., 1995; Shon et al., 1997)		
αA-EIVA	GCCGPYONAACHOCGCKVGROOYCDROSGG*	Competitive antagonist; muscle α/δ and α/γ subunit interface
αA-EIVB	GCCGKYONAACHOCGCTVGROOYCDROSGG*	Competitive antagonist; muscle α/δ and α/γ? subunit interface
αA-PIVA	GCCGSYONAACHOCSCKDROSYCGQ*	Competitive antagonist; muscle nAChR
ψ-PIIIE	HOOCCLYGKCRRYOGCSSASCCQR*	Noncompetitive antagonist; muscle nAChR
Conotoxins that inhibit neuronal ACh receptors (Cartier et al., 1996; Fainzilber et al., 1994; Johnson et al., 1995)		
Cys pattern: CC–C—C		
α-MII	GCCSNPVCHLEHSNLC*	Rat α3/β2 subunit interface
α-PnIA	GCCSLPPCAANNPDYC*	*Aplysia* neuronal nAChR
α-PnIB	GCCSLPPCALSNPDYC*	*Aplysia* neuronal nAChR
α-ImI	GCCSDPRCAWRC*	Rat α7 nAChR; *Aplysia* neuronal nAChR
Conantokins: *Conus* peptides targeted to glutamate receptors (McIntosh et al., 1984; Mena et al., 1990; Haack et al., 1990)		
Con-G	GEγγLQγNQγLIRγKSN*	Inhibits NMDA subtype
Con-T	GEγγYQKMLγNLRγAEVKKNA*	Inhibits NMDA subtype

‡ The capital letter or capital letter followed by a small letter after the name of the toxin class indicate the *Conus* species from which the toxin was isolated: G, *C. geographus*; M, *C. magus*; S, *C. striatus*; E, *C. ermineus*, P, *C. purpurascens;* Pn, *C. pennaceus*; T, *C. tulipa*.

* Peptides amidated at the C-terminus; ^peptides that have a free carboxyl end. Standard one-letter symbols are used to represent amino acid residues except the letter O, which represents hydroproline.

600 nm and fluorescence spectra ranging from ~400 to 700 nm (Haugland, 1996).

For applications in cell biology, a number of *Conus* peptides, radioiodinated ω-conotoxins, anti-ω-conotoxin GVIA, and fluorescent-labeled ω-conotoxin, and conantokin G are now available commercially (see Tab. 4). Radiolabeled derivatives of other *Conus* peptides can be prepared as described below for use in binding experiments, localization of ion channels/receptors, and cross-linking experiments with the macromolecular targets.

3 General Considerations in Handling and Storage of *Conus* Venom and Peptides

- The venom squeezed out of *Conus* ducts can be stored at –80°C and lyophilized venom at –20°C, although –80°C is better for long-term storage (up to several years). Lyophilized purified and synthetic peptides are quite stable and can stand a few days at room temperature during transport.
- At low humidities, static charge can cause lyophilized venom and purified peptides to "fly out" of test tubes or spatula, but this can be remedied by the use of an anti-static gun.
- A dust mask should be worn when taking lyophilized venom or conotoxins in and out of vials. Particularly after transport, it is wise to centrifuge the container to ensure that the material will not exit the vial upon opening.
- Significant losses of the more hydrophobic peptides can occur when the peptide is lyophilized to dryness particularly if left drying for a long time. This has been observed when drying the highly purified HPLC fractions during peptide isolation.
- Peptide solutions may be stored at 4°C, –20°C, or –80°C depending on the length of storage. For regular use they should be stored in small aliquots to avoid loss of activity due to repeated freezing and thawing.
- At very low concentrations, conopeptides tend to "stick" to glass or plastic, resulting in significant losses.
- To avoid nonspecific adsorption, carriers such as 0.1–0.2 mg/ml lysozyme or 0.1–1.0 mg/ml bovine serum albumin (BSA) can be added. Since lysozyme is basic (pI = 11), it serves as a good carrier for positively charged conotoxins.

Table 4 Commercially available ***Conus*** **peptides and related products**

Item	Suppliers**
***Conus* peptides**	
α-Conotoxin GI	a, c, d, h
MI	a, c, d, h
SI	a, c, d, h
SIA	a, c
EI	a, c
ImI	a, c, d, h
μ-Conotoxin GIIIA	a, c, d
GIIIB	a, c, d, h
PnIVB	a
μ-Conotoxin GS	a
ω-Conotoxin GVIA	a, c, d, h
MVIIA	a, b, c, d, h
MVIIC	a, b, c, d, h
SVIB	c, d, h
Conantokin G	a, c, d, h
Conantokin G (free acid)	a, d
($Glu^{3,4,7,10,14}$)-Conantokin G	a
(Tyr^{0})-conantokin G	a
Conantokin T	c
Conopressin G	a, c
(Arg^{8})-Conopressin G	a
Conopressin S	a, c
(Lys^{8})-Conopressin S	a
Radiolabeled peptides	
[^{125}I-*Tyr*^{22}]ω-Conotoxin GVIA	e, f,
[^{125}I-*Tyr*^{13}]ω-Conotoxin MVIIA	e
[^{125}I]ω-Conotoxin MVIIA	f
[^{125}I]ω-Conotoxin MVIIC	e, f
Fluorescent-labeled peptides	
ω-Conotoxin GVIA, fluorescein conjugate	g
ω-Conotoxin GVIA, tetramethylrhodamine conjugate	g
Conantokin-G, BODIPY FL conjugate	g
Conantokin-G, tetramethylrhodamine conjugate	g
Antibody	
Anti-ω-conotoxin GVIA	c

* The capital letter or capital letter followed by a small letter after the name of the toxin class indicates the *Conus* species from which the toxin was isolated: G, *C. geographus*; M, *C. magus*; S, *C. striatus*; E, *C. ermineus*; Im, *C. imperial*; Pn, *C. penaceus*; T, *C. tulipa*.

** a. Bachem (http://www.bachem.com)
b. Tocris (http://www.tocris.com);
c. albiochem (http://www.calbiochem.com)
d. Sigma (http://www.sigma-aldrich.com; http://www.calrbi.com)
e. Amersham (http://www.apbiotech.com)
f. NEN Life Science Products (http://www.nenlifesci.com)
g. Molecular Probes (http://www.probes.com)
h. Peptide International (http://www.pepnet.com)

4 Iodination as a General Labelling Technique

4.1 General aspects

The introduction of iodine is a convenient way to label peptides. Tyrosine and histidine residues can be directly iodinated, whereas amino groups can be modified to introduce an iodinatable phenolic group. Of the radioactive isotopes of iodine, ^{125}I is most widely used because its half-life ($t_{1/2}$ = 60.14 d) is relatively long compared with 13.02 days for ^{126}I and 8.04 days for ^{131}I. As a γ-ray emitter, ^{125}I does not require the use of a scintillant for activity measurements, and quenching is not a problem. The energy of the soft γ-emission of ^{125}I is in the range that can be readily detected and yet easily shielded off to protect personnel.

Iodine-125 is commercially available as $Na^{125}I$. The iodide itself will not react with Tyr, but must be converted to a reactive intermediate by oxidation (Seevers and Counsell, 1982). Although iodination can be done directly with I_2, it is usually done with a more reactive agent such as simple compounds of iodine with electronegative atoms such as O or Cl. Iodine monochloride may be formed by the reaction of iodide with chloramine-T or iodogen. Such reagents contain a stabilized form of chlorine attached to nitrogen, which reacts to generate the amide and ICl.

$$(CH_3\text{-}C_6H_4\text{-}SO_2\cdot NCl)^-Na^+ + I^- + 2H^+ \rightarrow CH_3\text{-}C_6H_4\text{-}SO_2\cdot NH_2 + ICl + Na^+$$

Chloramine-T

The iodous ion (I^+ or possibly H_2OI^+) from IC1 reacts with the ortho position of tyrosine through an electrophilic attack. Another method relies on the enzyme lactoperoxidase to convert iodide into a reactive species.

The possible iodinated derivatives of tyrosine and histidine residues are 3-iodotyrosine (MIY), 3,5-diiodotyrosine (DIY), 2-iodohistidine (MIH), and 2,4-diiodohistidine (DIH). At lower pH values (such as in the iodination of α-conotoxin GI at pH 5.3), iodination is almost exclusively at the tyrosine residue. Once the first iodine is attached to tyrosine, the second one reacts even more readily. At higher pH values (pH 8–9), the imidazole ring of histidine is also iodinated. In α-conotoxin GI, which contains His^{10} and Tyr^{11}, all peptide derivatives containing MIH and DIH are diiodinated at Tyr^{11} (Gray et al., 1984).

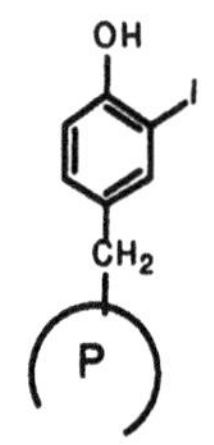

3-Iodotyrosine (MIT)

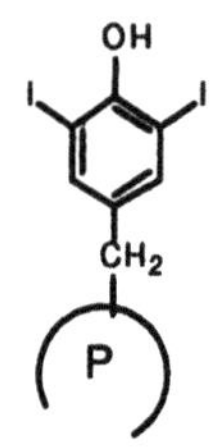

3,5-Diiodotyrosine (DIT)

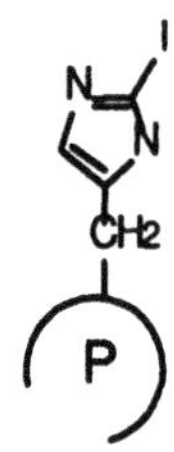

2-Iodohistidine (MIH)

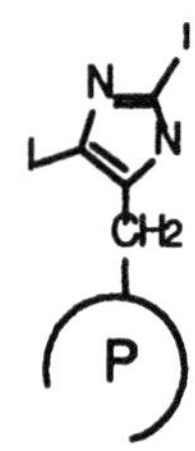

2,4-Diiodohistidine (DIH)

To minimize the effect of iodination on the biological activity of *Conus* peptides, it is usually best to limit iodination to the monoiodo derivative by using an excess of peptide over iodide in the reaction mixture. (Unmodified peptide can always be recycled, but diiodopeptide may just be a waste.)

4.2 Side reactions

Oxidation of amino acid residues such as Cys, Met, and Trp is an unwanted side reaction that may destroy biological activity.

Cysteine and any thiols (DTT, mercaptoethanol, glutathione) are extremely reactive, being oxidized to disulfides much faster than Tyr will be iodinated. In *Conus* peptides all Cys residues are present as disulfides, so this is not a problem. Sometimes, though, a synthetic peptide is made with an extra cysteine to allow it to be attached to a protein. It is best to oxidize the Cys first with a cheaper reagent than ^{125}I; iodinate the Tyr in a second step; reduce the disulfide when ready to do the coupling reaction onto protein.

For labelling of *Conus* peptides the bigger concern is that traces of thiol (e.g. mercaptoethanol from vapors in a refrigerator or speed-vac) may swamp the system. There are instances where certain batches of a peptide will simply not label, even though other batches behave perfectly well. Iodogen and chloramine-T themselves will react directly with thiols, so iodogen-coated tubes must be stored in such a way that there is no risk of exposure.

Methionine is normally less reactive than Tyr, but in some peptides this is not the case. Examples are conotoxins MVIIC and MVIID. Met is converted to the sulfoxide, Met (O), which is much more polar than Met itself. This usually makes the peptide elute earlier on HPLC, and also destroys its biological activity. So even if the Tyr does get iodinated, the peptide is not likely to bind to its target.

One solution in difficult cases is to synthesize an analog in which Met is replaced by norleucine (Nle), an amino acid which has much the same size, shape, and hydrophobicity. When this is not practical, it is also possible to reduce the Met (O) without affecting disulfides, using strong HCl and methyl sulfide.

Tryptophan reacts in several stages with the incorporation of one, two, or even three oxygens into the ring. Again, the typical result is loss of biological activity. This oxidation is *not* reversible, but Trp seems to be somewhat less reactive than Met.

4.3 Iodogen versus Chloramine T

Iodogen (1,3,4,6-tetrachloro-3α,6α-diphenylglycoluril) is a relatively insoluble compound with four chloroamide groups as oxidizing agents. It was developed as an attractive solid phase reagent for iodination of proteins and peptides (Fraker and Speck, 1978; Salacinski et al., 1981). With iodogen-coated tubes, one could simply put in the protein solution containing iodide, let it react, and then remove the protein solution leaving the iodogen behind. However, because of its insolubility, the reaction is harder to control. On the other hand, iodogen is a relatively mild reagent compared with chloramine T and it has been reported to produce less oxidative damage in proteins (Salacisnki et al., 1981). For labile peptides, the indirect method of Chizzonite et al. (1991, 1992) has been recommended. It involves preactivation of ^{125}I in the iodogen tube before the iodine is added to the sample. Since the peptide does not come in contact with iodogen, the potential for oxidative damage is minimized.

Chloramine T (sodium N-chlorobenzenesulfonamide) (Hunter, 1967) is water-soluble, and surprisingly stable in solution. Reactions are homogeneous, and more easily controlled by varying the chloramine T concentration. A disadvantage for work with proteins was that the excess reagent had to be removed, adding a complicated step. With peptides, putting the mix onto HPLC serves both to isolate the iodinated peptides and remove the chloramine T and excess iodide. Chloramine T or its degradation product runs very cleanly as a sharp peak, and it is not likely to interfere. Although the harsh oxidizing conditions may inactivate peptides containing methionine and tryptophan, the use of a relatively low concentration (~0.1 mM) of chloramine T can be tolerated as long as peptide exposure to the reagent is short (~ 10 min) and a reducing agent such as ascorbic acid is used to stop the reaction prior to chromatographic purification of the peptides. Chloramine T immobilized on polystyrene beads (Iodo-Beads from Pierce) is claimed to be more gentle than the soluble form (Markwell, 1982). The reaction is terminated by simply removing beads from the mixture.

4.4 Minimizing contamination and the risk of radiation exposure

Radioactive iodine (^{125}I) is a γ-emitter, and it requires lead shielding. It is usual to store and handle ^{125}I as iodide (I^-), which is nonvolatile, stable at basic pH (as $Na^{125}I$), and does not dissolve in plastic. Clear lead acrylic shields, flexible lead sheets or foil, lead-lined waste containers, and lead-vinyl gloves should be used. Gloves should be changed frequently especially after getting $Na^{125}I$ from the stock vial to avoid contamination of micropipets and the outside of the reaction tube. Before opening, vials containing radioactive material should be examined carefully and if necessary spun briefly to bring down droplets from the cap and sides of the container. Use disposable sleeve guards or clip the cuffs of lab coats to prevent contamination of clothing.

For safety reasons, stocks of $Na^{125}I$ should be kept in the dark to prevent photo conversion to free iodine and in the hood in case free iodine is liberated. The hood for iodination should have an airflow rate of 80–100 linear ft per min. Although iodide is nonvolatile, it will evaporate from acid solution as HI, and could be a real radiation hazard in HPLC effluent (pH 2). A reducing agent (STSI, see the section on radioiodination of *Conus* peptides) should be placed in the beaker for collecting HPLC waste to prevent volatilization of radioactive I_2. A bottle of the reducing agent should be kept handy in the work area in case of spills.

Radioiodination should be done only by trained personnel. The first-timer should do a test run using nonradioactive NaI instead of $Na^{125}I$ to become familiar with the steps and movements needed to avoid contamination of the experimenter and work area. The cold iodinated peptide obtained can then be used to test for biological activity and ensure that the iodinated peptide is active.

5 Radioiodination of Conotoxins

Iodination of peptides containing tyrosine but not histidine can be done at pH 7, whereas peptides that do not have tyrosine but contain histidine can be labeled at a basic pH (8–9). Selective iodination of tyrosine residues in the presence of histidine can be done at low pH. At ~5.3 when the imidazole ring is protonated, iodination is almost exclusively at tyrosine residues. Purification of the radioiodinated peptide is conveniently done using a reverse phase C18 HPLC column. Iodination makes the peptide more hydrophobic, and the magnitude of its shift to later elution times will depend on the residue iodinated and the number of iodine atoms introduced into the peptide. To ensure that the derivatives formed are mainly monoiodinated, use about 10-fold excess peptide over iodide.

The procedure to be presented is for radioiodination using chloramine T as the oxidizing agent. The reagent is cheap and the method is simple and relatively fast. Likewise, iodogen is widely used for the radioiodination of peptides. However, this method will not be presented here because it is very well discussed in the instructions for IODO-GEN precoated iodination tubes (Pierce Product No. 28601; see http://www.piercenet.com).

5.1 Materials and chemicals

Buffers and reagents

- **Buffer**: 250 mM Tris, pH 7
- **Alternative buffers**:
  For selective iodination of tyrosine residues: 0.25 M NH4Ac, pH 5.3
  For iodination of histidine residues: 0.25 M Tris buffer, pH 8.2
- **$Na^{125}I$**: Specific activity at 100% isotopic abundance is 17.4 mCi/μg iodine or 644 MBq/μg iodine (2.18 mCi/nmol); 1 mCi = 3.7×10^7 dps or Bq
- **Chloramine T** (2 mM)
- **Ascorbic acid** (0.2 M)
- **Lysozyme** (10 mg/ml)
- **Methanol** (40%)
- **Ethyl acetate**
- **HPLC solvent A** (0.1% trifluoroacetic acid)
- **HPLC solvent B** (60% acetonitrile (v/v) in 0.0.92% trifluoroacetic acid)
- **STSI solution**
  25 g sodium thiosulfate
  2 g NaI
  40 g NaOH per liter

Chemicals

Iodine-125, as $Na^{125}I$ in NaOH	Amersham*	IMS30
Chloramine T	Sigma**	C 9887
Ascorbic acid	Sigma	A 5960
Lysozyme	ICN***	100834
Trifluoroacetic acid (TFA, HPLC grade)	Fisher****	BP618-500
Acetonitrile (HPLC grade)	Fisher	A998-4

* Amersham http://www.apbiotech.com; *** ICN http://www.fishersci.com;
** Sigmahttp://www.sigma.sial.com; **** Fisher http://www.icnpharm.com

Equipment

- Hood with air flowrate 80 – 120 linear ft/min
- Portable gamma detector
- Gamma counter
- HPLC with reverse phase C18 column
- Table top centrifuge
- Vortex mixer
- Lyophilizer with vacuum centrifuge
- Clear lead acrylic shields and flexible lead sheets or foil, lead-lined waste containers, lead-vinyl gloves

5.2 Iodination of *Conus* peptides using Chloramine T

Prepare the HPLC and collection tubes prior to starting the iodination reaction. Line work area with plastic-backed absorbent paper. Shield the fraction collection area with clear lead acrylic barrier and other hot areas with flexible lead sheets. Place 2 ml of STSI in a 15-ml tube for collection of the very hot injection peak and place 50–100 ml of STSI in the beaker to be used for collecting radioactive HPLC waste. Do the iodination in the hood with an air flow of 80 – 120 linear ft/min.

Protocol 1 Carrier-free iodination of ω-conotoxin GVIA

1. In a 1.5-ml conical microtube, dissolve 7 nmol ω-GVIA in 50 μl of water (or 0.1% TFA).
2. To this add 50 μl 250 mM Tris, pH 7 + 5 μl of $Na^{125}I$ (~1.5 mCi, ~2175 Ci/mmol).
3. Close tube tightly, vortex gently, and centrifuge briefly to bring down droplets from the sides of tube.
4. Add 5 μl of 2 mM chloramine T, close tube tightly, vortex gently, spin down the droplets, and allow to stand at room temperature for 10 min.
5. Stop the reaction by adding 10 μl of freshly prepared 0.2 M ascorbic acid.
6. While still under the hood, load the reaction mixture in the syringe, wipe needle carefully, and transport the syringe on a tray.
7. Inject sample into the HPLC port and run on a reverse-phase C18 HPLC column preequilibrated with buffer A.

8 Elute with a gradient of acetonitrile in 0.1% TFA, collecting the injection peak in a 15-ml tube containing STSI and the peptides into tubes containing 20 µl of 10 mg/ml lysozyme as carrier.

9 Lyophilize in a vacuum centrifuge (e.g., Speed Vac) overnight and resuspend in 1 ml of 40% methanol.

10 Divide into small aliquots of about 20 x 10^6 cpm and store at –20°C.

Notes:
Methanol minimizes radiation damage by acting as a free radical scavenger.

Removal of free iodine/iodide from the reaction mixture

The amount of ^{125}I applied to the HPLC column can be minimized by extraction of free iodine from the reaction mixture with at least an equal volume of ethyl acetate. After addition of ethyl acetate, vortex gently and centrifuge to facilitate separation of layers. Pipette off the lower aqueous phase and apply on HPLC. The free iodine can also be removed by spinning the reaction mixture through a molecular sieve such as Sephadex G-15, but the method is more involved. Although these techniques are effective in reducing the volatile or unbound ^{125}I, the yield of peptide and derivatives is reduced. Removal of free iodine from the reaction mixture is not necessary if the syringe is loaded under the hood and the HPLC apparatus is well shielded and adjacent to the iodination area. Liberation of free iodine in the chloramine T method is prevented by the addition of ascorbic acid.

References

Albensi BC, Ryujin KT, McIntoxh JM et al (1993) Localization of [^{125}I]ω-conotoxin GVIA binding in human hippocampus and cerebellum. NeuroReport 4: 1331–1334

Bolton AE, Hunter WM (1973) The labelling of proteins to high specific radioactivities by conjugation to a ^{125}I-containing acylating agent. Biochem. J. 133: 529–539

Cartier GE, Yoshikami D, Gray WR et al (1996) A new α-conotoxin which targets α3β2 nicotinic acetylcholine receptors. J. Biol. Chem. 271: 7522–7528

Chizzonite R, Truitt T, Podlaski FJ et al (1991) IL-12: Monoclonal antibodies specific for the 40kDa subunit block receptor binding and biologic activity on activated human lymphoblasts. J. Immunol 147: 1548–1556

Craig AG, Norberg, T., Griffin D. et al. (1999) Contulakin-G, an O-glycosylated invertebrate neurotensin. J. Biol. Chem. 274: 13752–13759

Cruz LJ, Gray WR, Olivera BM (1978) Purification and properties of a myotoxin from *Conus geographus* venom. Arch. Biochem. Biophys. 190: 539–548

Cruz LJ, Gray WR, Olivera BM et al (1985) *Conus geographus* toxins that discriminate between neuronal and muscle sodium channels. J. Biol. Chem. 260: 9280–9288

Cruz LJ, Olivera BM (1986) Calcium channel antagonists. ω-Conotoxin defines a new high affinity site. J. Biol. Chem. 261: 6230–6233

Cruz LJ, Johnson DS, Olivera BM (1987) A characterization of the ω-conotoxin target. Evidence for tissue-specific heterogeneity in Ca channel types. Biochemistry 26: 820–824

Fainzilber M, Gordon D, Hasson A et al (1994a) A new neurotoxin receptor site on sodium channels is identified by a conotoxin that affects sodium channel inactivation in molluscs and acts as an antagonist in rat brain. J. Biol. Chem. 269: 2574–2580

Fainzilber M, Hasson A, Oren R et al (1994b) New mollusc-specific α-conotoxin block Aplysia neuronal acetylcholine receptor. Biochemistry 33: 9523–9529

Fainzilber M, van der Shors R, Lodder JC et al (1995) New sodium channel-blocking conotoxins also affect calcium currents in *Lymnea* neurons. Biochemistry 34: 5364–5371

Fraker PJ, Speck JC (1978) Protein and cell membrane iodinations with a sparingly soluble chloroamide, 1: 3: 4: 6-tetrachloro-3α,6α-diphenylglycouril. Biochem. Biophys. Res. Comm. 80: 849–857

Gately MK, Chizzonite R (1992) Measurement of human and mouse interleukin-12. Current Protocols in Immunology (Coligan J. E., Kruisbeek A. M., Margulies D. H., Shevach E. M. and Strober W., eds): New York: Wiley.

Gonoi T, Ohizumi Y, Nakamura H et al (1987) The *Conus geographutoxin* II distinguishes two functional sodium channel subtypes in rat muscle cells developing in vitro. J. Neurosci. 7: 1728–1731

Gray WR, Luque A, Olivera BM et al (1981) Peptide toxins from *Conus geographus* venom. J. Biol. Chem. 256: 4734–4740

Gray WR, Luque A, Galyean R et al (1984) Conotoxin GI: disulfide bridges, synthesis and preparation of iodinated derivatives. Biochemistry 23: 2796–2802

Gray WR, Olivera BM, Cruz LJ (1988) Peptide toxins from venomous *Conus* snails. Annu. Rev. Biochem. 57: 665–700

Haack JA, Rivier J, Parks TN et al (1990) Conantokin-T; a γ-carboxyglutamate-containing peptide with N-methyl-D-aspartate antagonist activity. J. Biol. Chem. 265: 6025–6029

Haugland RP (1996) Handbook of Fluorescent Probes and Research Chemicals. Eugene, Oregon: Molecular Probes

Hillyard DR, Monje VD, Mintz IM et al (1992) A new *Conus* ligand for mammalian presynaptic Ca channels. Neuron 9: 69–77

Hopkins C, Grilley M, Miller C et al (1995) A new family of *Conus* peptides targeted to the nACh receptor. J. Biol. Chem, 270: 22361–22367

Hunter WM (1967) The preparation of radioiodinated proteins of high activity, their reaction with antibody in vitro: the radioimmunoassay, Handbook of Experimental Immunology (Weir D. M., ed): Philadelphia: F. A. Davis

Jacobsen R, Yoshikami D, Ellison M et al (1997) Differential targeting of nicotinic acetylcholine receptors by novel αA-conotoxins. J. Biol. Chem 272: 22531–22537

Johnson DS, Martinez J, Elgoyhen AB et al (1995) α-Conotoxin ImI exhibits subtype-specific nicotinic acetylcholine receptor blockade: preferential inhibition of homomeric α7 and α9 receptors. Mol Pharmacol. 48: 194–199

Markwell MAK (1982) A new solid-state reagent to iodinate proteins: conditions for the efficient labelling of antiserum. Anal. Biochem 125: 427–432

Martinez JS, Olivera BM, Gray WR et al (1995) α-Conotoxin EI, a new nicotinic acetylcholine receptor antagonist with novel selectivity. Biochemistry 34: 14519–14526

McIntosh M, Cruz LJ, Hunkapiller MW et al (1982) A peptide toxin from *Conus magus* venom. Arch. Biochem. Biophys. 218: 329–334

McIntosh JM, Olivera BM, Cruz LJ, Gray WR (1984) γ-Carboxyglutamate in a neuroactive toxin. J. Biol. Chem. 259: 14343–14346

McIntosh JM, Adams ME, Olivera BM, Filloux F (1992) Autoradiographic localization of the binding of calcium channel antagonist, [^{125}I]ω-agatoxin IIIA, in rat brain. Brain Res 594: 109–114

McIntosh JM, Hasson A, Spira ME et al (1995) A new family of conotoxins that blocks voltage-gated sodium channels. J. Biol. Chem. 270: 16796–16802

Mena EE, Gullak MF, Pagnozzi MJ et al (1990) Conantokin-G: a novel peptide antagonist to the N-methyl-D-aspartic acid (NMDA) receptor. Neurosci Lett 118: 241–244

Mills LR, Niesen CE, So AT et al (1994) N-Type Ca^{2+} channels are located on somata, dendrites, and a subpopulation of dendritic pines on live hippocampal pyramidal neurons. J. Neurosci. 14: 6815–6824

Monje VD, Haack JA, Naisbitt SR et al (1993) A new *Conus* peptide ligand for Ca channel subtypes. Neuropharmacology 32: 1141–1149

Myers RA, Zafaralla GC, Gray WR et al (1991) α-Conotoxins, small peptide probes of nicotinic acetylcholine receptors. Biochemistry 30: 9370–9377

Myers, R.A, Cruz, L.J, Rivier, J.E, Olivera, B.M (1993) *Conus* peptides as chemical probes for receptors and ion channels. Chem. Rev. 93: 1923–1936

Olivera BM, Rivier J, Clark J, Ramilo CA et al (1990) Diversity of *Conus* neuropeptides. Science 249: 257–263.

Olivera BM, Miljanich GP, Ramachandran J, Adams ME (1994) Calcium channel diversity and neurotransmitter release: The ω-conotoxins and ω-agatoxins. Ann. Rev. Biochem. 63: 823–867

Ramilo CA, Zafaralla GC, Nadazdi L, Miljanich G et al (1992) Novel α- and ω-conotoxins from *Conus striatus* venom. Biochemistry 31: 9919–9926

Salacinski PRP, McLean C, Sykes JE et al (1981) Iodination of proteins, glycoproteins and peptides using a solid-phase oxidizing agent, 1: 3: 4: 6-tetrachloro-3α,6α-diphenyl glycouril. Anal. Biochem. 117: 136–146

Seevers RH, Counsell RE (1982) Radioiodination techniques for small organic molecules. Chem. Rev. 82: 575–590

Shon K-J, Hasson A, Spira ME et al (1994) δ-Conotoxin GmVIA, a Novel Peptide from the Venom of *Conus gloriamaris*. Biochemistry 33: 11420–11425

Shon K-J, Grilley M, Jacobsen R et al (1997) A noncompetitive peptide inhibitor of the nicotinic acetylcholine receptor from *Conus purpurascens* venom. Biochemistry 36: 9581–9587

Shon K-J, Stocker M, Terlau H et al (1998) κ-Conotoxin PVIIA is a peptide inhibiting the shaker K^+ channel. J. Biol. Chem. 273: 33–38

Shon K-J, Olivera BM, Watkins M et al (1998) μ-Conotoxin PIIIA, a new peptide for discriminating among tetrodotoxin-sensitive Na channel subtypes. J. Neurosci. 18: 4473–4481

Terlau H, Stocker M, Shon K-J et al (1996) μO-conotoxin MrVIA inhibits mammalian sodium channels, but not through Site I. J. Neurophys. 76: 1423–1429

Thompson JA, Lau AL, Cunningham DD (1987) Selective radiolabelling of cell surface proteins to a high specific activity. Biochem. 26: 743–750

Woodward SR, Cruz LJ, Olivera BM, Hillyard DR (1990) Constant and hypervariable regions in conotoxin propeptides. EMBO J 9: 1015–1020

Zafaralla GC, Ramilo C, Gray WR et al (1988) Phylogenetic specificity of cholinergic ligands: α-Conotoxin SI. Biochemistry 27: 7102–7105

6 Ant Polypeptide Toxins

Kirill A. Pluzhnikov, Ludmilla V. Shevchenko and Evgeniy V. Grishin

Contents

1 Introduction

Compared with other insects with social behaviour, ants have no peers. The number of the identified ant species is slightly less than 9,000, but an estimated 11,000 species are yet to be described, mostly from tropical habitats (Blum, 1992). Ants from most subfamilies are capable of stinging. Since stinging ants are particularly abundant in tropical regions, it may be anticipated that a large number of venom species are yet to be described, and the medical importance of these insects will be better appreciated. Ant venoms exhibit greater variability in composition and function than do venoms from any other group of arthropods. The radically different evolutionary and ecological histories of ants as compared with other arthropods would seem to argue against the similarity of the venoms produced in these diverse groups. Unlike social bees and wasps, which use their stings solely for defence, ant stings serve for a variety of additional functions, including prey capture and the elaboration of trail, sex, aggre-

Methods and Tools in Biosciences and Medicine
Animal Toxins, ed. by H. Rochat and M.-F. Martin-Eauclaire

gation, and alarm pheromones (Schmidt, 1986). Ant venoms, in general, cause immediate, often intense locomotor activity followed by clonic convulsions, peripheral vasodilitation, and sedation. This is usually followed by cyanosis, shallow breathing and a gradual loss of activity until death (Schmidt *et al.*, 1980).

The chemistry and pharmacology of ant venoms have been reviewed by Schmidt (1986) and Blum (1992). The most important components of ant venoms are fairly volatile organic compounds and proteins. Ants of several myrmicine genera produce venoms containing piperidines, pyrrolidines, and indolizidines. The protein-rich ant venoms are found in the majority of ant species. Ant venoms are not as rich in glycoproteins as those of wasps or bees. Enzymes are the major components of the stinging ant venom (Schmidt *et al.*, 1986). Typically, the predominant *Ponerinae* venom ingredients are phospholipase A, lipase, hyaluronidase, and acidic phosphatase. Hemolytic activity is also common for these venoms (Schmidt *et al.*, 1984). Neurotoxic polypeptide components were studied in greater detail in *Paraponera clavata*. A neurotoxic polypeptide of 25 amino acid residues, called poneratoxin, has been isolated from *P. clavata* venom (Piek *et al.*, 1991a,b).

2 Materials and Reagents

Reagents

Endoproteinase Glu C	Boehringer Mannheim	791156
Endoproteinase Arg C	Boehringer Mannheim	269590
Trypsin	Boehringer Mannheim	109819
1,4-Dithiothreitol	Sigma	D9760
4-Vinylpyridine	Sigma	V3877

Solutions

- **Solution 1**
  0.2 M Tris-HCl, pH 7.5
  7.5 M guanidine hydrochloride
  2 mM EDTA

3 Methods

3.1 Venom preparation

Ants were collected in the Amazonian rain forest (Department of Ucauali, Peru). To collect the venom, ants were held with forceps to a sponge, and the venom release was induced by tactile stimulation of the abdomen. For large-scale preparations, 30–40 ants may be treated simultaneously. The procedure is as follows:

Protocol 1 Venom preparation

1. The sponge is prewashed sequentially with the extraction buffer, water, ethanol, and dried.
2. Ants are placed into the jar with the bottom completely lined with sponge, and covered with the exact same size sponge. The top sponge is gradually lifted and lowered in order to stimulate stinging.
3. Venom extraction from the sponge is done with a 20–50-ml syringe in ice-cold ammonium bicarbonate or acetate buffer.

   Notes:
   This procedure allows treatment of more than a thousand ants a day and causes them practically no damage.

3.2 Isolation of the toxin

Protocol 2 Isolation of the toxin

1. 40 mg of crude venom from the *Ectatomma tuberculatum* ant were dissolved in 1 ml of 0.1 M NH_4HCO_3 buffer (pH 8.1) and applied to a Sephadex G-50F column (2.5 x 90 cm) in the same buffer.
2. The toxic fraction was further purified by HPLC using a reverse-phase Ultrapore RPMC column (0.46 x 25 cm) Beckman in a linear gradient 0–60% of acetonitrile in 0.1% trifluoroacetic acid.

3.3 Reduction of the disulfide bonds and modification of the SH groups

Prior to sequencing Et-1, disulfide bonds were reduced with dithiothreitol, and SH groups were modified with vinylpyridine.

Protocol 3 Reduction of the disulfide bonds and modification of the SH groups

1. 1 nmol of ectatomin was dissolved in 0.1 ml of **Solution 1** predegassed with nitrogen.
2. The sample was heated at 60°C for 1 h, cooled to room temperature, and a 150-fold molar excess of dithiothreitol solution in the same buffer was added.
3. The reduction reaction was performed at 37°C for 18 h.
4. Further alkylation was done after addition to the reaction mixture of 4-vinylpyridine in three-fold excess with respect to dithiothreitol in 10 µl of methanol.
5. The modification reaction was performed at room temperature for 15 min in the dark.

3.4 Separation of the chains and proteolytic fragmentation of Et-1

Protocol 4 Separation of the chains and proteolytic fragmentation of Et-1

1. The modified A and B chains of Et-1 were separated by reverse-phase chromatography on an Ultrapore RPSC column in a linear gradient of 0–50% of acetonitrile in 0.1% trifluoroacetic acid.
2. Each chain was digested with *Staphylococcus aureus* V8 endoproteinase in 50 mM ammonium bicarbonate pH 8 at 37°C for 6 h, (enzyme:protein ratio 1:15, by weight) or with trypsin in the same buffer for 16 h at 37°C (enzyme:protein ratio 1:25).
3. In some cases, in order to obtain additional structural information, chains were digested with endoproteinase Arg-C in 50 mM ammonium bicarbonate pH 8 at 37°C for 12 h (enzyme:protein ratio 1:30).
4. The hydrolysates were separated by reverse-phase HPLC on an Ultrapore RPSC column using a linear gradient of 0–50% of acetonitrile in 0.1% trifluoroacetic acid.

5. Each chain and all the overlapping fragments were sequenced on a gas-phase automatic sequencer 470A (Applied Biosystems, USA).

   Notes:
   Isolation, modification and determination of primary amino acid sequences of Et-2, Eq-1, and Eq-2 were performed according to the same procedure.

3.5 Bioassay

The toxic activity of the purified venom fractions and ectatomins was assayed in BALB/C mice (body weight 15–20 g) and cockroaches *Blata orientalis* (body weight 60–130 mg). 3–5 μl of the toxin water samples with concentrations of 0.03–2 μg/μl were injected intracerebroventricularly into mice or into the second thoracic segment of cockroaches.

4 Collection and Composition of Ant Venom

The practical problem of obtaining reasonable quantities of ant venom of purities sufficient for scientific studies severely hinders research into these venoms. Ants generally produce 10 μg of dry venom or less per individual, whereas many spiders, scorpions, and snakes produce from 1 to 300 mg per individual (Schmidt, 1986). Ants are usually collected wearing latex gloves or with forceps, particularly the most aggressive types such as *Paraponera clavata*. In a number of cases, special bites are used. Most often, in order to obtain the sufficient amounts of venom, it is necessary to draw it from dozens or hundreds of ants. A number of methods have been described. The most frequently used are the following: dissection of the poison sack from the venom gland, venom extraction with buffer solutions from the whole ants or their abdominal segments, venom collection by electrostimulation or by tactile stimulation with forceps of stinging the sponge (Hermann and Blum, 1966; Schultz and Arnold, 1984).

5 Double-stranded Polypeptide Toxins

A novel class of polypeptide toxins (ectatomins) has been discovered in the venom of tropical stinging ants *Ectatomma tuberculatum* and *E. qudridens*. The venom of the *E. tuberculatum* ant is one of the most toxic venoms. Its toxicity exceeds that of solitary wasps and bees. The LD_{50} value of the crude *Ectatomma tuberculatum* venom in mice is 1.7 mg/kg as determined by intraperitoneal

(Schmidt *et al.*, 1980) or 31 μg/kg by intracerebroventricular injections, respectively. The venom is also highly toxic for cockroaches (LD_{50} = 7 mg/kg). The major toxic component of the venom was isolated by chromatography on Sephadex G-50f followed by reversed-phase HPLC of the toxic fraction. The toxin was named ectatomin (Et-1), bearing in mind the species from which it was isolated and the abundance of amino groups within the molecule (the calculated pI = 9.95). The molecular weight of ectatomin was estimated to be 7928 Da. The protein (Fig. 1) consists of two highly homologous polypeptide chains (47% identity) with a disulfide bridge in each, linked together by a disulfide bond (Arseniev *et al.*, 1994; Pluzhnikov *et al.*, 1994).

The affinity-purified polyclonal antibodies against ectatomin helped identify a novel structural homologue of ectatomin (Et-2) in *E. tuberculatum* ant venom. Its complete amino acid sequence has been determined. Et-2, like ectatomin, consists of two highly homologous chains (38% identity), linked by a disulfide bond. The mobility of each chain is limited by a single intrachain disulfide (Fig. 1).

A-chain 37
GVIPKKIWETVCPTVEPWAKKCSGDIATYIKRECGKL
WSTIVKLTICPTLKSMAKKCEGSIATMIKKKCDK
B-chain 34

Ectatomin Et-1, Mw 7928Da, pI 9,95.

A-chain 40
GVKEKLLAVLRKACPVIEKLAKKCHGTVATLVKKACSLIH
GLKEWAMKKMCPIAVRLAKKCDGTLATKIKEICDNL
B-chain 36

Ectatomin Et-2, Mw 8314Da, pI 9,78 .

Figure 1 Amino acid sequences of the ectatomins Et-1 and Et-2 from the venom of *Ectatomma tuberculatum* ant.

Investigation into venoms of other *Ponerinae* ants with the use of affinity-purified antibodies against ectatomin revealed its structural homologues in the venom of the closely related ant species *E. qudridens*. Five of them were isolated. The complete amino acid sequences of the two most abundant ectatomin homologues (Fig. 2) and partial sequences of the other three have been determined.

The amino acid sequences of some ectatomins were confirmed by cloning and sequencing of the corresponding complementary DNAs (cDNAs). It was shown that the main structural moiety of this toxin class (Et-1, Et-2, Eq-1, Eq-2) is two highly homologous polypeptide chains linked by a disulfide bond.

A-chain 37

FAIGAIIKKFGKKLVEYAVKHKDLYAPYIKKHLCEKL

INFGAIKAILKKWGKKLVEYALKHKDLYAPYIKKHLCEKL

B-chain 40

Ectatomin Eq-1, Mi 9016 Da,pI 9,79.

A-chain 40

INFGAIKAILKKWGKKLVEYALKHKDLYAPYIKKHLCEKL

INFGAIKAILKKWGKKLVEYALKHKDLYAPYIKKHLCEKL

B-chain 40

Ectatomin Eq-2, Mi 9429 Da,pI 9,89.

Figure 2 Amino acid sequences of the ectatomins Eq-1 and Eq-2 from the venom of *Ectatomma qudridens* ant.

Each chain consists of 34–40 amino acids with clearly presented clusters of positively charged lysine residues. All these toxins have molecular weights of 8000–9500 Da and isoelectric points of ~10. A high content of α-helices (~60%) and a low percentage of unfolded regions are typical of the secondary structures of these toxins. This novel class of toxins may be divided into two structural subgroups, Et and Eq. These subgroups are distinguished primarily by either the presence or absence of the intrachain disulfide bonds and the location of the interchain disulfide bridge. Toxin molecules of the first subgroup consist of two homologous polypeptide chains and have an interchain disulfide bridge in the central part. Provided the intrachain disulfides are absent, the molecule may comprise either highly homologous or identical polypeptides connected by a disulfide bond in the C-terminal region.

The purified ectatomin (Et-1) is responsible for the major toxic effect of the venom in both mammals and insects. Ectatomin content in crude venom is >10%. Et-1 has an LD_{50} of 6.8 μg/kg when injected intracerebroventricularly into mice. Its toxic effect in cockroaches was estimated as an LD_{50} value of 2.1 mg/kg. The high doses of the toxin caused a quick development of complete paralysis that was irreversible and lasted for more than 7 days. Studies of toxic effects of ectatomin homologues showed that, despite the same type of molecular organization, their toxicity is lower than that of Et-1: more than 40-fold in vertebrates and 3–7-fold in insects.

6 Spatial Structure

Ectatomin spatial structure was determined by two-dimensional NMR spectroscopy techniques (Nodle *et al.*, 1995). Et-1 in aqueous solution forms a bundle of four α-helices (Fig. 3). The spatial structure of the two ectatomin chains are similar, which accords with their fairly high sequence homology. Each chain consists of two antiparallel α-helices connected by a hinge region of four residues. Two intrachain disulfide bridges connect the termini of the antiparallel α-helices and stabilize their relative positioning. The third disulfide bridge connects the two hinge regions. A substantial part of the molecule surface is positively charged, due to the large number of surface lysine side chains which are mobile and exposed to the solvent. Nonpolar amino acid residues form a hydrophobic core of the protein. The prevalence of positively charged residues at the molecule surface facilitates its interaction with negatively charged molecules (e.g., lipids).

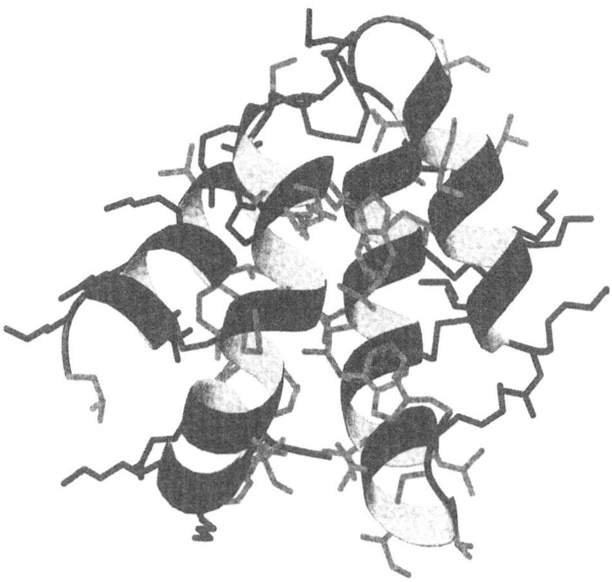

Figure 3 Schematic representation of the Et-1 structure.

7 Membrane Active Properties

The structure of ectatomin suggests the possibility of its interaction with cellular membranes. Patch-clamp experiments revealed the channel-forming activity of Et-1. At a concentration of 0.05–0.1 μM it formed channels in *Xenopus laevis* oocyte membranes. The conductance of a single channel was 8 pS. The results obtained for oocytes were confirmed in model membrane systems. The channel formation depended on the membrane potential and occurred only at positive cis potential. Analysis of the correlation between the membrane permeability and the ectatomin concentration revealed that each pore is formed by two ectatomin molecules (Pluzhnikov *et al.*, 1994).

The channel-forming properties of ectatomins may account for their toxic activity. On the other hand, a high level of Et-1 toxic activity may have effects other than pore formation in plasma membrane.

8 Troubleshooting

No separation of A and B chains of ectatomins after reduction and modification

- Low molecular weight proteins containing several disulfide bonds are highly stable. For reduction of disulfide bonds it is essential to use special conditions: high concentration of chaotropic agents (urea or guanidine hydrochloride) followed by heating for at least 1–2 h; dithiothreitol should be added in a hundredfold or greater excess per disulfide bond.

Poor separation of peptides after enzymatic digestion:

- The acetonitrile gradient is too steep.
- The chromatography buffer system should be changed. Rechromatography of mixed fractions should be done in a buffer with higher pH, for example, 10 mM ammonium acetate, pH 5.7.
- The matrix phase is not suitable for separation of the peptides.

References

Arseniev AS, Pluzhnikov KA, Nolde DE et al (1994) Toxic principle of selva ant venom is a pore-forming protein transformer. FEBS Lett. 347: 112–6

Blum MS (1992) Ant venoms: chemical and pharmacological properties. J. Toxicol. – Toxin reviews 11(2): 115–64

Hermann HR Jr, Blum M (1966) The morphology and histology of the hymenopterous poison apparatus. Annals of the Enthomological Society of America 59(2): 397–405

Nolde DE, Sobol AG, Pluzhnikov KA et al (1995) Three-dimensional structure of ectatomin from Ectatomma tuberculatum ant venom. J.Bio NMR 219(5): 1–13

Piek T, Hue B, Mantel P et al (1991a) Pharmacological characterization and chemical fractionation of the venom of the Panerinae ant, Paraponera clavata (F.). Comp. Biochem. Physiol. 99C(3): 481–6

Piek T, Duval A, Hue B et al (1991b) Poneratoxin, a novel peptide neurotoxin from the venom of the ant Paraponera clavata. Comp. Biochem. Physiol. 99C(3): 487–95

Pluznikov KA, Nolde DE, Tertishnikova SM et al (1994) Sructural and funtional studies of toxic principle of Ectatomma tuberculatum ant venom. Bioorg. Khim (Rus.) 20(8–9): 857–871

Schmidt JO, Blum MS, Overal WL (1980) Comparative lethality of venoms from stinging hymenoptera. Toxicon 18: 469–74

Schmidt JO, Blum MS, Overal WL (1984) Hemolytic activities of stinging insect venoms. Arch. Biochem. Physiol. 2: 155–60

Schmidt JO (1986) Chemistry, Pharmacology, and Chemical Ecology of Ant Venoms. Venoms of the hymenoptera (Piek T, ed) pp 425–508: London: Academic Press

Schmidt JO, Blum MS, Overal WL (1986) Comparative enzymology of venoms from stinging hymenoptera. Toxicon 24(9): 907–21

Schults DR, Arnold PI (1984) Immunochemical and clinical studies of venom from the ant Pseudomyrmex sp. Insects poisons, alergens and other invertabrate venoms (Tu AT, ed) pp 243–90: New York: Marcel Dekker, Inc

7 Wasp Kinins and Kinin Analogues

Tom Piek

Contents

1 Introduction

Due to the powerful venoms secreted by bees, social wasps, and ants, all belonging to the order of the Hymenoptera, these insects appear early in man's history and prehistory (Piek, 1986). According to Westwood (1840) the Hymenoptera are distinguished as an order from other insects by the number, the comparative size and the structure of the wings, the mandibulated mouth enclosing the labium ensheated by the maxillae, the ovipositor, or the sting, and the nature of their metamorphosis. In the modern literature the Hymenoptera are normally divided into the suborders Symphyta and Apocrita. The Symphyta are phytophagous Hymenoptera, the females of which deposit their eggs in plant tissues. Haviland (1922) suggested that the evolutionary ancestors of Apocrita might have laid eggs in or nearby other plant parasitic larvae and may have become in this way parasitic on these insects. This would then represent the first evolutionary step in the development of the hymenopteran venom system. The Apocrita are divided into two major groups:

- The *Terebrantia*, which have an ovipositor (terebra or drill) which is also used as a ductus venatus and

Methods and Tools in Biosciences and Medicine
Animal Toxins, ed. by H. Rochat and M.-F. Martin-Eauclaire

- The Aculeata, which have an ovipositor (aculeus or sting) fully modified for injection venom into a host, prey, or enemy.

To the aculeate Hymenoptera belong the bees, the wasps, and the ants. The wasps are divided into the solitary wasps and the social wasps (Vespoidea). The ants are social. A single group of solitary wasps (the Scoliidae), all the social wasps (Vespidae), and the ants (Formicidae) are regarded to be related. The suggestion that the formicid-scoliid-vespid group is associated by a number of synapomorphies (Brothers, 1975) is supported by the fact that in the venoms of these Hymenoptera kinins or kinin–like activities have been found (Piek, 1992), while in the venoms isolated from bees and two groups of bee-related solitary wasps no kinin–like activity is detected (Tab. 1).

1.1 Vespid venoms and toxins

Stings by social wasps (Vespidae) can be painful (repellent function) and occasionally lethal. There are two main causes for adverse reactions and lethality:
- Mass attack, as sometimes occurs in the vicinity of the nest and
- dysfunction of the victims, body immune system.

For practical reasons we divide the Vespidae into three well-known groups:
- paper wasps (*Polistes*)
- yellow jackets (*Vespula, Paravespula*)
- hornets (*Vespa*)

The venoms of these wasps contain proteins such as acid phosphatase, hyaluronidase and phospholipase. These enzymes and a few nonenzymatic proteins possess antigenic properties, which can become an allergen to human beings. Moreover, wasp venoms contain a number of low molecular weight compounds, for example acetylcholine, histamine, sertonin, catecholamines (Tab. 1), and a number of polypeptides of moderate molecular weight:
- histamine-releasing peptides (Mastoparans)
- chemotactic peptides
- pain-producing peptides or wasp kinins (Nakajima, 1986)

1.2 Wasp kinins

Kinin-like activity was first demonstrated by the presence of a hypertensive substance in human urine (Frey and Kraut, 1928). Rocha e Silva *et al.* (1949) found that trypsinization of mammalian blood serum released a substance that

Table 1 Neurotoxins in venoms of Hymenoptera.

Family and genus	Ach	Hist	Ser	Glu	Cat	Kin	O
Vespidae(social wasps)							
Dolichovespula	-	+++	+++		+	++	
Vespula	-	+++	+++		+	++	
Vespa	+++	+++	+++		++	+	
Polites	-	+++	+++			+++	
Polybia		++	+			++	
Parapolybia						+	
Ropalidia		++	++		+		
Synoeca			+++			++	
Brachygastra		++	+			++	
Parachartergus			++	+			++
Scoliidae(solitary wasps)							
Megascolia	-	++	-			++	P
Scolia	-	+++		-		+	P
Colpa	-	++	-			+	P
Formicidae(ants)							
Pogonomyrmex	-	+?	-?			++	P
Paraponera	-	-?	-?			++	Poneratoxin
Dinoponera	-	-?	-?			++	P
Ectatomma	-	-?	-?			?	P
Diacamma	-	-?	-?			?	P
Myrmecia	-	-?	-?			++	P
Pompilidae (solitary wasps)							
Anoplius	-	+	-			-	
Batozonellus	-	+	-			-	
Episyron	++	-	-			-	
Pepsis	+++	-	-			-	
Sphecidae(solitary wasps)							
Bembix	-	++	-			-	
Cerceris	-	+	-			-	
Mellinus	-	++	-			-	
Philanthus	++	-	-	++		-	Philanthotoxin
Sceliphron	-	++	-			-	
Palmodes	-	+++	-			-	
Apidae(bees)							
Apis	-	+++	-		+++	-	Apamin, Melittin
Bombus	+++	+++	-			-	Bombolitin
Xylocopa	-	++	-			-	P

Ach = acetylcholine, Hist = histamine, Ser = serotonin, Glu = glutamate, Cat = catecholamines, Kin = kinins, O = other toxins;
- = <1 ng; + = 1–100 ng; ++ = 100–1000 ng; +++ >1000 ng, per venom resevoir, per milliliter or per milligram, which is roughly equivalent. No entry means «not tested for», P means present or probably present without quantification. ? = doubtful, the activity is masked by unknown agonists. For references see Blum and Hermann (1978), Piek and Spanjer (1986), Piek (1986), Nakajima (1986), Piek et al. (1989) and Toki et al. (1988).

caused a slow contraction of the guinea-pig ileum. They called the substance bradykinin because of its slow action compared with that of acetycholine or histamine. This bradykinin is a nonapeptide of which the amino acid sequence is presented in Table 2 (see also Elliot, 1970).

Bradykinin and two other mammalian kinins (Lys-bradykinin, or kallidin and Met-Lys-bradykinin) act on a variety of smooth muscles with contractions or relaxations. In the peripheral nervous system of vertebrates, bradykinin depolarizes nerve cells in sympathetic ganglia, and it may evoke release of catecholamines by depolarizing nerve terminals. Although in the mammalian CNS the function and mode of action of bradykinin are still obscure, nevertheless bradykinin is considered to be a brain peptide (Iversen, 1983).

A bradykinin-like substance was discovered as a venom constituent of the wasp *Paravespula vulgaris*, named wasp kinin (Schachter and Thain, 1954). Although the chemical characterization of this wasp kinin is still unknown, other wasp kinins have now been characterized chemically (Tab. 2). All wasp kinins show remarkable similarities to bradykinin. All of them contain a bradykinin-like sequence. Most of them are longer than bradykinin, either at one end or the other. The bradykinin-like part is either identical to bradykinin, or differs at position 3 or 6 where proline is replaced by hydroxyproline or serine is replaced by threonine, which means a single extra hydroxyl or methyl group, respectively.

It was not before the discovery of threonine^6-bradykinin (Thr^{6}BK) in solitary wasp venoms (Piek *et al.* 1983, 1987, 1990) that research on the effects of kinins on the synaptic transmission started. The solitary wasp, *Megascolia flavifrons*, stings its prey in all nerve ganglia involved in locomotion, and this results in an irreversible paralysis.

Intraganglionic injection or ganglionic perfusion of the venom from *Megascolia flavifrons* or of the venom from a related wasp, *Colpa interrupta*, as well as of the kinin (Thr^{6}BK, see Tab. 2) present in these fractions resulted in an irreversible block of transmission from the cockroach cercal nerve to a giant interneuron in the sixth abdominal ganglion of the cockroach (Piek *et al.* 1983, 1990; Hue and Piek 1988, 1989).

Kinins are neurotoxic components of venoms of several groups of Hymenoptera, including ants and social wasps. Kinins in the venom of solitary wasps may contribute to the permanent paralysis of the prey. However, social wasps do not sting the prey, but kill them by biting. They probably use the kinin activity for producing pain in vertebrates, which may function as an effective repellent against vertebrate intruders of the colony. Wasp intruders (interspecies competition) can be paralyzed or killed by stinging, and kinins may play an essential role here.

Studies on the mode of action of natural kinins may contribute to a better understanding of the role of (brady)kinin in the mammalian CNS.

Table 2 Wasp kinins in venoms of the Hymenoptera.

Species	Amino acid sequence[a]	Name of kinin	Ref.[b]
	1 2 3 4 5 6 7 8 9 Arg-Pro-Pro-Gly-Phe-Ser-Pro-Phe-Arg	Bradykinin (BK)	
Vespidae (Vespinae)			
Vespa mandarinia	Gly-Arg-Pro-Hyp-Gly-Phe-Ser-Pro-Phe-Arg-Ile-Asp	Vespakinin-M	1
Vespa xanthoptera	Ala-Arg-Pro-Pro-Gly-Phe-Ser-Pro-Phe-Arg-Ile-Val	Vespakinin-X	2
Vespa analis	Gly-Arg-Pro-Pro-Gly-Phe-Ser-Pro-Phe-Arg-Val-Ile	Vespakinin-A	3
Vespa tropica	Gly-Arg-Pro-Hyp-Gly-Phe-Ser-Pro-Phe-Arg-Val-Val	Vespakinin-T	4
Paravespula maculifrons	C C Thr-Ala-Thr-Thr-Arg-Arg-Arg-Gly-Arg-Pro-Pro-Gly-Phe-Ser-Pro-Phe-Arg	Vespulakinin	5
Paravespula lewisii	Thr-Ala-Thr-Thr-Lys-Arg-Arg-Gly-Arg-Pro-Pro-Gly-Phe-Ser-Pro-Phe-Arg NAcGal Gal	Vespulakinin-L	3
Vespidae (Polistinae)			
Polistes fuscatus *Polistes exclamans* *Polistes annularis*	Pyr-Thr-Asn-Lys-Lys-Lys-Leu-Arg-Gly-Arg-Pro-Pro-Gly-Phe-Ser-Pro-Phe-Arg	Polisteskinin-3	6
Polistes rothneyi	Arg-Pro-Pro-Gly-Phe-Thr-Pro-Phe-Arg	Thr6BK	7
	Ala-Arg-Arg-Pro-Pro-Gly-Phe-Thr-Pro-Phe-Arg	Polisteskinin-R	7
Polistes jadwigae	Arg-Arg-Arg-Pro-Pro-Gly-Phe-Thr-Pro-Phe-Arg	Polisteskinin-J	3
	Arg-Thr-Arg-Pro-Pro-Gly-Phe-Ser-Pro-Phe-Arg		
Polistes chinensis	Ser-Lys-Arg-Pro-Pro-Gly-Phe-Ser-Pro-Phe-Arg	Polisteskinin-C	3
Parapolybia indica	pGlu-Glx-Lys-Arg-Pro-Pro-Gly-Phe-Ser-Pro-Phe-Arg-Lys		9
	Gly-Glx-Glx-Lys-Arg-Pro-Pro-Gly-Phe-Ser-Pro-Phe-Arg-Lys-Arg		
	Glx-Glx-Lys-Arg-Pro-Pro-Gly-Phe-Ser-Pro-Phe-Arg-Lys-Arg		
	Glx-Glx-Lys-Arg-Pro-Pro-Gly-Phe-Ser-Pro-Phe-Arg-Lys-Lys		
Scoliidae			
Megascolia flavifrons	Arg-Pro-Pro-Gly-Phe-Thr-Pro-Phe-Arg	Thr6BK	8
	Arg-Pro-Pro-Gly-Phe-Thr-Pro-Phe-Arg-Lys-Ala	Megascoliakinin	8
Colpa interrupta	Arg-Pro-Pro-Gly-Phe-Thr-Pro-Phe-Arg	Thr6BK	10

[a] C = carbohydrate

[b] Key to references: 1, Kishimura et al. (1976); 2,Yasuhara et al. (1977); 3, Nakajima et al. (1984); 4, Yasuhara et al. (1983); 5, Yohida et al. (1976); 6, Udenfriend et al. (1967); 7, Watanabe et al. (1987); 9, Toki et al. (1988); 10, Piek et al. (1990)

2 Materials and Chemicals

Venom preparations

Hymenoptera venoms collected from dissected reservoirs according to the method of Schmidt (1986) are available from the Southwest Venoms (1961 West Brichta, AZ 85745, Fax: 520-884-9345).

Chemicals

NAdamantane acetyl-D-Arg-[Hyp^3,Thi^{5,8},DPhe^7] bradykinin	Sigma	B6029
D-Arg[Hyp^3, Thi^5, D-Tic^7, Oic^8] bradykinin	Hoechst	HOE-140
des Arg^9 bradykinin	Sigma	B4397
des Arg^9 [Leu^8] bradykinin	Sigma	6769
Bradykinin	Sigma	B3256
BPP_{5a}	Sigma	P8772
Captopril	Sigma	C-4042
Kallidin	Sigma	B4889
Sodium tetraborate	Aldrich	22,173-2

Physiological salt solutions

These solutions must be carefully prepared with pure materials, e.g., "Analar" grade, and glass-distilled or deionized water.

- **Solution for smooth muscle**
  119 mM NaCl
  4.7 mM KCl
  2.5 mM $CaCl_2$
  1.2 mM KH_2PO_4
  1.2 mM $MgCl_2$
  25 mM $NaHCO_3$
  11.1 mM glucose
- **Insect saline,** pH 7.2
  208 mM NaCl
  3.1 mM KCl
  >10.0 mM $CaCl_2$
  2.1 mM $NaHCO_3$
  >26 mM sucrose

Equipment

- vacuum-insulated container (Union Carbide, type LR-10C)
- transducer for isometric contractions (Grass Force-Displacement Transducer FT 03 C) connected to a bridge amplifier

- displacement transducers for isotonic contractions: older convenient transducers are Hewlett-Packard (7DCDT-1000 and 7DCDT-500); a modern transducer is the Lever Transducer Type B (368) from Hugo Sachs Elektronik (D7806 March-Hugstetten, F.R. Germany) connected to a Bridge Amplifier Type 301 (H. Sachs).

3 Methods

3.1 Methods for the collection of venoms

Social Hymenoptera usually can be collected in the field in large quantities; solitary wasps must be collected one by one in the field. When collected in the field, the wasps have to be in good condition before storing them in a freezer. A convenient method for collecting wasps in the field is to kill and transport them in liquid nitrogen. A very convenient cryogenic equipment is the vacuum-insulated container of about 10 l liquid nitrogen made by Union Carbide. Later the venom reservoirs can be removed from freshly thawed abdomens and extracted in distilled water in ice. The extracts can be filtered and freeze-dried. Extracts prepared from venom reservoirs are contaminated with soluble substances originating from the tissues of the reservoir and/or the venom gland. These venom preparations can only be used in physiological studies if the venoms are very potent and thus are applied in extremely diluted solutions. Purification of the active principles of the venom is normally required.

3.2 Microbioassays using smooth muscle

In the past, large amounts of wasp venom reservoirs were needed for chemical analyses. Nakajima (1986) describes how in the 1960s he required more than 6000 wasps to obtain the entire sequence of polisteskinin in 3 years, whereas in 1982 one venom reservoir was sufficient to isolate and characterize three kinds of peptides, including vespakinin-T, within a year. As a logical consequence, the amount of toxins in the fractions is so low that specialized microbioassays are requested.

The pharmacological activity of venoms or their neurotoxic fractions can be screened using a cascade of isolated vertebrate smooth muscle preparations (Ferreira and Vane, 1967). A useful variant of the cascade consists of an isolated rat fundus strip, a rat duodenum, a guinea pig ileum, a rat colon, and a guinea pig rectum (Piek, 1991).

Each muscle preparation is mounted in a bath, surrounded by a warming-up tube. Between the saline storage container and the first warming-up tube a continuous perfusion apparatus guarantees the continuous flow of saline, which is administered dropwise in order to superfuse the preparations in cascade. The Krebs saline is aerated with carbogen.

Contractions of the fundus strip can be recorded using a transducer for isometric contractions (see the section on equipment) connected to a bridge amplifier. Contractions and/or relaxations of the other muscles can be recorded using displacement transducers for isotonic contractions (see the section on equipment).

Examples of cascade registrations are presented in Figures 1 and 2, where the contractions and relaxations evoked by two wasp venom preparations, respectively, of two polybine wasps are compared with the reactions to 10-ng doses of acetylcholine, histamine, serotonin, and bradykinin. The rat fundus strip is extremely sensitive to serotonin. The rat duodenum shows a specific relaxation (often followed by a very slow contraction) with bradykinin. The rat colon is sensitive to angiotensin and often shows a weak relaxation, or only a decrease in spontaneous activity with bradykinin. Finally, the guinea pig rectum is sensitive to bradykinin and histamine.

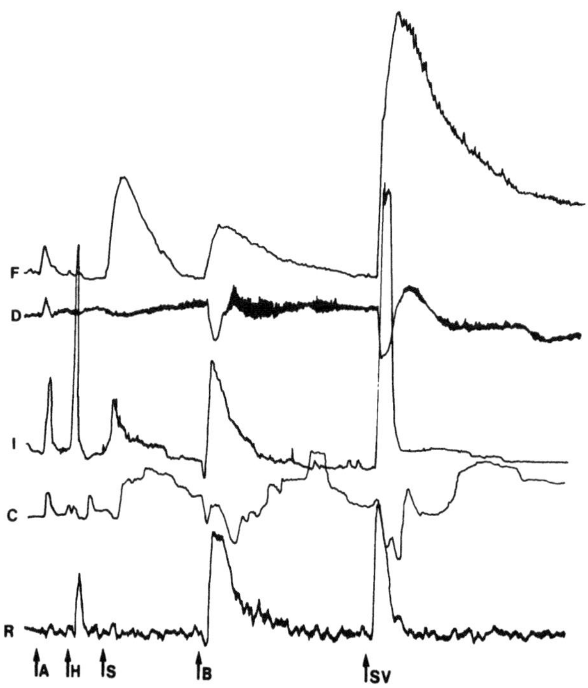

Figure 1 Effects of 10 ng acetylcholine (A), 10 ng histamine (H), 10 ng serotonin (S), 10 ng bradykinin (B), and 25 µg dried venom preparation from the wasp *Synoeca septentrionalis* (SV) on five different mammalian smooth muscle preparations, suspended in a cascade setup, and superfused by a Krebs solution at 37°C. (F) Rat fundus strip, (D) rat duodenem, (I) guinea-pig ileum, (C) rat colon, and (R) guinea-pig ileum.

Note the characteristic relaxation of (D) and (C) caused by bradykinin and by the venom.

The combination of contractions and/or relaxations of the various muscle preparations in response to an unknown agonist can be compared with the response to known agonists. Sometimes the results are clear; sometimes interactions mask some of the effects. For example, Figures 1 shows that the venom of the wasp *Synoeca septentrionalis*, applied at a dose of 25 µg in 50 µl saline at the top of the superfusion stream of the cascade, causes relaxation of the duodenum and colon and contraction of the fundus, the ileum, and the rectum, as is also seen in response to bradykinin (10 pmol). Quantitative differences can be caused by different active compounds present in the venom. The presence of a bradykinin-like toxin is evident. Figure 2 shows that the venom of the wasp *Polybia sericea* (25 μg) causes comparable effects. However, the initial relaxation of the duodenum is followed by a large contraction. Although such subsequent contractions are often present with kinins (see Fig. 1), interruption of the relaxation in a very early stage indicates the presence of another active component in the venom. In all cases superfusion with antagonists like the cholinergic antagonist atropine, the histaminergic antagonist mepiramine, the serotonergic antagonist methysergide, or the bradykinin antagonists Nα-adamantane acetyl-D.Arg (Hyp^3, Thi^{5,8}, D.Phe^7) bradykinin (Lammek *et al.*, 1990) or HOE-140 (D-Arg [Hyp^3, Thi^5, D.Tic^7, Oic^8) bradykinin (Hock *et al.*, 1991) may provide conclusive information.

The vertebrate smooth muscles described here contain a variety of chemically mediated transmission types, all present in the nervous networks as well

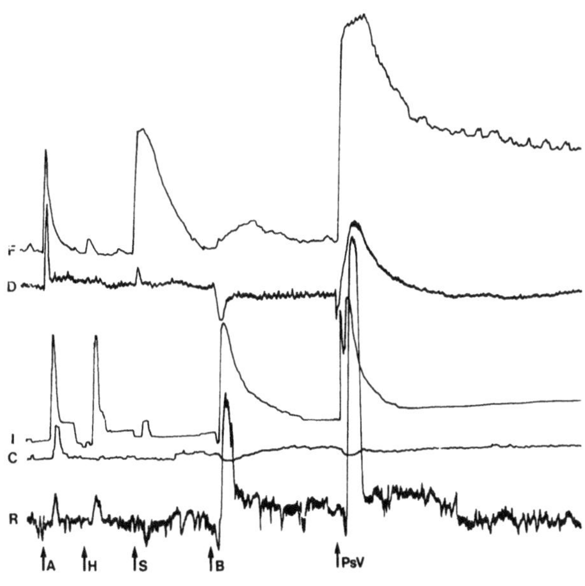

Figure 2 Effect of known agonists (see legend Fig. 1) and 25 μg of the venom of the wasp *Polybia sericea* (PSV) on mammalian smooth muscle preparations in cascade (see legend Fig. 1).

Note that the initial relaxation of (D) is antagonized by a contraction, which might be caused by an unknown agonist present in the crude venom preparation.

as in the neuromuscular synapses. They form a suitable combination of preparations providing a first indication of the presence of neuroactive components in the venom serve as a useful microbioassay guiding the process of chemical separation of toxins.

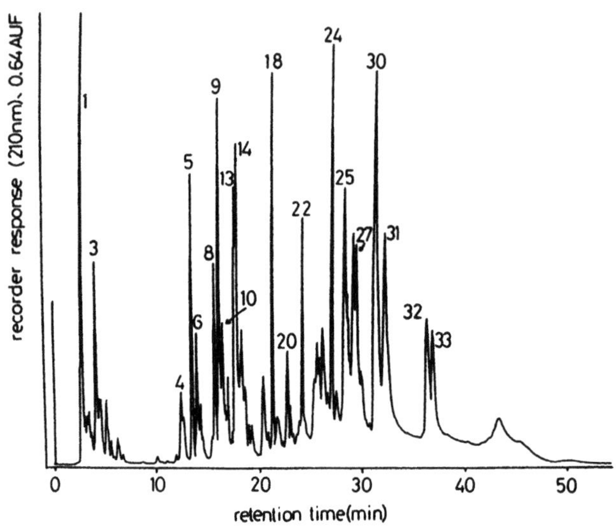

Figure 3 Chromatogram of 176 µg of the crude venom preparation of the wasp *Megascolia flavifrons*.

HPLC using a reverse phase column (ODS-LS-120T, Toyo-Soda). Linear gradient in 30 minutes from elute A (CH_3CN: 0.05% Ch_3 COOH, 5:95) to elute B (Ch_3CN: 0.05% CH_3 COOH, 3:7). Flow rate (ml/min) column temperature 40°C. Detection: UV at 210 nm. Reprinted by permission of the publisher from Yasuhara *et al.* (1987), *Toxicon* vol 25, pp 530, Elsevier Science Inc, New York.

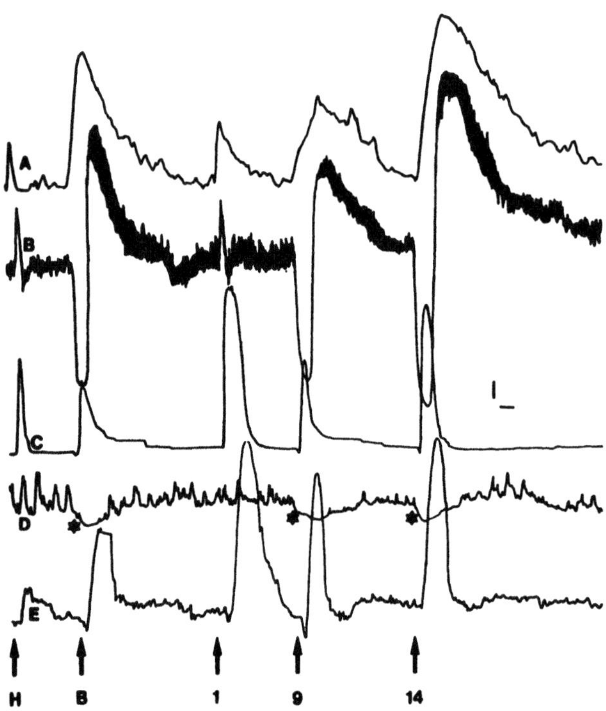

Figure 4 Effects 10 m ng histamine (H), 10 ng bradykinin (B), and half a venom reservoir equivalent of the fractions 1, 9, and 14 of Figure 3 on (A) rat fundus strip, (B) rat duodenum, (C) guinea pig ileum, (D) rat colon, and (E) guinea-pig rectum.

Note the contraction of all muscles, except the colon, with histamine and fraction 1 and the contraction of the fundus, the relaxation followed by a contraction of the duodenum, the ileum and the rectum, and the relaxation of the colon (asterisks) with bradykinin and fractions 9 and 14. All other fractions were inactive. Calibration 1 mm, 1 min. Reprinted by permission of the publisher from Yasuhara *et al.* (1987), *Toxicon,* vol. 25, pp 532, Elsevier Science Inc., New York.

An example of a cascade-guided isolation of kinins from the venom of the solitary scoliid wasp *Megascolia flavifrons* is illustrated in Figures 3 and 4.

Figure 3 shows a chromatogram of the crude venom preparation with more than 30 peaks. Figure 4 shows a cascade registration that includes the effects of fractions 1, 9, and 14, indicating that fractions 9 and 14 may contain a kinin. Of course, preceding this registration all fractions were tested, beginning with those containing small peaks, followed by fractions corresponding to relatively low molecular weight components and finally by the rest of the fractions. By following these sequences Yasuhara *et al.* (1987) were able to bioassay more than 30 fractions using only two cascade setups. In both cases the presence of a toxic fraction terminated in both cases the cascade experiments.

The amino acid sequences of the two kinins of the above fractions 9 and 14 are presented in Table 2.

Although cascade screening is a useful microbioassay, some imperfections are evident. The technique does not provide the researcher with quantitative results. It is possible to compare three doses of a known agonist (bradykinin) with three doses of the kinin-containing fraction to roughly indicate the kinin activity expressed as bradykinin equivalents. Another shortcoming may be that bioactive products produced by the upper muscle preparation as a result of the treatment by a venom or a fraction may cause contractions and/or relaxations of muscle preparation lower in the system. This is not the case with the known (reference) agonists, but is possible with unknown mixtures of bioactive products present in venoms or fractions. At suspicion of such effects a simple technique is to admit the samples not only at the top of the superfusion flow, but also above each of the individual muscle preparations.

To overcome the absence of a reliable quantification by the cascade experiments, bath experiments are indicated. For a simple comparison of kininergic activity, Figure 5 shows the relaxation of the isolated rat duodenum preparation in a bath by bradykinin (BK) and the crude venom of the wasp *Megascolia flavifrons* (MV) expressed as the number of individuals (venom reservoirs). The conclusion that here the venom preparation contains a kinin-like activity per wasp, which is comparable to about 7 ng bradykinin, is more reliable than the semiquantitative result obtained with the cascade technique.

3.3 Dose-response curves

For more reliable comparison of the potency of kinins, several mammalian smooth muscle preparation can be used. In our hands the isolated rat duodenum preparation is the most stable and reliable preparation. The kinins are tested using the duodenum in a 10-ml bath filled with aerated Krebs solution at 34°C. Kinins dissolved in physiological salt solution are added to the bath at doses of 100 μl resulting in different final concentrations. The duodenum preparation shows a characteristic concentration-dependent relaxation (Fig. 6). The amplitudes of relaxation can be plotted as the percentages of the maximal relaxation to bradykinin on a probit scale against log doses (Fig. 6,la). EC_{50} values are found by interpolation (±SEM). The 5% significance limits can be calculated using the t-test.

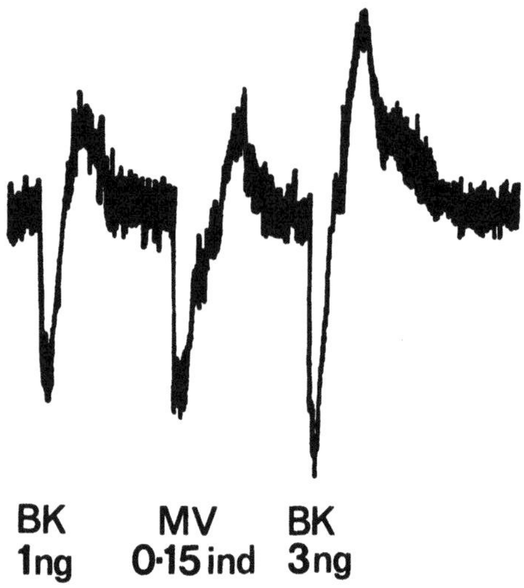

Figure 5 Relaxation of the isolated rat duodenum preparation by bradykinin (BK) and the crude venom of the wasp *Megascolia flavifrons* (MV), expressed as the number of individuals (venom reservoirs).

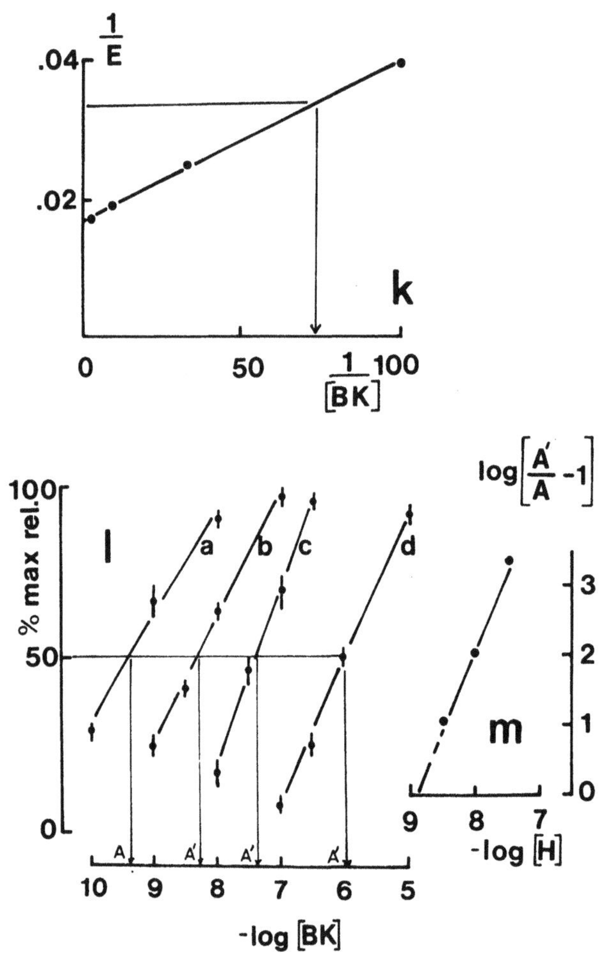

Figure 6 Dose-response relationships of the relaxation of the rat duodenum by bradykinin (BK) in the absence (a) and the presence of the B2-antagonist HOE-140 (H): (b) 10^{-8} M, (c) 10^{-7} M and (d) 10^{-6} M.

(K) Double reciprocal dose-response curve (Lineweaver-Burk plot) of bradykinin indicating an E_{max} at 1/[BK] = 0 and a K_d corresponding to 2x E_{max} (arrow). (l) Log dose-response curves of bradykinin in the absence and presence of HOE-140. (A, A^1) indicate the EC_{50}, respectively. Vertical bases are SEM (n = 4). (m) Schild plot indicating that the K_d (H) is about 10^{-9} M.

Besides the EC_{50} values it can be useful to determine the dissociation constant (K_d). If we accept the assumptions of the classical receptor theory, K_d can be estimated by plotting a number of concentration-effect values in a double reciprocal plot (Lineweaver-Burk plot). Figure 6k shows such a plot for the same data as presented at Figure 6,la. It is obvious that the EC_{50} and the K_d values for bradykinin are not exactly identical. However, they do not differ significantly.

Besides kininergic agonists, like the wasp kinins, there a number of (brady)-kinin antagonists have been developed. An example is HOE-140, which probably is a specific and competitive B2-antagonist (see below). This competitive antagonist shifts the dose-response curve to the right side (to higher antagonist concentration), in a concentration-dependent way (Fig. 6l). In the presence of the different concentrations of the antagonist (H) the EC_{50} of the agonist (B^1) can be divided by that in the absence of antagonist (B). In a Schild plot log (B/H-1) is plotted against log HV (Fig. 6m). This plot is based on the theoretical relationship between the K_d (H) the concentration of H and that op B with or without antagonist ($B^1$and B, respectively).

$$B^1/B = B/K_d(H)$$

For the derivation of this equation see Tallarida and Jacob (1979).

Figure 6m shows that the K_d value for the antagonist HOE-140 in the duodenum of the rat is $1{,}6 \times 10^{-9}$ M.

3.4 Receptor specification

Two different bradykinin receptors are distinguished in mammalian smooth muscle. B1 receptors mediate the action of bradykinin in blood vessels. The most potent agonist is des Arg^9 bradykinin; bradykinin is less potent at these sites. Intestinal muscles contain B2 receptors and sometimes also B1 receptors. Compared bradykinin, des Arg^9 bradykinin is less potent, in most cases about by two orders of magnitude.

Figure 7 shows relaxations of an isolated preparation of the duodenum of a rat, caused by bradykinin (A), des Arg^9 BK (B), kallidin, and cyclo-Lys-Lys-BK (see Section 5). Figure 7 also shows that anti-B2 antagonizes bradykinin and cyclo-Lys-Lys-BK and that anti-B1 antagonizes des Arg^9BK and kallidin.

3.5 Synaptic assays

Studies of the role of kinins in the mammalian CNS have demonstrated that these peptides significantly affect various aspects of behaviour (Capek, 1962;

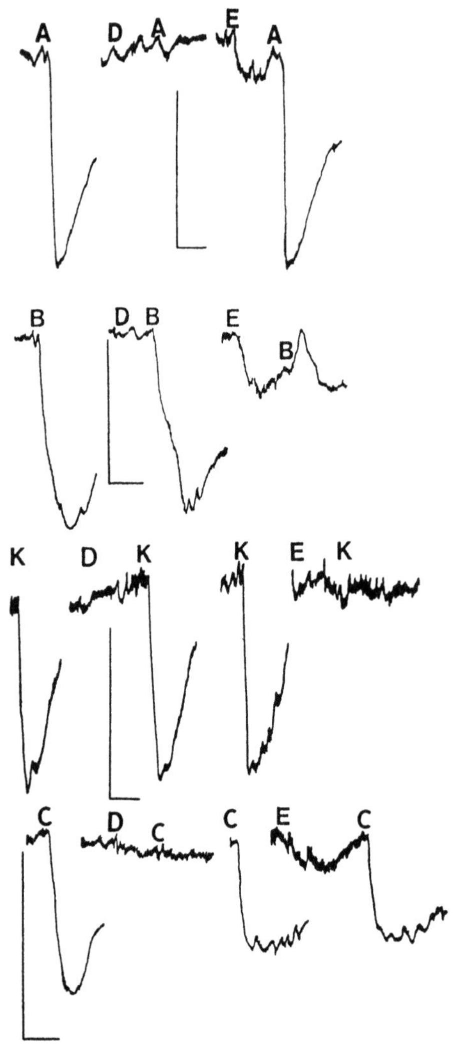

Figure 7 Relaxations of the isolated rat duodenum preparations in the presence of the antagonists (A) 10^{-9} M bradykinin (BK), (B) 10^{-7} M des Arg^{9}BK, (C) 10^{-9} M cyclo-Lys-Lys-BK, and (K) 10^{-9} M Lys-BK, all with and without a preceding administration of the antagonists (D) HOE-140 (anti-B2) 10^{-8}M or (E) des Arg^9 [Leu^8]-BK (anti-B1) 10^{-6}M. Note that E at this high concentration has its own agonistic effect. The recording shows that the rat duodenum may contain both B1 and B2 receptors, that BK and cyclo-Lys-Lys-BK are B2 agonists, and des Arg^9 BK and Lys-BK are B1 agonists. Calibration, 1 min and 50% of the maximal relaxation.

Moniuszko-Jakonick *et al*, 1986) and the effects of catecholamines or acetylcholine (Moniuszko-Jakonick *et al.*, 1976; Wisniewski *et al.*, 1974; Wisniewski and Bodzenta, 1975).

Bradykinin excites dorsal root ganglion cells, leading to the sensation of pain (Dray and Perkins, 1993). A nonpathological (natural) role of bradykinin in permeability of the blood-brain barrier has been suggested (Elliot *et al.*, 1996).

Hue developed a technique by means of which the sixth abdominal ganglion of the cockroach could perfused by pressure with toxins (Hue and Piek, 1988). The desheated A6 ganglion could be stimulated via the cercal nerve XI and to one of the identified isolated giant axons (GI_2, according to the nomenclature by Harris and Smyth, 1971). For recording electrical activity a single-fiber oil method was used (Callec *et al.*, 1980).

Bradykinin and threonine^{6}-bradykinin irreversibly block synaptic transmission from the cercal nerve to the giant interneuron. This block is frequency-dependent in a way which is similar to that of the block by hemicholineum-3. Inophoric applied carbachol causes postsynaptic potentials which are not affected by the kinins (Hue and Piek, *1989*). It was concluded that the kinins, like hemicholineum-3, block transmission at a presynaptic site, probably by depletion of the readily releasable store of transmitter substance. In contradiction to what is described in a next section on cyclic kinins, which in the mammalian smooth muscle preparation are less potent than their linear analogues, in the insect CNS cyclic bradykinin and cyclic vespulakinin 1 are equally potent or even more potent than linear kinins (Piek *et al.*, submitted)

4 Troubleshooting

Isolated mammalian smooth muscle preparations are generally studied in a bath containing Krebs saline solution and having an inlet for the inflow of fresh saline and an outlet, often controlled by a valve. The container (bath), the valve and all tubes or canules must be cleaned carefully after the experiments. If this is not done, microorganisms will grow in the remaining solution, which also contains organic materials released from the muscle. Cleaning with distilled water is not sufficient. Residue of detergent or disinfectant may also poisonignore the next preparation, but invisible microorganisms developing in the flow system are the worst. They can destroy the activity of a new muscle preparation very quickly, often within 20 min. A convenient desinfectant which is removed quickly with distilled water and/or saline solution is a 5% solution of sodium tetraborate (borax).

5 Stable Kinin Mimics

Since bradykinin may be an example of a short loop-forming bioactive peptide (Milner-White, 1989), cyclic kinin could also show kininergic potency, and as a matter of fact this is the case (Gobbo *et al.*, 1994; 1995,a,b;1996; Piek *et al.*, 1996). When tested on mammalian muscles, cyclic kinins are of low potency compared with their linear analogues.

However, their actions are potentiated by the bradykinin potentiating peptide BPP_{5a} isolated from the snake venom and inhibited by the bradykinin antagonist N-adamantanacetyl D-Arg-[Hyp^{3}, Thi^{5,8}, D-Phe^{7}] bradykinin (Gobbo *et al.*, 1995b) and by the specific B2-receptor antagonist HOE-140 (Piek *et al.*, 1995).

Although cyclic kinins are less potent than linear kinins, their potency depends on the dimensions of the cycle, i.e., on the number of backbone atoms.

There are 27 backbone atoms in cyclo-bradykinin, 30 in cyclo-kallidin, and 33 in cyclo-Lys-kallidin (cyclo-Lys-Lys-BK). The latter cyclic kinin is the most potent one; at a higher backbone atom number the potency declines progressively. Cyclo-Lys-Lys-BK is only one order of magnitude less potent than bradykinin (Gobbo *et al.*, submitted). It is not broken down quickly by kininase. Therefore, we can conclude that the first step in the direction of designing a stable active mimic of bradykinin is successful, which encourages further investigation.

Cyclic kinins are a lead to the development of a (nonpeptide) mimic of (bradykinin, both as a pharmacon affecting blood pressure or affecting permeability of the blood-brain barrier and as a new type of pesticide.

References

Blum MS, Hermann HR (1978) Venoms and venom apparatus of the Formicidae, Arthropod Venoms (Bettini S, ed) pp 801–869: Berlin: Springer

Brothers DJ (1975) Phylogeny and classification of the aculeate Hymenoptera, with special reference to Mutillide. Univ Kans Sc Bull 50: 483–648

Callec JJ, Sattelle DB, Hue B et al (1980) Central synaptic actions of pharmacological agents in insects: oil-gap and manitol-gap studies. Neurotox 19 (Sherwood M, ed) pp 93–100: New York: Plenum Press

Dray A, Perkins M (1993) Bradykinin and infammatory pain. Trends Neurosci 16: 99–104

Capek R (1962) Some effects of bradykinin on the central nervous system. Biochem Pharmacol. 10: 61–64

Elliot DF (1970) The discovery and characterization of bradykinin. Handbook exp. Pharmacol. 25: 7–13

Elliot PJ, Hayward NJ, Huft MR et al (1996) Unlocking the blood-brain barrier: a role for RMP-7 in brain tumor therapy. Exp. Neurol. 141: 204–224

Ferreira SH, Vane JR (1967) The disappearance of bradykinin and eledoisin in the circulation and vascular beds of the cat. Brit. J. Pharmacol. 30: 417–424

Frey EK, Kraut H (1928) Ein neues Kreislaufhormon und seine Wirking. Naunyn-Schmiedeberg's Arch. Exp. Pathol. Pharamacol. 133: 1–56

Gobbo M, Biondi L, Filira F et al (1994) Synthesis and biological activity of some linear and cyclic kinin analogues. Int J Peptide Protein Res. 44: 1–9

Gobbo M, Biondi L, Filira F et al (1995a) Cyclic analogues of wasp kinins from Vespa analis and Vespa tropica. Int. J. Peptide Protein Res. 45: 282–289

Gobbo M, Biondi L, Filira F et al (1995b) Cyclic analogues of Thr^{6}-bradykinin, N'-Lys-bradykinin and endo -Lys^{8a} vespulakinin 1. Int. J. Peptide Protein Res. 45: 459–465

Gobbo M, Biondi L, Filira F et al (1996) Linear and cyclic Tyr^{6}- and Trp^{6}-kallidin (Lys-bradykinin) analogues. Gazzetta Chimica Italiana 126: 359–363

Harris CL, Smyth T (1971) Structural details of cockroach grant axons revealed by injection dye. Comp. Biochem. Physiol. 40: 295–304

Haviland MD (1922) On the post-embryonic development of certain chalchids, hyperparasites of Aphids. Q. J. Micros. Sci. 66: 321–338

Hock FJ, Wirth U, Albus W et al (1991) HOE 140 a new potent and long acting bradykinin antagonist. Brit. J. Pharmacol 102: 769–773

Hue B, Piek T (1988) Effects of kinins and related peptides on synaptic transmission in the insect CNS. Neurotox 88: Molecular Basis of Drugs and Pesticide Action. (Lunt G.G, ed) pp 27–33: Amsterdam: Elsevier

Hue B, Piek T (1989) Irreversible presynaptic activation-induced block of transmission in the insect CNS by hemicholineum-3 and threonine^{6} bradykinin. Comp. Biochem. Physiol. 93C: 87–89

Jaques R, Schachter M (1954) The presence of histamine, 5-hydroxytryptamine and a potent. Slow contracting substance in wasp venom. Br. J. Pharmacol. 9: 53–57

Lammek B, Wang Yi-Xin, Gravas I et al (1990) A new highly potent antagonist of bradykinin. Peptides 11: 1041–1043

Milner-White EJ (1989) Predicting the biologically active conformations of short polypeptides. TIPS 10: 70–74

Moniusko-Jakonick J, Wisniewski K, Koscielak M (1976) Investigations of the mechanism of central actions of kinins. Psychopharmacol. 50: 181–186

Nakajima T (1986) Pharmalogical biochemistry of vespid venoms. Venoms of the Hymenoptera (Piek T., ed) pp 309–327: London: Academic Press

Piek T (1986) Venoms of the bumble bees and carpenter bees. Venoms of the Hymenoptera (Piek T, ed) pp 417–424: London: Academic Press

Piek T (1986) Historical introduction. Venoms of the Hymenoptera (Piek T, ed) pp 1–16: London: Academic Press

Piek T (1991) Chemistry, Isolation and Microbioassays of Arthropod Neurotoxins. Pesticide Chemistry (Frehse H, ed.) pp 75–85: Weinheim, VCH

Piek T (1992) A toxinological argument in favour of the close relationship of Vespidae, Scoliidae, Tiphiidae, Mutillidae, and Formicidae (Hymenoptera). Proc. Exper. Appl. Entomol. 3: 99–104

Piek T, Buitenhuis A, Simon Thomas RT et al (1983) Smooth muscle contracting compounds in the venom of Megascolia flavifrons (Hym: Scoliidae). Comp. Biochem. Physiol. 75C: 145–152

Piek T, Hue B, Mony L et al (1987) Block of synaptic transmission in insect CNS by toxins from the venom of the wasp Megascolia flavifrons (Fab.) Comp. Biochem. Physiol. 87C: 287–295

Piek T, Hue B, Mantel P et al (1990) Threonine^{6}-bradykinin in the venom of the wasp Colpa interrupta (F) presynaptically blocks nicotinic transmission in the insect CNS. Comp. Biochem. Physiol. 96C: 157–162

Piek T, Gobbo M, Mantel P et al (1996) B2-kininergic action of linear and cyclic tryptophan^{6}- and tyrosine^{6}-kallidin. Life Sci 59: 391–397

Schachter M, Thain EM (1954) Chemical and pharmacological properties of the potent slow contracting substance (kinin) in wasp venom. Br. J. Pharmacol. 9: 352–359

Schmidt JO (1986) Chemistry, pharmacology, and chemical ecology of ant venom. Venoms of the Hymenoptera. (Piek T, ed) pp 425–507: London: Academic Press

Tallarida RJ, Jacob LS (1979) The Dose-Response Relation in Pharmacology. Berlin: Springer

Toki T, Nakajima T, Yasuhara T (1988) Isolation and sequential analysis of peptides on the venom sac of Parapolybia indica. Jpn. J. San. Zool. 39: 105–111

Westwood JO (1840) An Introduction to the Modern Classification of Insects. London: Longman, Orme, Brown, Green, and Longmans

Wisniewski K, Bodzenta A (1975) Kinins and central effects on the acetylcholine. Acta Neurobiol. Exp. 35: 85–92

Yasuhara T, Mantel P, Nakajima T et al (1987) Two kinins isolated from an extract of the venom reservoirs of the solitary wasp Megascolia flavifrons. Toxicon 25: 527–535

8 Mastoparan as a G Protein Activator

Terumi Nakajima, Kaori Wakamatsu and Hidehito Mukai

Contents

1 Introduction

Mastoparan was isolated from the venom of wasp, *Vespula lewisii*, as a peptide that induces histamine release from rat peritoneal mast cells (Hirai *et al.*, 1979). It is a C-terminally amidated tetradecapeptide comprising primarily basic and hydrophobic amino acid residues but no acidic residues (Tab. 1). Structurally similar peptides that also induce the histamine release have been isolated from venom of other wasps and hornets, and are collectively termed mastoparans (Nakajima *et al.*, 1985). Mastoparans were also found to induce exocytosis from a variety of cells: serotonin from platelets (Ozaki *et al.*, 1990), insulin from pancreatic β cells (Yokokawa *et al.*, 1989), surfactant from pulmonary alveolar type 2 epithelial cells (Joyce-Brady *et al.*, 1991), and catecholamine from chromaffin cells (Kuroda *et al.*, 1980; Vitale *et al.*, 1993). This peptide also elevates intracellular Ca^{2+} concentration in neutrophils (Perianin and Snyderman, 1989) and promotes proliferation of Swiss 3T3 cells (Gil *et al.*, 1991).

Investigations directed toward elucidation of the regulatory mechanism of mastoparan have shown that this peptide, directly or indirectly, modulates activities of proteins as diverse as trimeric GTP-binding regulatory proteins (G

Methods and Tools in Biosciences and Medicine
Animal Toxins, ed. by H. Rochat and M.-F. Martin-Eauclaire

Table 1 Primary structure of mastoparan and its derivatives.

Peptide	Primary structure[a]	Commercial source[b]	notes
mastoparan	Ile Asn Leu Lys Ala Leu Ala Ala Leu Ala Lys Lys Ile Leu NH2	*1,2,3,4,5,6*	*c*
Mas7	Ile Asn Leu Lys Ala Leu Ala Ala Leu Ala Lys <u>Ala</u> <u>Leu</u> Leu NH2	*1,2,3,4*	*d*
Mas8	Ile Asn Leu Lys Ala Leu Ala Ala Leu Ala Lys <u>Arg</u> <u>Leu</u> Leu NH2	*1,3*	*d*
Mas11	Ile Asn Leu Lys Ala Leu Ala Ala Leu <u>Lys</u> Lys Lys <u>Leu</u> Leu NH2	none	*d*
Mas17	Ile Asn Leu Lys Ala <u>Ala</u> Ala Ala Leu Ala Lys Lys <u>Leu</u> Leu NH2	*1,2,3,4*	*d*

[a] Amino acid residues that differ from those of mastoparan are shown underlined.
[b] Commercial source: *1*, Peninsula Laboratories; *2*, Bachem; *3*, Phoenix Pharmaceuticals; *4*, Biomolecular Research Laboratories; *5*, Peptide Institute; *6*, Sigma.
[c] from (Hirai *et al.*, 1979)
[d] from (Higashijima *et al.*, 1990)

proteins) of the G_i/G_o family (Higashijima *et al.*, 1988, 1990; Ross and Higashijima, 1994), calmodulin (Busse and Mülsch, 1990), phospholipases (Argiolas and Pisano, 1983), protein kinase C (Raynor *et al.*, 1991), etc. (Tab. 2). Mastoparan is also known to increase permeability of ions and small molecules through membranes by forming pores at a high peptide concentration (Matsuzaki *et al.*, 1996).

Mastoparan can be used to activate G proteins bound to cytoplasmic membranes in cells where no G protein-coupled receptors are known (e.g., peritoneal mast cells) and G proteins bound to intracellular organelles (Ktistakis *et al.*, 1992). And, in fact, mastoparan has been used as an activator of G proteins in more than 60% of applications at the time of preparing this manuscript (January, 1997). Researchers, however, should keep in mind that the mastoparans act on various proteins and do appropriate control experiments to exclude the possibility that the mastoparan-induced responses are mediated by proteins other than G proteins or through the damage of cell membranes. In this chapter we describe general precautions to be taken in applying mastoparan as a G protein activator and present protocols to assay exocytosis and determine the

Table 2 Some proteins known to be modulated by mastoparan.

Protein	Action	Effective Conc. [μM]	Reference
Gi/Go	↑	10	(Vitale *et al.*, 1993) (Mukai and Higashijima, 1994)
Rho	↑	20	(Koch *et al.*, 1991)
Calmodulin	↓	0.3	(Busse and Mülsch, 1990)
Phospholipase A2	↑	5	(Argiolas and Pisano, 1983)
Phospholipase D	↑	10	(Mizuno *et al.*, 1995)
Protein kinase C	↓	10	(Raynor *et al.*, 1991)
G protein coupled Receptor kinase	↑/↓	10/100[a]	(Haga *et al.*, 1994)

[a] Activation at a low peptide concentration and inhibition at a high concentration.

change in intracellular free Ca^{2+} concentration ($[Ca^{2+}]_i$) in rat peritoneal mast cells when stimulated by mastoparan, as examples for the application of mastoparan in cell biology.

2 Precautions in Applying Mastoparan as a G Protein Activator

Cell damage: First of all, for a particular cell type used, the mastoparan concentration range must be determined where membrane damage can be neglected, e.g., by determining the activity of lactate dehydrogenase (LDH, a maker enzyme of cytosol which is not released upon physiological exocytosis) leaked from the cells. In this respect, an analog Mas11 which is *ca.* 10 times more potent than mastoparan in stimulating exocytosis from mast cells yet causes no membrane damage even at 100 μM can be a promising substitute for mastoparan (Figure1 and Tab. 3; Mukai *et al.*, submitted for publication). Concentration dependence of mastoparan-induced response also gives some insight into the pathway(s) involved, since effective concentration of this peptide varies with target proteins regulated (Tab. 2).

Blockade by PTX and GPAnt: The most convincing evidence for the involvement of G_i/G_o in mastoparan's action would come from the blockade of the mastoparan-induced response by pertussis toxin (PTX) (Higashijima *et al.*, 1987), which specifically ADP-ribosylates G proteins of the G_i/G_o family to make them insensitive to mastoparan as well as liganded receptors. Use of G protein antagonists (GPAnt; Mukai *et al.*, 1992), which reverses the activation of G_i/G_o by mastoparan (Vitale *et al.*, 1993), would also be helpful.

Other activators: The use of mastoparan analogs which are more potent (Mas7) or less potent (Mas17) in activating G proteins (Tab. 3; refer also to Hi-

Table 3 Effects of mastoparan and its analogs on rat peritoneal mast cells and on the activity of G_i *in vitro*.

Peptide	Exocytosis[a]	Cell Damage[b]	Fold G_i Activation[c]	
	EC_{50} (μM)	EC_{20} (μM)	10 μM	100 μM
Mastoparan	7.1	39.8	10.2	17.0
Mas7	3.8	18.0	22.8	75.6
Mas8	3.6	33.0	n.t.[d]	
Mas11	0.5	–[e]	3.0	9.8
Mas17	16.9	–	n.t.	

[a] Determined from β-hexosaminidase release.
[b] Determined from leakage of LDH.
[c] Activation (in fold) of rabbit liver G_i reconstituted into phospholipid vesicles.
[d] n.t.: not tested.
[e] –: Activity was not detected at 100 μM.

gashijima *et al.*, 1990) will provide additional support for the involvement of G proteins. It should be noted, however, that the potencies of mastoparan analogs vary depending on the particular cell type used: e.g., although mastoparan potently stimulates exocytosis from the HL-60 cells differentiated to neutrophil-like ones, Mas11 has no effects on these cells. When permeabilized cells are used, application of GTPγS or aluminum fluoride that directly activates G proteins is expected to elicit responses similar to those of mastoparan (Vitale *et al.*, 1993).

3 Materials and Chemicals

Buffers and solutions

- **Incubation buffer I** (IB1, for exocytosis assays), pH 7.4 **(Solution 1)**
  - 135 mM NaCl
  - 4.7 mM KCl
  - 1.2 mM KH_2PO_4
  - 1.0 mM $CaCl_2$
  - 0.6 mM $MgSO_4$
  - 3.0 mM $NaHCO_3$
  - 5.6 mM glucose
  - 5.0 mM HepesNa
  - 0.1% bovine serum albumin (BSA)
  - 0.1% gelatin
  - 10 U/ml heparin
- **Incubation buffer II** (IB2, for $[Ca^{2+}]_i$ measurements), pH 7.4 **(Solution 2)**
  - 140 mM NaCl
  - 4 mM KCl
  - 1 mM $MgCl_2$
  - 1.25 mM $CaCl_2$
  - 1 mM NaH_2PO_4
  - 11 mM glucose
  - 0.2% BSA
- **70% Percoll-IB1**, pH 7.4 **(Solution 3)**
  - 70% Percoll
  - 135 mM NaCl
  - 4.7 mM KCl
  - 1.2 mM KH_2PO_4
  - 1.0 mM $CaCl_2$
  - 0.6 mM $MgSO_4$
  - 3.0 mM $NaHCO_3$
  - 5.0 mM HepesNa

- **85% Percoll-IB1**, pH 7.4 **(Solution 4)**
  85% Percoll
    135 mM NaCl
    4.7 mM KCl
    1.2 mM KH_2PO_4
    1.0 mM $CaCl_2$
    0.6 mM $MgSO_4$
    3.0 mM $NaHCO_3$
    5.0 mM HepesNa
- **Stop buffer I** (for termination of cell-stimulation), pH 7.4 **(Solution 5)**
    25 mM Tris/Cl
    123 mM NaCl
    2.7 mM KCl
- **Stop buffer II** (for termination of β-hexosaminidase reaction), pH 10.7 **(Solution 6)**
    0.4 M glycine
- **Stop buffer III** (for termination of LDH reaction), pH 8.5 **(Solution 7)**
    0.2 M Tris/Cl
    5 mM $NaHCO_3$
    50 mM Na_2CO_3
- **Substrate solution I** (for β-hexosaminidase reaction) **(Solution 8)**
    10 mM of *p*-nitrophenyl-*N*-acetyl-β-D-glucosaminide in
    40 mM citric acid and
    70 mM Na_2HPO_4, pH 4.5
- **Substrate solution II** (for LDH reaction) **(Solution 9)**
    2 mM NAD and
    10 mM lactic acid in
    0.2 M Tris/Cl, pH 8.5
- **Peptide solution**
    Peptides are dissolved in phosphate-buffered saline (PBS).
    Notes:
    If they are insoluble in PBS, peptide solutions can contain dimethyl sulfoxide (DMSO) up to 0.1% (v/v). If needed, higher DMSO concentrations can be adopted when experimentally confirmed to be permissive; the maximal concentration of DMSO where cell functions are not inhibited varies greatly depending on the particular cell types used.

Lyophilized peptides and peptide solutions should be stored below –20°C to avoid oxidation and adsorption onto tubes. Peptides listed in Table 1 and GPAnt's are stable over months under these conditions.

Chemicals

Mastoparan-related peptides	Listed in Table 1	
Compound 48/80	Sigma	C2313
p-Nitrophenyl-*N*-acetyl-β-D-glucosaminide	Sigma	N9376
Lipid-poor BSA	Calbiochem	126575
Percoll	Pharmacia Biotech	17-0891
NAD	Sigma	N7129
lactic acid	Sigma	L1750
Fura-2/acetoxymethyl ester (fura-2/AM)	Dojin	343-05401
	Molecular Probe	F-1201
Toluidine blue O	Sigma	T0394
Trypan blue	Wako Chemical	207-03252

Equipment

- Water bath with humidified chamber
- 96-well microplate reader
- Spectrofluorometer for $[Ca^{2+}]_i$ measurements: CAF-100 or CAF-110 (Japan Spectroscopy Inc.) are strongly recommended
- Spectrofluorometer for cell damage estimation: any standard spectrofluorometer

4 Methods

4.1 Preparation of rat peritoneal mast cells

In contrast to receptor agonists that specifically stimulate a certain set of cells possessing their target receptor molecules, mastoparans can stimulate a wide variety of cells. Purification of cells of interest is hence essential when investigating the effects of mastoparan.

Protocol 1 Preparation of rat peritoneal mast cells

1. Sprague Dawley or Wister rats (400–500 g) are sacrificed under anesthesia with ether, and cells from the peritoneal cavity are collected in incubation buffer I (**Solution 1**) and washed twice with **Solution 1** by centrifugation at 200 × g for 5 min.
2. The cells are suspended in 1.5 ml of 85% Percoll-IB1 (**Solution 4**) in a 15-ml conical polypropylene tube, overlaid with 1.5 ml of 70% Percoll-IB1 (**Solution 3**), and then centrifuged at 225 × g for 10 min at 4°C.
3. The 70% Percoll-IB1 layer and the upper half of the 85% Percoll-IB1 layer are discarded, and the cell pellet is resuspended in IB1 (**Solution 1**).
4. The last step is repeated once again, and the purified cells are washed three times with IB1 (**Solution 1**) and kept on ice until use.

Notes:
About 2 × 10^6 cells per rat will be obtained at over 95% purity (determined by toluidine blue staining) and 98% viability (determined by trypan blue exclusion).

When the involvement of G proteins of the G_i/G_o family in the $[Ca^{2+}]_i$ increase or exocytosis is to be confirmed by blockade with PTX, purified cells are incubated with or without 100 ng/ml PTX for 2 h at 37°C, washed three times with IB1, then kept on ice until use.

The involvement of the G proteins can also be confirmed by using G protein antagonists (GPAnt's, Mukai *et al.*, 1992), which antagonize the stimulatory effect of liganded receptors or mastoparan when applied during preincubation. Among the GPAnt's examined so far, GPAnt-2 is the most potent. The low solubility of GPAnt-2 in an aqueous medium, however, necessitates inclusion of DMSO at 1% in the assay medium to attain 30 µM of this antagonist such that responses via G_i/G_o are completely inhibited. When such a high concentration of DMSO inhibits cellular functions as in mast cells, the second choice is the use of GPAnt-1. Although GPAnt-1 partially activates G_i/G_o, it inhibits at 30 µM the exocytosis from mast cells almost completely (refer also to Mukai and Higashijima, 1993; Mukai *et al.*, 1992).

4.2 β-Hexosaminidase release assay

When rat peritoneal mast cells are challenged by mastoparan or compound 48/80, exocytosis of secretory vesicles containing histamine occurs. Since it takes *ca.* half an hour per sample to quantify secreted histamine by HPLC, we recommend assay of exocytosis by determining secreted β-hexosaminidase, which is known to colocalize with histamine in the vesicles and can be quantified in multiples simultaneously in a 96-well plate. This assay ,however, necessitates, purification of mast cells as described in the previous section, because other cells in the rat peritoneal cavity, neutrophils and macrophages, also contain this enzyme in secretory vesicles. Figure 1A shows β-hexosaminidase release induced by mastoparan and its analogs.

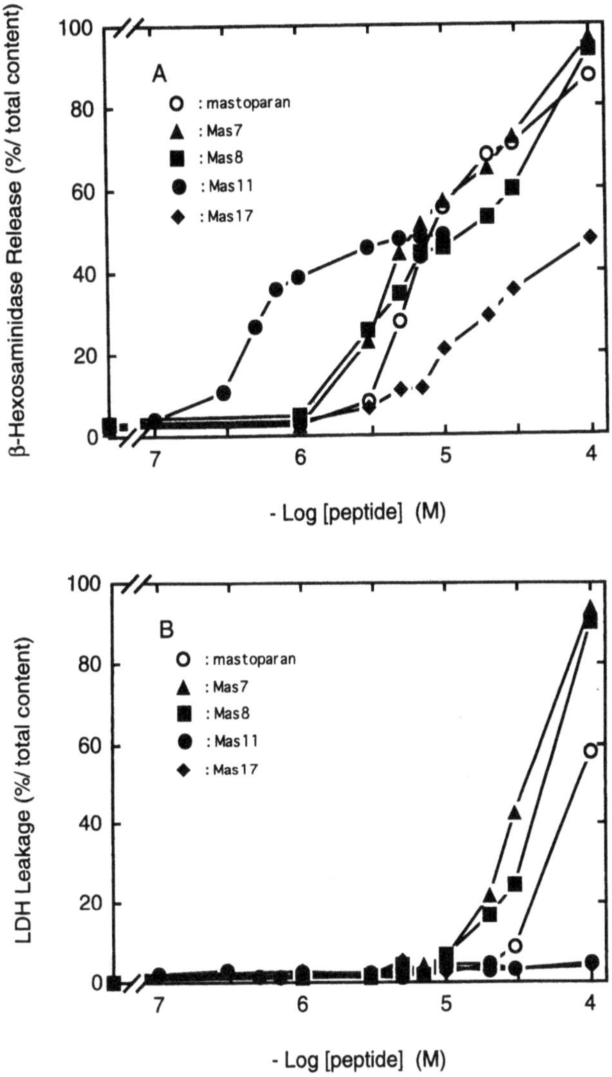

Figure 1 Effects of mastoparan and its analogs on rat peritoneal mast cells (Mukai *et al.*, submitted for publication).

Purified rat peritoneal mast cells were stimulated by each concentration of mastoparan (○), Mas7 (▲), Mas8 (■), Mas11 (●), and Mas17 (◆) for 10 min at 37°C, and β-hexosaminidase released (panel A) and LDH leaked (panel B) were determined as described in the text.

Protocol 2 β-Hexosaminidase release assay

1. Rat peritoneal mast cells (5×10^4 cells) in 90 μl of IB1 are preincubated for 10 min at 37°C.
2. The secretion of β-hexosaminidase is initiated by adding 10 μl of peptide solution or compound 48/80 solution (100 μg/ml). The cell suspension is incubated for 10 min at 37°C, then terminated with 200 μl of ice cold stop buffer I (**Solution 5**).
3. The released enzyme is separated from cells by immediate centrifugation (1 min, Fisher model 235C).
4. The cell-free supernatant (90 μl) is incubated at 37° C for 2 h with 60 μl of substrate solution I (**Solution 8**) in a 96-well microplate.
5. The enzyme reaction is stopped by adding 100 μl of stop buffer II (**Solution 6**).
6. Absorbance at 410 nm (of *p*-nitrophenol produced) is determined on a microplate reader.
7. The total content of β-hexosaminidase in the cells is determined with cells that are incubated with 0.05% Triton X-100 for 10 min at 37°C.

Notes:
The extent of the enzyme secretion by each peptide sample is expressed as the percentage versus the total content or vs. the amount released in the presence of 10 μg/ml compound 48/80.

4.3 $[Ca^{2+}]_i$ Measurements (Mukai *et al.*, in preparation)

Protocol 3 $[Ca^{2+}]_i$ Measurements

1. Purified rat peritoneal mast cells are washed twice with IB2 (**Solution 2**), incubated with 4 μM fura-2/AM for 1 h at 20°C to load fura-2 into cytosol.
2. The cells are then washed twice, resuspended in fresh IB2 without fura-2/AM, and stored on ice until use.
3. Cell suspension (1 ml, 10^6 cells/ml) in a cuvette is constantly stirred, and the ratio of fluorescence (λ_{em} = 500 nm) at two excitation wavelengths, F(λ_{ex} = 350 nm)/F(λ_{ex} = 340 nm) is monitored on a CAF-100 spectrofluorometer.
4. The cells are stimulated by adding 10 μl of peptide solutions, and the $[Ca^{2+}]_i$ is calculated as described (Grynkiewicz *et al.*, 1985).

Note:
Since fura-2 loaded in cytosol of rat peritoneal mast cells easily leaks from the cells or translocates from cytosol to organelles, the cells should be kept on ice until use and warmed just prior to the measurement.

Figure 2 shows the elevation of $[Ca^{2+}]_i$ induced by mastoparan and Mas11.

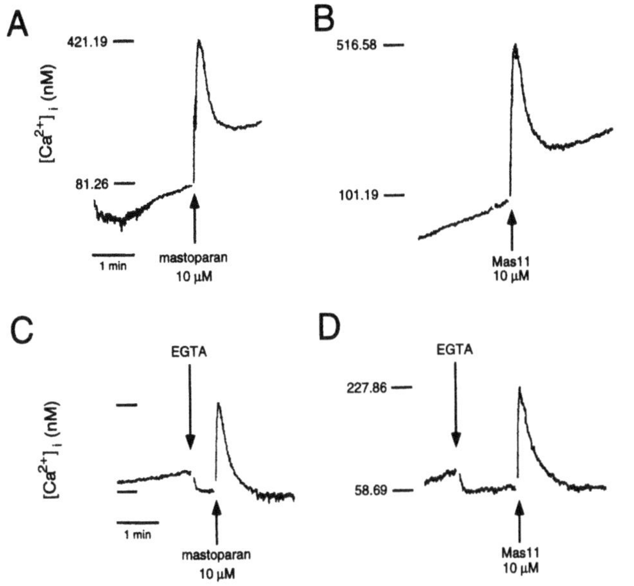

Figure 2 Effects of mastoparan and Mas11 on $[Ca^{2+}]_i$ in rat peritoneal mast cells (Mukai *et al.*, in preparation).

Purified mast cells were stimulated by 10 μM mastoparan (panels A and C) or Mas11 (panels B and D) in the presence of 1.25 mM Ca^{2+} (panels A and B) or 4 mM EGTA (panels C and D) at 37°C. The attenuated but definite rise in $[Ca^{2+}]_i$ in panels C and D suggests that these peptides induce both mobilization of Ca^{2+} from intracellular Ca^{2+} stores and influx of extracellular Ca^{2+}.

4.4 Estimation of cell damage

Figure 1B shows leakage of LDH induced by mastoparan and its analogs.

Protocol 4 Estimation of cell damage

1. Aliquots of cell-free supernatants (20 μl) obtained in the above experiments are incubated with 100 μl of substrate solution II (**Solution 9**) for 15 min at 37°C in a glass tube.
2. The reaction is stopped with 1 ml of ice-cold stop buffer III (**Solution 7**), and the fluorescence of each tube (λ_{ex} = 475 nm, λ_{em} = 345 nm) is determined on a spectrofluorometer.
3. The LDH released is expressed as the percentage *versus* the total content of LDH in cells that is determined using the cell-free supernatant solubilized with 0.05% Triton X-100.

References

Argiolas A, Pisano JJ (1983) Facilitation of phospholipase A_2 activity by mastoparans, a new class of mast cell degranulating peptides from wasp venom. J. Biol. Chem. 258: 13697–702

Busse R, Mülsch A (1990) Calcium-dependent nitric oxide synthesis in endothelial cytosol is mediated by calmodulin. FEBS Lett. 265: 133–6

Gil J, Higgins T, Rozengurt E (1991) Mastoparan, a novel mitogen for Swiss 3T3 cells, stimulates pertussis toxin-sensitive arachidonic acid release without inositol phosphate accumulation. J. Cell. Biol. 113: 943–50

Grynkiewicz G, Poenie M, Tsien RY (1985) A new generation of Ca^{2+} indicators with greatly improved fluorescence properties. J. Biol. Chem. 260: 3440–50

Haga K, Kameyama K, Haga T (1994) Synergistic activation of a G protein-coupled receptor kinase by G protein βγ subunits and mastoparan or related peptides. J. Biol. Chem. 269: 12594–9

Higashijima T, Uzu S, Nakajima T et al (1987) Mechanism of histamine-release by mastoparan. Peptide Chemistry 1986 (Miyazawa T, ed) pp 75–8: Osaka: Protein Research Foundation

Higashijima T, Uzu S, Nakajima T et al (1988) Mastoparan, a peptide toxin from wasp venom, mimics receptors by activating GTP-binding regulatory proteins (G proteins). J. Biol. Chem. 263: 6491–4

Higashijima T, Burnier J, Ross EM (1990) Regulation of G_i and G_i by mastoparan, related amphiphilic peptides, and hydrophobic amines. J. Biol. Chem. 265: 14176–86

Hirai Y, Yasuhara T, Yoshida H et al (1979) A new mast cell degranulating peptide "mastoparan" in the venom of Vespula lewisii. Chem. Pharm. Bull (Tokyo) 27: 1942–4

Joyce-Brady M, Rubins JB, Panchenko MP et al (1991) Mechanisms of mastoparan-stimulated surfactant secretion from isolated pulmonary alveolar type 2 cells. J. Biol. Chem. 266: 6859–65

Koch G, Haberman B, Mohr C et al (1991) Interaction of mastoparan with the low molecular mass GTP-binding proteins rho/rac. FEBS Lett. 291: 336–40

Ktistakis NT, Linder ME, Roth MG (1992) Action of brefeldin A blocked by activation of a pertussis-toxin-sensitive G protein. Nature 356: 344–6

Kuroda Y, Yoshida M, Kumakura K et al (1980) Effects of peptides on the release of catecholamines and adenine nucleotides from cultured adrenal chromaffin cells. Proc. Japan Acad. [B] 56: 660–4

Matsuzaki K, Yoneyama S, Murase O et al (1996) Transbilayer transport of ions and lipids coupled with mastoparan X translocation. Biochemistry 35: 8450–6

Mizuno K, Nakahata N, Ohizumi Y (1995) Mastoparan-induced phosphatidylcholine hydrolysis by phospholipase D activation in human astrocytoma cells. Br. J. Pharmacol. 116: 2090–6

Mukai H, Higashijima T (1993) Inhibitory effects of substance P analogs on the activation of GTP-binding regulatory proteins by receptors. Regul. Pept. 46: 338–9

Mukai H, Higashijima T (1994) The regulatory mechanisms of GTP-binding regulatory proteins by mastoparan and its derivatives. Peptide Chemistry 1993 (Okada Y, ed) pp 309–12: Osaka: Protein Research Foundation

Mukai H, Munekata E, Higashijima T (1992) G protein antagonists. J. Biol. Chem. 267: 16237–43

Nakajima T, Yasuhara T, Uzu S et al (1985) Wasp venom peptides; wasp kinins, new cytotrophic peptide families and their physico-chemical properties. Peptides 6 Suppl. 3: 425–30

Ozaki Y, Matsumoto Y, Yatomi Y et al (1990) Mastoparan, a wasp venom, activates platelets via pertussis toxin-sensitive GTP-binding proteins. Biochem. Biophys. Res. Commun. 170: 779–85

Perianin A, Snyderman R (1989) Mastoparan, a wasp venom peptide, identifies two discrete mechanisms for elevating cytosolic calcium and inositol trisphosphates in human polymorphonuclear leukocytes. J. Immunol. 143: 1669–73

Raynor RL, Zheng B, Kuo JF (1991) Membrane interactions of amphiphilic polypeptides mastoparan, melittin, polymyxin B, and cardiotoxin. J. Biol. Chem. 266: 2753–8

Ross EM, Higashijima T (1994) Regulation of G-protein activation by mastoparans and other cationic peptides. Methods Enzymol. 237: 26–37

Vitale N, Mukai H, Rouot B et al (1993) Exocytosis in chromaffin cells. Possible involvement of the heterotrimeric GTP-binding protein G_i. J. Biol. Chem. 268: 14715–23

Yokokawa N, Komatsu M, Takeda T et al (1989) Mastoparan, a wasp venom, stimulates insulin release by pancreatic islets through pertussis toxin sensitive GTP-binding protein. Biochem. Biophys. Res. Commun. 158: 712–6

9 Pharmacologically Active Peptides and Proteins from Bee Venom*

Peter N. Strong and Jonathan D.F. Wadsworth

* Dedicated to Professor Barbara Banks and the late Professor Charles Vernon, whose kindness and generosity stimulated the interest of one of the authors (PNS) as well as many others, into the fascinating secrets of the venom of the honeybee.

Contents

Methods and Tools in Biosciences and Medicine
Animal Toxins, ed. by H. Rochat and M.-F. Martin-Eauclaire

1 Introduction

In comparison with the venoms of snakes, scorpions, and marine snails, honeybee (*Apis mellifera*) venom appears to contain a relatively small number of peptides and proteins, which are estimated to make up ~. 70% of the venom dry weight. Early studies on the pharmacology and biochemistry of identified components of bee venom were due to the pioneering efforts of Habermann (for a review, see also Habermann, 1972). Apart from several biogenic amines (primarily histamine), and two identified hydrolytic enzymes (phospholipase A_2 and hyaluronidase), the majority of the venom comprises of a series of low molecular weight, basic peptides ranging from 18 to 26 amino acids. The most abundant of these is melittin, a pharmacologically unique peptide surfactant, comprising 50% of the venom dry weight. The other peptides are all neurotoxins (the best characterised being apamin and MCD peptide) and they share a similar structure, rich in disulphide bridges.

2 Bee Venom Neurotoxins

2.1 Apamin

Apamin is a basic 18-amino acid peptide (M_r ~2000) (Fig. 1). (For reviews, see Habermann, 1984; Strong, 1990). It possesses an amidated C-terminal carboxyl group and two intramolecular disulphide bridges. It is a very stable molecule, and its structure is remarkably insensitive to temperature and solvent environment. Synthetic apamin has identical structural characteristics and biolo-

Melittin*	2846.5	G I G A V L K V L T T G L P A L I S W I K R K R Q Q-NH_2
Apamin	2027.3	C N C K - - - A P E T A L C A R R C Q Q H-NH_2
MCD peptide	2587.2	I K C N C K R H V I K P H I C R K I C G K N-NH_2
Tertiapin	2455.1	A L C N C - N R - I I I P H M C W K K C G K K-NH_2
Secapin	2866.5	Y I I D V P P R C P P G S K F I K N R C R V I V P

Figure 1 Amino acid sequences of bee venom peptides.

The above sequences have been verified using a BLASTP search via GenomeNet (http://www.blast.genome.ad.jp/) using SwissProt and PRF (for tertiapin) protein data bases. Molecular masses were verified using ExPASy sequences analysis tools (http://www.expasy.ch/tools/). Molecular masses were corrected for disulphide bridges and carboxy terminal amidation.

* Other species of the genus Apis show slight variations in amino acid sequence (Kreil, 1975).

gical properties to the native toxin, and structure-activity studies (with both native and synthetic derivatives) demonstrate that a pair of adjacent arginine residues (Arg_{13} and Arg_{14}) together with maintenance of the disulphide-bonded structure are the only real critical determinants of biological activity.

NMR studies suggest that the toxin possesses an α-helical core with the arginine residues protruding from the surface (Dempsey, 1986; Wemmer and Kallenbach, 1983). Although early immunological studies also identified the same arginine residues as the major contributor to the dominant epitope of apamin, more detailed investigation has subsequently shown that epitopes can be found over the whole toxin surface, depending on how it is presented as an antigen (Defendini *et al.*, 1990).

Although apamin was the first potassium channel toxin to be isolated and characterised, its precise mechanism of action was unknown for many years and it was simply described as a centrally acting neurotoxin, producing motor hyperactivity and convulsions. The first evidence to show that apamin specifically targeted potassium channels came from experiments demonstrating that the toxin blocked increases in potassium permeability resulting from receptor activation (Banks *et al.*, 1979). Subsequent studies have shown that apamin blocks receptor-mediated calcium-activated potassium permeabilities in many cells. Apamin-sensitive fluxes can either be measured with potassium-sensitive electrodes or with radioisotopic tracers ($^{42}K^+$ or $^{86}Rb^+$) (Burgess *et al.*, 1981; Maas *et al.*, 1980; Schmid Antomarchi *et al.*, 1986). Another means of activating potassium flow through the same channel is to induce efflux artificially by the application of ionophores such as A_{23187} (Burgess *et al.*, 1981). For specific details of the effects of apamin on a given tissue or cell, the reader is referred to the appropriate references in Table 1.

The slow after-hyperpolarisation (AHP) which follows bursts of action potentials often results from the activation of calcium-dependent potassium currents, and apamin has been shown to reduce the amplitude of many different AHP currents, without affecting their decay rate (for reviews, see Sah, 1996; Strong, 1990; Strong and Brewster, 1992). Although the effects of apamin are rapid, blockade is often incomplete, and washout times of 1 h are usually required for full recovery. Single-channel conductance studies on guinea-pig hepatocytes and rat myotubes have shown that apamin specifically blocks small conductance, calcium-activated potassium currents (SK_{Ca} currents), with single-channel conductance of 12–20 pS in symmetrical (*ca.* 140 mM) potassium solutions (Blatz and Magleby, 1986; Capiod & Ogden, 1989). Thus block by apamin has come to be regarded as *prima facie* evidence for an unknown current being of the SK_{Ca} subtype (however, it should be remembered that the converse is not true – there are many examples of SK_{Ca} channels that are not blocked by apamin).

Apamin radiolabelled with ^{125}I can be used as a pharmacolgical probe in various tissues, to quantify apamin-binding sites and to examine toxin distribution using autoradiographic techniques (for reviews, see Strong, 1990; Strong & Brewster, 1992). ^{125}I is incorporated into the C-terminal His_{18} residue of the

Table 1 Apamin interaction with tissues and cells: key references.

Tissue or cell type	^{125}I-apamin binding	Inhibition of K_{Ca} permeabilities/currents
Synaptic plasma membranes	1–4	
Neurons	5	6–10
Neuroblastoma cells	11	11
PC12 cells	12	
Chromaffin cells	13	14
Smooth muscle	15, 16	15, 17
Skeletal muscle	18–20	18, 29, 21
Liver	4,16,22,23	23-25
Heart	16	
Astrocytes	26	

1. Hugues et al. (1982) J. Biol. Chem. 257, 2762–2769
2. Wu et al. (1985) Brain Res. 360, 183-194
3. Seagar et al. (1986) Biochemistry 25, 4051-4057
4. Wadsworth et al. (1994) J. Biol. Chem. 269, 18053-18061
5. Seagar et al. (1984) J. Biol. Chem. 259, 1491-1495
6. Pennefather et al. (1985) Proc. Natl. Acad. Sci. USA 82, 3040-3044
7. Kawai & Watanabe (1986) Br. J. Pharmacol. 87, 225-232
8. Zang & Krnjevic (1987) Neurosci. Lett. 74, 58-62
9. Bourque & Brown (1987) Neurosci. Lett. 82, 185-190
10. Selyanko (1996) J. Physiol. (Lond.) 494P, P52
11. Hugues et al. (1982) Proc. Natl. Acad. Sci. USA 79, 1308-1312
12. Schmid-Antomarchi et al. (1986) J. Biol. Chem. 261, 8633-8637
13. Lara et al. (1995) Biochem. Pharmacol. 49, 1459-1468
14. Park (1994) J. Physiol. (Lond.) 481, 555-570
15. Hugues et al. (1982) Life Sci. 31, 437-443
16. Marqueze et al. (1987) Eur. J. Biochem. 169, 295-298
17. Gater et al. (1985) Br. J. Pharmacol. 86, 861-868
18. Hugues et al. (1982) EMBO J. 1, 1039-1042
19. Schmid-Antomarchi et al. (1985) Proc. Natl. Acad. Sci. USA 82, 2188-2191
20. Behrens & Vergara (1992) Am. J. Physiol. 263, C794-C802
21. Blatz & Magleby (1986) Nature 323, 718-720
22. Cook et al. (1983) FEBS Lett. 152, 265-269
23. Cook & Haylett (1985) J. Physiol. (Lond.) 253, 373-394
24. Banks et al. (1979) Nature 282, 415-417
25. Capiod & Ogden (1989) J. Physiol. (Lond.) 409, 285-295
26. Seagar et al. (1987) Brain Res. 411, 226-230

toxin with negligible change in toxin affinity (Hugues *et al.*, 1982). Apamin can be selectively modified at either Cys_1 or Lys_4 to produce photolabile toxin derivatives that also show only slightly reduced binding affinity with respect to native toxin (Seagar *et al.*, 1986; Wadsworth *et al.*, 1996). These photolabile toxin derivatives can be used to label and identify apamin-binding polypeptides, which are thought to be subunits of SK_{Ca} channels. Radiolabelled toxin can also be used to monitor the solubilization and purification of these ion channel components (Schmid Antomarchi *et al.*, 1984; Seagar *et al.*, 1987).

Both monovalent and divalent cations (e.g., Na^+, Ca^{2+}) inhibit toxin binding (Hugues *et al.*, 1982) and therefore both the potency of apamin as an SK_{Ca} channel blocker and the value of dissociation constants obtained in binding studies are critically dependent on the ionic composition of the medium used. In low ionic strength media (e.g., 5 mM K^+ 10 mM Tris buffer), picomolar dissociation constants are obtained (typically 2–70 pM). However, in physiological (150 mM Na^+) buffers, the affinity of apamin for SK_{Ca} channels is reduced, and dissociation constants an order of magnitude greater are observed (typically 60–400 pM; see for example Cook & Haylett, 1985).

An apamin precursor has been identified from a worker bee complementary DNA (cDNA) library (Gmachl & Kreil, 1995). Preproapamin (46 amino acids) consists of a 19-amino acid signal sequence and an eight-residue proregion preceding the sequence of the mature peptide. The precursor ends in a Gly residue, necessary for the formation of the C-terminal amide. The extremely small size of preproapamin suggests that it is transported across the endoplasmic reticulum via a posttranslational route that is independent of signal recognition particles and docking proteins.

2.2 MCD peptide

MCD peptide (M_r ~2600) (Gauldie *et al.*, 1978; Habermann, 1972) has an even more pronounced basic character than apamin. Seven out of its 22 amino acids are either arginine or lysine, there are no acidic residues, and the peptide has an amidated carboxyl terminus. NMR studies have demonstrated that the structural scaffold on which MCD peptide is built is very similar to apamin, with two analogously positioned disulphide bridges defining an α-helical core (Kumar *et al.*, 1988).

MCD peptide was originally found to be a potent antiinflammatory agent (IC_{50} = 1 μg/g rat), triggering histamine release from mast cells [hence its original name, mast cell degranulating (MCD) peptide], although it was later shown that much lower concentrations caused convulsions upon direct injection (10 ng/mouse) into the brain (for a review see Strong, 1990). The antiinflammatory properties of MCD peptide are most likely nonspecific, probably related to its polycationic structure; however, there is some evidence to suggest that MCD peptide's mast-cell degranulating activity may result from a direct interaction with pertussus toxin-sensitive G proteins, in turn leading to the activation of phospholipase C (Mousli *et al.*, 1991). In brain, MCD peptide initially produces long-lasting hippocampal theta rhythms, which are associated with long-term potentiation and can be observed *in vitro* by application of the peptide to the CA1 region of hippocampal brain slices (Bidard *et al.*, 1987a). The similar effects of convulsions and hyperactivity upon injection with either MCD peptide or the snake venom toxin dendrotoxin I (an inhibitor of A-type potassium channels with highest affinity for Kv1.1 and Kv1.2) initially suggested that the two toxins might have similar cellular targets, and both electrophysiological and autoradiographic studies show this to be the case (Bidard *et al.*, 1987b; Mourre *et al.*, 1988). The most elegant proof that MCD peptide blocks a noninactivating delayed rectifier K^+ current has been provided by data demonstrating that MCD peptide inhibits outward currents generated by injection of the $K_v1.1$ gene into *Xenopus* oocytes (Stuhmer *et al.*, 1988).

MCD peptide can be labelled with ^{125}I. Two monoiodopeptides are obtained, and it is reasonable to assume that these correspond to His_8- and His_{13}-labelled peptides (Taylor *et al.*, 1984). Both iodopeptides bind with high-affinity binding to brain membranes, although there is a twofold difference between the two. The affinity of the more readily (His_8?) labelled peptide (K_d = 150 pM) is very similar to that of the native toxin (K_d = 140 pM). On the other hand, the less reactive site (His_{13}?) is probably less accessible, and this results in greater disruption of the peptide conformation, leading to a reduced affinity (K_d = 300 pM). Iodinated toxin can also be used to identify an MCD peptide-binding polypeptide on rat brain synaptosomes, and studies would suggest that it is the same polypeptide (subsequently shown to be the $K_v1.1$ α-subunit) that binds α-dendrotoxin and dendrotoxin-I (Rehm *et al.*, 1988).

MCD peptide is synthesised as a prepropeptide, with an identical length signal sequence and proregion as found in preproapamin. At the cDNA level, the 5' untranslated region and segments encoding the precursor peptides are 85% identical. Even more remarkably, the two toxin genes are completely identical at the 3' ends, and this has led to the discovery that MCD peptide and apamin are encoded by two genes in tandem, with a common 3' exon (Fig. 2) (Gmachl & Kreil, 1995). Exons encoding the bulk of the apamin sequence reside in the third and final intron of MCD peptide. The final, common exon of both peptides encodes the last six amino acids of apamin but is completely within the 3' untranslated region of MCD peptide.

2.3 Tertiapin

Tertiapin (M_r ~2400) is a 21-amino acid peptide with two disulphide bridges and other secondary structural similarities to both apamin and MCD peptide (Gauldie *et al.*, 1976; Xu & Nelson, 1994). It represents less than 1% of the total venom protein. Although little is known of the biological properties of tertiapin, it is included as a neurotoxin because of evidence suggesting that it has presynaptic activity (Ovchinnikov *et al.*, 1980). At a cellular level, tertiapin has been shown to inhibit Ca^{2+}/calmodulin activation of cAMP phosphodiesterase in PC12 cells by binding to calmodulin (2:1 mole ratio, tertiapin:calmodulin), and it has been suggested that the presynaptic effects of tertiapin may result from interfering with calmodulin function (Dudkin *et al.*, 1983) . Tertiapin also competes for ^{125}I-MCD peptide binding sites in the central nervous system (Taylor *et al.*, 1984), suggesting a role for the toxin in blocking voltage-activated K^+ channels.

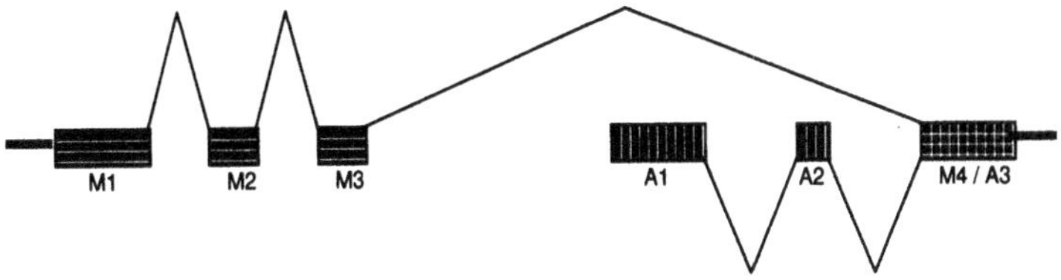

Figure 2 MCP peptide and apamin tandem gene structure.
Boxes labelled M1-M4 and A1-A3 represent exons encoding MCD-peptide and apamin genes respectively. Lines joining boxes represent introns.

2.4 Secapin

Secapin (M_r ~2900) represents ~ 1% of worker bee venom protein (Gauldie *et al.*, 1978), although its abundance in the venom of queen bees has been estimated to be ~. 10-fold higher (Vlasak & Kreil, 1984). Using a queen bee venom cDNA library, clones with inserts coding for secapin precursors have been isolated and sequenced. The mature peptide is produced from a 77-amino acid preprosecapin species, consisting of a 21-amino acid signal sequence and a 20-residue proregion, followed by 25 amino acids which correspond to the sequence of the purified peptide isolated from the venom.

Secapin has a single disulphide bridge with little structural similarities to the other basic peptides in bee venom. Secapin has some central activity since intravenous injection causes sedation and marked hypothermia (Gauldie *et al.*, 1976). Apart from this observation *en passant,* there are no further clues as to its biological properties or its functional role within the venom, which is nevertheless intriguing, given that it is predominantly found in queen bees.

3 Other Pharmacologically Active Peptides and Enzymes Present in Bee Venom

3.1 Melittin

Melittin is a small 26-amino acid peptide (M_r ~2800) (Habermann, 1972). Melittin is an extremely basic molecule; however, since it has no acidic residues and four of the five positively charged amino acids are clustered in a contiguous sequence (Lys_{21}-Arg-Lys-Arg) at the amidated C-terminus, this results in an amphipathic molecule with surfactant, detergent-like properties. Not surprisingly perhaps, melittin has an extremely broad spectrum of rather nonspecific pharmacological activities (e.g., on the heart, both smooth and skeletal muscle, and both peripheral and central synapses). It is also a powerful direct haemolytic, agent and many of its diverse cellular properties are the result of the liberation of pharmacologically active molecules during haemolysis. A model for melittin's surfactant properties suggests that the basic C-terminal hexapeptide associates with negatively charged membrane phospholipids, followed by insertion of the N-terminal apolar portion of the molecule into the bilayer structure, causing disruption of membrane lipids (Dawson *et al.*, 1978). Melittin has also been shown to inhibit Ca^{2+}/calmodulin-activated phosphodiesterase, by binding to calmodulin (Dudkin *et al.*, 1983). Whether this is a specific effect of melittin or is simply another example of its affinity for negatively charged (acidic) molecules is unclear but certainly worthy of further attention.

In a manner similar to the other venom peptides, melittin is generated from a prepromelittin precursor. After cleavage of a 21-amino acid signal peptide, promelittin is secreted into the venom sac before being fully processed, thus protecting the bee from the lytic effects of the mature peptide (Kindas-Mugge, 1979). Unlike proapamin, prosecapin, and pro MCD peptide, however, promelittin is cleaved in a multistep process, catalysed by a dipeptidylpeptidase IV, which sequentially removes two amino acid residues at a time from the N-terminus (Kreil *et al.*, 1980). Therefore unusually, at least two different propeptide activation mechanisms are found in bee venom. 10% of melittin in bee venom appears to be formylated at the N-terminal Gly residue, and the venom also contains a partially degraded fragment, missing the first seven amino terminal residues, known as melittin F (Gauldie *et al.*, 1978) comprising <1% of the total venom protein. Slight variations in melittin sequence have been reported for different species of the genus *Apis* (Kreil, 1975).

3.2 Phospholipase A_2

Phospholipase A_2 enzymes are major components of most animal venoms and represents ~ 12% of bee venom by weight. These enzymes specifically hydrolyse fatty acids from the *sn*2 position of diacylglycerophospholipids. When bee venom phospholipase A_2 (M_r ~15,800) was purified and the amino acid sequence reported (Shipolini *et al.*, 1974), it appeared that the bee enzyme was unrelated to vertebrate and other venom phospholipases A_2. However, this conclusion needs to be reconsidered in light of the more recent sequence data deduced from a cloned cDNA (Kuchler *et al.*, 1989), which indicates the presence of an extra pair of Cys residues. This suggests a revised arrangement of three key disulphide bridges which does in fact maintain the common three-dimensional structure surrounding the active site found in all phospholipase A_2 enzymes. Nevertheless, the bee venom enzyme is still unique among venom phospholipase A_2 enzymes in being a glycoprotein, and although it retains a conserved Ca^{2+} binding site (Ca^{2+} is an essential cofactor for the enzyme), it lacks the proton-relay system proposed for the active site of pancreatic and other venom A_2 enzymes.

In contrast to secapin, phospholipase A_2 is virtually absent in queen bee venom and is found exclusively in the venom of worker bees (Kuchler *et al.*, 1989). The enzyme is derived from a precursor starting with a typical signal peptide, although the resultant proenzyme has a proregion that is twice as long as pancreatic phospholipases. Activation of the proenzyme results from cleavage after a single Arg residue in the C-terminus of the propeptide. This is analogous to the processing of secapin (and other mammalian phospholipases A_2) but unlike reptilian enzymes.

The most common cause of fatal accidents from bee venom stings occurs as a result of anaphylactic reactions in hypersensitive patients. The most important allergen in bee venom is generally recognised to be phospholipase A_2, although for a small minority of patients, hyaluronidase (see below) is also allergenic (King *et al.*, 1976). Melittin is weakly allergenic, most probably as a result of its surfactant properties.

3.3 Hyaluronidase

Hyaluronidases are glycoproteins which hydrolyse long chains of alternating units of N-acetylglucosamine and D-gluronic acid present in hyaluronic acid. The enzymes are present in the venoms of snakes, scorpions, bees, and wasps. The bee enzyme represents 2% of the venom by weight and has typical hyaluronidase activity, although it also hydrolyzes chrondroitin sulphates. It was originally known as a spreading factor (Habermann, 1972), facilitating the diffusion of other venom components. It is inhibited by polysaccharide sulphates like heparin. The enzyme has been purified to homogeneity (Kemeny *et al.*, 1984) and has more recently been cloned (Gmachl & Kreil, 1993). The deduced amino acid sequence is a polypeptide consisting of 349 amino acids (M_r 40,700), with three potential glycosylation sites.

4 Purification of Bee Venom Peptides

It is convenient to consider purification of all bee venom peptides together, as they can be separated in one rational purification scheme.

Protocol 1 Purification of bee venom peptides

1. Crude bee venom (25 g) is dissolved in 200 ml formic acid (0.1 M) and dialysed (Visking tubing M_r cutoff 12,000) against four changes of the same buffer for 48 h at 4°C.
2. The dialysate is lyophilised.
3. 5 g of lyophilised powder is dissolved in 10 ml ammonium phosphate (0.5 M) adjusted to pH 7.4. This solution is applied to a column (100 × 5 cm diam.) of Sephadex G50 (Pharmacia) equilibrated in the same buffer at room temperature.
4. The column is eluted with 0.5 M ammonium phosphate pH 7.4 (3 l) and the eluate monitored at 229 or 254 nm.
5. Fractions are run on cellulose thin-layer chromatography (tlc) plates (Merck) and the chromatogram developed in a solvent of butanol:pyridine:acetic acid:water (3:3:1:3, v/v).
6. Peptide spots are detected using ninhydrin and the R_f values for different peptides is given in Table 2.

   Notes:
   Although fractions containing different peptides can be distinguished, many fractions will contain more than one peptide at this stage.
7. Peptide-rich fractions of interest are combined and desalted on a G10 column (100 x 2.6 cm diam.), eluting with 2% acetic acid. Column fractions are monitored as before, at 229 nm or 254 nm.
8. Eluates can be lyophilised and stored at –20°C indefinitely.
9. Pooled fractions enriched in each peptide (up to 0.5 g) are each separately chromatographed on an SP-Sephadex (Pharmacia) ion-exchange column (16 cm x 1 cm diam), equilibrated with 0.1 M sodium acetate (pH 4.7).

   Notes:
   On application of a linear salt gradient (0–1.2 M NaCl), individual peptides elute as indicated in Table 4.
10. Peptides are desalted as before, on a G10 column (100 x 1.6 cm diam), eluting with 2% acetic acid.

Traces of melittin (~ 2–3%) may still contaminate many of the other peptides, and these can be removed using either reverse-phase HPLC (Dotimas *et al.*, 1987) or, on a larger scale, if preparative columns are not available, using an immobilised heparin affinity support (Banks *et al.*, 1981). Reverse-phase HPLC of all bee venom peptides is best carried out with ion-pair reagents (Table 3).

Table 2 Chromatographic properties of bee venom peptide toxins.

Peptide	R_f value	Elution on SP-Sephadex (NaCl, M)
MCD peptide	0.15	0.9
Apamin	0.30	0.3
Secapin	0.75	0.4
Tertiapin	0.50	0.7
Mellitin	0.95	1.2

R_f values determined from tlc plates run in butanol:pyridine:acetic acid:water (3:3:1:3). Ion-exchange chromatography: peptides eluted with a gradient of 0–1.2 M NaCl in 0.1 M sodium acetate, pH 4.7

Table 3 Elution characteristics of bee venom peptides.

Peptide	Ion pair reagent* (mM)	pH	1-propanol (%)	Retention time (min)
Apamin	methanesulphonate (2)	2.5	3	8
Secapin	methanesulphonate (10)	3.0	8	7
MCD peptide	ethanesulphonate (15)	3.0	10	4

* in the presence of triethylamine (10 mM)
Peptides are chromatographed on a Vydac C_{18} reverse-phase column (250 x 4.1 mm diam.), flow rate 1 ml/min. The columns are eluted under isocratic conditions, with the various ion-pair reagents indicated, in the presence of 1-propanol. The eluate is monitored at 229 nm and peptides elute with the approximate retention times as shown.

Protocol 2 Removing melittin by heparin-affinnity and reverse-phase chromatography

1. Peptides (~ 5 mg) are dissolved in 3 ml sodium phophate buffer (10 mM, pH 6.8) and applied to a column (12 x 0.9 cm diam) of heparin-Sepharose CL-6B (Pharmacia) equilibrated in the same buffer.
2. The column is eluted with a linear salt gradient (0.02–1.5 M NaCl, total volume 600 ml) in the same buffer.

   Notes:
   Apamin elutes at a salt concentration of 0.1 M, and MCD peptide elutes at a concentration of 0.3 M.
3. Residual traces of melittin can be finally removed from the column by eluting with 1.5 M salt.
4. Purified peptides are desalted on a 1-ml C_{18} reverse-phase column (Bond-Elut: Analytichem) eluting with methanol:acetonitrile:water (3:1:30 v/v) containing 1% (v/v) phosphoric acid.
5. Peptides can be lyophilised and stored dessicated at –20°C for at least 5 years.

   Notes:
   Melittin contamination of apamin and MCD peptide preparations can be checked fluorimetrically. The tryptophan fluorescence of suspected samples (excitation 295 nm, emission 354 nm) can be measured and compared with a standard linear calibration curve of synthetic melittin (Bachem) in the range 0.1–50 μM.

Alternatively, melittin (<0.1% phospholipase contamination) can be specifically and rapidly isolated in bulk from crude bee venom by extraction and gel filtration.

Protocol 3 Removing melittin by extraction and ion exchange chromatography

1. 1 g of venom is dissolved in 10 ml distilled water, and 75 mg sodium sulphite is added; the pH is adjusted to 8.0 with 1 M ammonium hydroxide.
2. After incubation for 10 min (RT) and centrifugation (10,000 × g, 10 min, RT), the supernatant is extracted twice with n-butanol (5 ml) and evaporated to dryness.
3. After dissolving in 5 ml of water, the sample is then chromatographed on SP Sephadex as described above (Protocol 1, Step 9).
4. Pure melittin (~350 mg), free of contamination with either enzymes or other peptides, elutes at 1.2 M NaCl and is desalted in the usual way, using Sephadex G10.

Notes:

- These methods, presently in use in our laboratory, are based upon and refined from previous original studies (Banks *et al.*, 1981; Dotimas *et al.*, 1987; Gauldie *et al.*, 1976).
- Although apamin and MCD peptide, isolated from bee venom, commercially available (Sigma; Latoxan), the **quality** is variable because of contamination with melittin (see below). Synthetic apamin and synthetic MCD peptide are now comercially available (Alexis Biochemicals). Crude bee venom can be obtained from Sigma or Latoxan.
- **WARNING:** Lyopholised bee venom can cause severe **respiratory allergies**. It is recommended that a face mask be worn at all times when working with any lyopholised samples of bee venom origin. Both MCD peptide and apamin are neurotoxic and can be fatal if they enter the bloodstream.
- The **hydrophobic properties of monomeric melittin** cause it to bind tightly to both column matrices and other peptides. Melittin can be more easily separated from apamin and MCD peptide on gel exclusion columns run in the presence of salts that promote α-helical structure (e.g., oxyanions such as phosphate and sulphate) (Dotimas *et al.*, 1987). Under these conditions, melittin associates as a stable tetramer (M_r ~12,000) with considerably reduced hydrophobic properties. Association is also promoted at high peptide concentrations and high ionic strength.
- **Tlc** represents a quick and simple way to determine the peptide profiles of multiple samples eluted from the initial gel filtration column.
- **Heparin-Sepharose chromatography** provides a convenient affinity support for separating apamin and MCD peptide from final traces of melittin on a milligram scale (Banks *et al.*, 1981). It is especially recommended as a simple, one-step cleanup for small amounts of commercial samples of apamin. Reverse-phase radial compression C_{18} columns can also be used on a preparative basis. In order to distinguish between the various closely related, extremely basic peptides, it is essential to use ion-pair solvent reagents with all types of reverse-phase columns.
- **Contamination of other peptides with melittin** is a serious problem and one of the main causes of difficulties with commercial samples of native (as opposed to synthetic) apamin and MCD peptide. The detergent properties of melittin rapidly cause cell lysis, and even 1% levels of contamination can compromise the use of apamin and MCD peptide as pharmacological and biochemical probes. Minor contamination with phospholipase A_2 can also cause severe complications in any cell-based assays. The procedures described here result in peptides that are essentially free of phopholipase A_2 (< 0.1% contamination).
- The large-scale, rapid, and selective **purification of phospholipase-free melittin** makes use of the fact that melittin (in contrast to all other identified peptides and proteins present in bee venom) does not contain any disulphide bridges. After treatment of crude venom with the reducing agent so-

dium sulphite and subsequent centrifugation to remove insoluble, denatured products, the resultant supernatant is completely devoid of enzymatic or channel-blocking activities and is >95% melittin.

- **Melittin impurities** (at the 0.1% level) in apamin and MCD peptide can be detected in a simple and sensitive fluorimetric assay. The assay is based on the fact that tryptophan-containing peptides (such as melittin) can easily be identified in peptides such as MCD peptide and apamin (but not tertiapin), which contain no aromatic amino acid residues.

5 Preparation of Radiolabelled Apamin and MCD Peptide

Solutions

- **Equilibration buffer 1 (Solution 1)**
  160 mM NaCl
  50 mM sodium phosphate buffer, pH 6.0
- **Equilibration buffer 2 (Solution 2)**
  500 mM NaCl
  50 mM sodium phosphate buffer, pH 5.6
- **High salt buffer 1 (Solution 3)**
  300 mM NaCl
  50 mM sodium phosphate, pH 6.0
- **High salt buffer 2 (Solution 4)**
  720 mM NaCl
  50 mM sodium phosphate, pH 5.6

5.1 Preparation of ^{125}I-monoiodoapamin

Apamin can be conveniently iodinated using the Iodogen procedure.

Protocol 4 Preparation of ^{125}I-monoiodoapamin

1. 10 mg of Iodogen (Pierce) is dissolved in dichloromethane (25 ml), and an 80-µl aliquot is dispensed to a flat-bottomed tube and evaporated to dryness under a gentle stream of nitrogen.
2. 100 µg (50 nmol) apamin, dissolved in 10 mM Tris/HCl buffer pH 8.6 (80 µl), is added to the Iodogen tube, followed by 2 mCi Na^{125}I (Amersham) and a further aliquot of 10 mM Tris/HCl buffer, to bring the final reaction volume to 160 µl.
3. Following incubation (with intermittent shaking) at room temperature for 20 min, the contents of the Iodogen tube is decanted into a clean plastic vial.
4. After 10 min the reaction mixture is applied to an S-Sepharose (Pharmacia) column (14 × 0.9 cm) equilibrated in **Solution 1** and eluted (8 ml/h) with the same buffer, collecting 1-ml fractions.

   Notes:
   Two peaks of radioactivity are obtained, the first being unincorporated Na^{125}I, the second being ^{125}I-monoiodoapamin.
5. After approximately six column volumes of equilibration buffer (**Solution 1**), unlabelled apamin is eluted with high salt buffer 1 (**Solution 3**). ^{125}I-monoiodoapamin is stored at 4°C in the presence of bovine serum albumin (2 mg/ml) and retains full binding activity for at least 2 months.

5.2 Preparation of ^{125}I-monoiodo MCD peptide

MCD peptide can also be iodinated using the Iodogen procedure, and the protocol for apamin (Protocol 4) can be used effectively, with the following modifications:

- for the **Iodogen reaction**: 50 mM sodium phosphate buffer, pH 6.0, used instead of 10 mM Tris/HCl pH 8.6;
- for **SP Sephadex chromatography**: column equilibrated in **Solution 2** (500 mM NaCl, pH 5.6) instead of **Solution 1** (160 mM NaCl, pH 6.0). For the high salt buffer, **Solution 4** (720 mM NaCl, pH 5.6) is used instead of **Solution 3** (160 mM NaCl, pH 6.0).

Two, partially resolved monoiodopeptides are eluted, both of which are active in subsequent binding assays. The second eluting peptide, obtained in higher yield is assumed to be monoiodo ^{125}I-[His$_8$]MCD peptide (see above). ^{125}I-Monoiodo MCD peptides are stored at 4°C in the presence of bovine serum albumin (2 mg/ml) and retain full binding activity for at least 2 months.

Notes:

- These procedures are a refinement of the previously published work of ourselves and others (Hugues *et al.*, 1982; Strong & Brewster, 1992; Taylor *et al.*, 1984).
- **Iodogen tubes** are best prepared fresh, immediately prior to use. The reaction mixture is decanted from the Iododgen reaction tube and allowed to sit for 10 min in order to permit unincorporated $^{125}I^{+}$ to revert to molecular iodine. This greatly reduces nonspecific labelling of the gel bed and permits the same column to be used many times.
- Although this method avoids **contamination of HPLC apparatus** or having a dedicated apparatus for use with radioactive samples, ^{125}I-monoiodopeptides can be separated and purified on a C_{18} reverse-phase column as described for the preparation of the native peptides.
- A **minor radioactive peak**, sometimes occurring between free iodine and ^{125}I-monoiodopeptide, is due to ^{125}I-diiodopeptide and should be discarded. Unlabelled peptides can be detected in the high salt eluates by HPLC or by radioimmunoassay.

6 Preparation of Photolabile Derivatives of ^{125}I-Monoiodoapamin for Identifying Apamin-binding Polypeptides

6.1 Photolabile ^{125}I-[ε-formyl-lys$_4$]apamin derivatives

Protocol 5 Preparation of [ε-formyl-lys$_4$]apamin

1. 4.3 mg (~ 2 µmol) apamin is dissolved in 0.75 ml 0.1 M sodium borate buffer pH 9.2 and cooled to 0°C in an ice bath.
2. 1.7 mg (10 µmol, 5-fold molar excess) 4-nitrophenylformate in 10 µl dry tetrahydrofuran is added in two equal aliquots (at time zero and after 5 min).
3. The reaction is allowed to proceed for a further 7 min (12 min total) and then quenched with 4 x 20 µl 2 M formic acid (until the yellow colour of the nitrophenyl anion disappears).
4. After extraction of the reaction mixture with diethylether (3 x 750 µl), the aqueous phase is directly applied to an S-Sepharose (Pharmacia) cation-exchange column (8 x 1 cm diam.) equilibrated in 50 mM sodium acetate pH 4.8.
5. The column is eluted (flow rate 15 ml/h) with three column volumes of equilibration buffer and then with a linear gradient (120 ml) of 0–0.5 M NaCl developed in the same buffer, monitoring the eluate at 254 nm. [ε-Formyl-lys$_4$]apamin elutes at 0.35 M NaCl.

Protocol 6 Iodination of [ε-formyl-lys$_4$]apamin

[ε-Formyl-lys$_4$]apamin is iodinated using the identical Iodogen procedure that is used for labelling apamin.

1. Following iodination, the reaction mixture is applied to an S-Sepharose cation exchange column (5 x 1 cm diam.) equilibrated with 10 mM sodium phosphate pH 6.0.
2. After removal of free ^{125}I with equilibration buffer (35 ml), the column is eluted with a linear gradient of 0–200 mM NaCl developed in the same buffer.

   Notes:
   ^{125}I-[ε-formyl-lys$_4$]apamin is eluted as a single peak between 140–160 mM NaCl.
3. Subsequent elution of the column with 300 mM NaCl elutes unmodified [ε-formyl-lys$_4$]apamin from the column.
4. Fractions containing ^{125}I-[ε-formyl-lys$_4$]apamin (2200 Ci/mmol) are pooled to ~20 nM final concentration and are stored at 4°C in the presence of 2 mg/ml BSA.

Protocol 7 Preparation of photolabile derivatives of ^{125}I-[ε-formyl-lys$_4$]apamin

1. An aliquot of ^{125}I-[ε-formyl-lys$_4$]apamin (in ~0.15 M NaCl, 10 mM sodium phosphate, pH 6.0, containing 2 mg/ml BSA) is diluted with an equal volume of 50 mM sodium phosphate, pH 6.0, and incubated (at pH 6.0) with a 500-fold molar (including BSA) excess of succinimidyl ester photoprobe (Pierce).
2. The reaction is allowed to proceed to completion by incubation in the dark for 3 h at 20°C, after which time photolabile products are stored in the dark at 4°C and used within 24 h of preparation.

6.2. Photolabile ^{125}I-[α-formyl-cys$_1$]apamin derivatives

Protocol 8 Preparation of [α-formyl-cys$_1$]apamin

1. 6 mg (3 μmol) apamin is dissolved in 120 μl formic acid (98%).
2. 10 x 2 μl aliquots of acetic anhydride are added at 30 s intervals with brief agitation between each addition [final ratio of formic acid: acetic anhydride = 6:1 (v/v)].
3. Following gentle agitation at room temperature for a further 90 min, the reaction is diluted to 1 ml with distilled water and applied to an S-Sepharose cation-exchange column (8 x 1 cm diameter) equilibrated in 0.1 M sodium acetate, pH 4.7.
4. The column is eluted (flow rate 15 ml/h) with three column volumes of equilibration buffer and then with a linear gradient (120 ml) of 0–0.5 M NaCl developed in the same buffer.

   Notes:
   Eluate fractions are monitored at 254 nm, and [α-formyl-cys$_1$]apamin elutes at 0.24 M NaCl.

[α-Formyl-cys$_1$]apamin is iodinated and purified using the identical procedure employed for the iodination of [ε-formyl-lys$_4$]apamin. ^{125}I-[α-formyl-cys$_1$]apamin (2200 Ci/mmol) elutes from the S-Sepharose column at 140–160 mM NaCl and fractions (pooled to ~20 nM final concentration) ae stored at 4°C in the presence of 2 mg/ml BSA.

Protocol 9 Preparation of photolabile derivatives of ^{125}I-[α-formyl-cys$_1$]apamin

1. An aliquot of ^{125}I-[α-formyl-cys$_1$]apamin (in ~0.15 M NaCl, 10 mM sodium phosphate, pH 6.0, containing 2 mg/ml BSA) is diluted with an equal volume of 100 mM borax, pH 8.5, and incubated (at pH 8.5) with a 500-fold molar excess (including BSA) of succinimidyl ester photoprobe (Pierce).
2. The reaction is allowed to proceed to completion by incubation in the dark for 3 h at 20°C, after which time photolabile products are stored in the dark at 4°C and used within 24 h of preparation.

Notes

- **Apamin** has two primary amino groups (Cys1 α-NH_2 and Lys$_4$ ε-NH_2) to which a variety of photolabile succinimidyl ester probes can be selectively attached. Three stages are involved in their preparation:
  (a) introduction of a formyl protecting group to block the activity of the "interfering" amino group;
  (b) radiolabelling the formylapamin derivative at His$_{18}$;
  (c) reaction of a photolabile succinimidyl ester with the unique free amino group of the radiolabelled formylapamin derivative.
- A large (and continually expanding) range of **succinimdyl ester photoprobes** is available from Pierce, and those that have been used successfully are described (Wadsworth *et al.*, 1996).
- The **amount of photoprobe** used is determined as a 10-fold excess of the ratio of reagent to the total primary reactive amines present in ^{125}I-apamin plus BSA (assuming 50 lysine groups present in BSA available to react with succinimidyl esters).
- 100% **acetonitrile** is used as a carrier for stock solutions of succinimidyl ester photoprobes, whereas 75% (v/v) acetonitrile prepared in H_2O is used for sulfo-succinimidyl ester photoprobes.
- All **manipulations with azido-succinimidyl ester photoprobes** are performed in a darkroom under a 25-watt photographic red lamp.

7 Binding Assays

Solutions

- **Incubation buffer (Solution 5)**
  10 mM KCl
  25 mM Tris pH 8.4 containing 0.1% (w/v) BSA

7.1 ^{125}I-Apamin

^{125}I-apamin binding to isolated membranes can be conveniently measured using a filtration binding assay (Hugues *et al.*, 1982; Wadsworth *et al.*, 1994). Additionally, the measurement of apamin binding sites under physiological conditions using cells in suspension (e.g., heptocytes) may be measured using a rapid centrifugation assay (Cook *et al.*, 1983). The incubation conditions in these two assays described below (or close variants of them) have been used to characterise ^{125}I-apamin acceptor binding sites in a variety of preparations (Tab. 1).

Protocol 10 Filtration assay

1. Membranes (50 µg protein) are incubated with ^{125}I-apamin (0.2–150 pM) in a final volume of 1 ml, all dilutions being made in incubation buffer **(Solution 5)**.
2. Samples are incubated in triplicate for 1 h at 0–4°C.
3. At the end of the incubation, samples are filtered through cellulose acetate filters (Sartorius SM, 0.45 µm pore size) that have been soaked in incubation buffer for at least 1 h at 0°C and then rinsed with a 5-ml aliquot of ice-cold buffer immediately prior to application of the membrane sample.
4. After filtration of the membranes, the filter is washed rapidly (<25 s) with two further 5-ml aliquots of ice-cold buffer **(Solution 5)**.
5. Wet filters can be counted directly in a calibrated γ counter.

   Notes:
   Nonsaturable binding is determined in parallel experiments by the inclusion of an excess of unlabelled apamin (0.1 µM). Binding data is analysed using a nonlinear curve-fitting computer program, available in a variety of commercial software packages (e.g., EBDA / LIGAND, Biosoft Cambridge, UK).

Protocol 11 Whole-cell centrifugation assay

1. A suspension (8×10^6 cells/ml) of isolated hepatocytes (Cook & Haylett, 1985) is incubated at 37°C in Eagle's minimum essential medium, supplemented with 10% (v/v) newborn calf serum and 2% (w/v) BSA.
2. An aliquot (0.3 ml) of this suspension is layered above 0.25 ml di-n-butyl phthalate (oil layer) in a 1.5-ml centrifuge tube.
3. 0.2-ml aliquots of cell incubation medium containing a range of concentrations of ^{125}I-apamin are then added to the cells to produce final concentrations of 1–600 pM ^{125}I-apamin.
4. Samples are incubated with labelled toxin for 2 min at 37°C.
5. Incubations are performed in quadruplicate, and at the end of the incubation period the cells are rapidly centrifuged (15 s, 6000 x g) through the oil layer to separate bound from unbound toxin.
6. After removing the aqueous layer (top) and the oil layer by careful aspiration, the bottom of the tubes, containing the cell pellets, are cut off with a guillotine.
7. Nonsaturable binding is measured in the presence of unlabelled apamin (1 μM). Pellets are counted in a calibrated γ counter.

Notes:

- **Binding site density** is extremely low on most cells and membranes, and the use of purified ^{125}I-monoiodopeptides, specific activity 2200 Ci/mmol, is vital for reliable and accurate quantification.
- **Whatman GF/B glass fibre filters**, presoaked in aqueous 0.5% (v/v) polyethyleneimine, for 1 h at 4°C, can be used equally satisfactorily as cellulose acetate filters and have the advantage of being considerably cheaper.
- **Inorganic cations** inhibit apamin binding ($k_{0.5}$ for Na^+ is 35 mM); the affinity of apamin for its binding protein is therefore considerably reduced in physiological saline as compared with the low ionic strength buffer typically used for membrane assays.
- Attainment of equilibrium is extremely rapid, approaching the diffusion-limited rate. Dissociation is equally fast – the half-time for dissociation at 37°C is 40 s (Castle, 1987), and therefore the speed and efficiency at which labelled cells are separated from unbound ligand is crucial.
- **[^{3}H]Inulin** can be used to determine the amount of extracellular fluid carried through with the cells after centrifugation. This carryover is the predominant contributor to noninhibitable binding.

7.2 ^{125}I-MCD peptide

Both filtration and centrifugation procedures can be used for measuring ^{125}I-MCD peptide binding to isolated rat brain synaptic membranes (Taylor *et al.*, 1984).

The **incubation medium** for either assay consists of
140 mM NaCl
5 mM KCl
2.8 mM $CaCl_2$
1.3 mM $MgSO_4$
20 mM Tris, pH 7.4, containing 0.1% (w/v) BSA.

Binding experiments are terminated either by rapid centifugation followed by immediate aspiration of membrane pellets or by rapid filtration, using cellulose acetate filters (Millipore EHWP 0.5-µm pore size).

Notes:
- The inclusion of 10 µg/ml compound 48/80 in the binding medium reduces levels of **nonsaturable binding** (Bidard *et al.*, 1987b).
- **Whatman GF/C glass fibre filters**, presoaked in aqueous 0.5% (v/v) polyethyleneimine can be used in place of cellulose acetate filters [see Section 7.1, note (ii)].

8 Conclusion

The value of neurotoxins as molecular probes to characterise the structure and function of receptors and ion channels stems from their high affinity and exquisite sensitivity for the macromolecular structures with which they interact. Nowadays, new toxins are nearly always identified through highly sensitive biological and pharmacological screening assays. (Indeed it would be a herculean effort to isolate toxins such as charybdotoxin and other scorpion venom potassium channel toxins without being able to monitor function during the purification process.) What makes the early isolation of toxins like apamin and MCD peptide so remarkable is that they were isolated and structurally characterised without any indication of their ion channel-blocking properties. Indeed, the painstaking purification schemes and structural characterisation of these peptides both by Habermann's group and by Vernon, Shipolini, Banks, and colleagues has provided a firm foundation for subsequent studies which have brought these initially esoteric toxins to the attention of a wide range of neuroscientists.

Only relatively high abundance (>1%) peptide neurotoxins have been characterised in bee venom to date. Since all of these toxins have very unusual structural features when compared with other peptides possessing similar biological properties, the use of sophisticated and sensitive pharmacological screening assays (as mentioned earlier, employed with great success to isolate low-abundance toxins in scorpion venoms), might reveal the presence of many more structurally novel toxins in honeybee venom.

Note added in proof: Tertiapin has recently been shown to block inward rectifier, G-protein gated K^+ Channels (GIRK 1/4 and ROMIK 1) with nanomolar affinities: Jin W and Lu Z (1998) Biochemistry 37: 13291–9

Acknowledgements

We thank the Wellcome Trust and the Muscular Dystrophy Group of Great Britain for financial support.

References

Banks BE, Brown C, Burgess GM et al (1979) Apamin blocks certain neurotransmitter-induced increases in potassium permeability. Nature 282: 415–7

Banks BE, Dempsey CE, Pearce FL et al (1981) New methods of isolating been venom peptides. Anal Biochem. 116: 48–52

Bidard JN, Gandolfo G, Mourre C et al (1987a) The brain response to the bee venom peptide MCD. Activation and desensitization of a hippocampal target. Brain Res. 418: 235–44

Bidard JN, Mourre C, Lazdunski M (1987b) Two potent central convulsant peptides, a bee venom toxin, the MCD peptide, and a snake venom toxin, dendrotoxin I, known to block K+ channels, have interacting receptor sites. Biochem. Biophys. Res. Commun. 143: 383–9

Blatz AL, Magleby KL (1986) Single apamin-blocked Ca-activated K+ channels of small conductance in cultured rat skeletal muscle. Nature 323: 718–20

Burgess GM, Claret M, Jenkinson DH (1981) Effects of quinine and apamin on the calcium-dependent potassium permeability of mammalian hepatocytes and red cells. J. Physiol (Lond) 317: 67–90

Capiod T, Ogden DC (1989) The properties of calcium-activated potassium ion channels in guinea-pig isolated hepatocytes. J. Physiol (Lond) 409: 285–95

Castle NA (1987) Ph.D. Thesis University of London

Cook NS, Haylett DG (1985) Effects of apamin, quinine and neuromuscular blockers on calcium-activated potassium channels in guinea-pig hepatocytes. J. Physiol (Lond) 358: 373–94

Cook NS, Haylett DG, Strong PN (1983) High affinity binding of [^{125}I]monoiodoapamin to isolated guinea-pig hepatocytes. FEBS Lett. 152: 265–9

Dawson CR, Drake AF, Helliwell J, Hider RC (1978) The interaction of bee melittin with lipid bilayer membranes Biochem. Biophys. Acta 510: 75–86

Defendini ML, Pierres M, Regnier Vigouroux A et al (1990) Epitope mapping of apamin by means of monoclonal antibodies raised against free or carrier-coupled peptide. Mol Immunol. 27: 551–8

Dempsey CE (1986) pH dependence of hydrogen exchange from backbone peptide amides in apamin. Biochemistry 25: 3904–11

Dotimas EM, Hamid KR, Hider RC, Ragnarsson U (1987) Isolation and structure analysis of bee venom mast cell degranulating peptide. Biochim. Biophys. Acta 911: 285–93

Dudkin SM, Mikchaylova LI, Severin ES Jr (1983) Mechanisms of cyclic AMP phosphodiesterase regulation. Adv. Enzyme Regul. 21: 333–52

Gauldie JA, Hanson JM, Rumjanek FD et al (1976) The peptide components of bee venom. Eur. J. Biochem. 61: 369–376

Gauldie JA, Hanson JM, Shipolini RA, Vernon CA (1976) The structures of some peptides from bee venom. Eur. J. Biochem. 83: 405–410

Gauldie J, Hanson JM, Shipolini RA, Vernon CA (1978) The structures of some peptides from bee venom. Eur. J. Biochem. 83: 405–10

Gmachl M, Kreil G (1993) Bee venom hyaluronidase is homologous to a membrane protein of mammalian sperm. Proc. Natl. Acad. Sci. USA 90: 3569–73

Gmachl M, Kreil G (1995) The precursors of the bee venom constituents apamin and MCD peptide are encoded by two genes in tandem which share the same 3'-exon. J. Biol. Chem. 270: 12704–8

Habermann E (1972) Bee and wasp venoms. Science 177: 314–322

Habermann E (1984) Apamin. Pharmacol. Ther. 25: 255–70

Hugues M, Duval D, Kitabgi P et al (1982) Preparation of a pure monoiodo derivative of the bee venom neurotoxin apamin and its binding properties to rat brain synaptosomes. J. Biol. Chem. 257: 2762–9

Kemeny DM, Dalton N, Lawrence AJ et al (1984) The purification and characterisation of hyaluronidase from the venom of the honey bee, Apis mellifera. Eur. J. Biochem. 139: 217–23

King TP, Sobotka AK, Kochoumian L, Lichtenstein LM (1976) Allergens of honey bee venom. Arch. Biochem. Biophys. 172: 661–71

Kindas-Mugge I (1979) Isolation and translation of melittin mRNA. Trends Biochem. Sci. 4: 135–137

Kreil G (1975) The structure of Apis dorsata melittin: phylogenetic relationships between honeybees as deduced from sequence data. FEBS Lett 54: 100–102

Kreil G, Haiml L, Suchanek G (1980) Stepwise cleavage of the pro part of promellitin by dipeptidylpeptidase IV. Eur. J. Biochem. 111: 49–58

Kuchler K, Gmachl M, Sippl MJ, Kreil G (1989) Analysis of the cDNA for phospholipase A2 from honeybee venom glands. The deduced amino acid sequence reveals homology to the corresponding vertebrate enzymes. Eur. J. Biochem. 184: 249–54

Kumar NV, Wemmer DE, Kallenbach NR (1988) Structure of P401 (mast cell degranulating peptide) in solution. Biophys. Chem. 31: 113–9

Maas A.J, Den Hertog A, Ras R. Van den Akker J (1980) The action of apamin on guinea-pig taenia caeci. Eur. J. Pharmacol. 67: 265–74

Mourre C, Nehlig A, Lazdunski M (1988) Cerebral glucose utilization after administration of apamin, a toxin active on Ca^{2+}-dependent K^+ channels. Brain Res. 451: 274–84

Mousli M, Bronner C, Bueb JL, Landry Y (1991) Evidence for the interaction of mast cell-degranulating peptide with pertussis toxin-sensitive G proteins in mast cells. Eur. J. Pharmacol. 207: 249–255

Ovchinnikov YA, Miroshnikov AI, Kudelin IA et al (1980) Structure and presynaptic activity of tertiapin-neurotoxin from bee venom Apis mellifera. Bioorg. Khim. 6: 359–365

Rehm H, Bidard JN, Schweitz H, Lazdunski M (1988) The receptor site for the bee venom mast cell degranulating peptide. Affinity labelling and evidence for a common molecular target for mast cell degranulating peptide and dendrotoxin I, a snake toxin active on K^+ channels. Biochemistry 27: 1827–32

Sah P (1996) Ca^{2+}-activated K^+ currents in neurones: types, physiological roles and modulation. Trends Neurosci 19: 150–154

Schmid Antomarchi H, Hugues M, Norman R et al (1984) Molecular properties of the apamin-binding component of the Ca^{2+}-dependent K+ channel. Radiation-inactivation, affinity labelling and solubilization. Eur. J. Biochem. 142: 1–6

Schmid Antomarchi H, Hugues M, Lazdunski, M (1986) Properties of the apamin-sensitive Ca^{2+}-activated K^+

channel in PC12 pheochromocytoma cells which hyper-produce the apamin receptor. J. Biol. Chem. 261: 8633–7

Seagar MJ, Labbé-Jullié C, Granier C et al (1986) Molecular structure of rat brain apamin receptor: differential photoaffinity labelling of putative K^+ channel subunits and target size analysis. Biochemistry 25: 4051–4057

Seagar MJ, Marqueze B, Couraud F (1987) Solubilization of the apamin receptor associated with a calcium-activated potassium channel from rat brain. J. Neurosci. 7: 565–70

Shipolini RA, Callewaert GL, Cottrell RC, Vernon CA (1974) The amino-acid sequence and carbohydrate content of phospholipase A2 from bee venom. Eur. J. Biochem. 48: 465–76

Strong PN (1990) Potassium channel toxins. Pharmacol. Ther. 46: 137–162

Strong PN, Brewster BS (1992) Apamin: a probe for small conductance, calcium-activated potassium channels. Methods Neurosci. 8: 15–24

Stuhmer W, Stocker M, Sakman B et al (1988) Potassium channels expressed from rat brain cDNA have delayed rectifier properties. FEBS Lett 242: 199–206

Taylor JW, Bidard JN, Lazdunski M (1984) The characterization of high-affinity binding sites in rat brain for the mast cell-degranulating peptide from bee venom using the purified monoiodinated peptide. J. Biol. Chem. 259: 13957–67

Vlasak R, Kreil G (1984) Nucleotide sequence of cloned cDNAs coding for pre-prosecapin, a major product of queen-bee venom glands. Eur. J. Biochem. 145: 279–82

Wadsworth JDF, Doorty KB, Strong PN (1994) Comparable 30kDa apamin-binding polypeptides may fulfill equivalent roles within putative subtypes of small conductance, Ca^{2+}-activated K^+ channels. J. Biol. Chem. 269: 18053–180561

Wadsworth JDF, Doorty KB, Ganellin CR, Strong PN (1996) Photolabile derivatives of ^{125}I-apamin defining the structural criteria required for labelling high and low molecular mass polypeptides associated with small conductance Ca^{2+}-activated K^+ channels. Biochemistry 35: 7917–7927

Wemmer D, Kallenbach NR (1983) Structure of apamin in solution a two-dimensional nuclear magnetic resonance study. Biochemistry 22: 1901–6

Xu X, Nelson JW (1994) One-disulfide intermediates of apamin exhibit native-like structure. Biochemistry 33: 5253–61

10 Purification and Characterization of Scorpion Toxins Acting on Voltage-Sensitive Na^+ Channels

Marie-France Martin-Eauclaire and Hervé Rochat

Contents

1 Introduction

Scorpions produce some of the most potent venoms in the animal kingdom. The toxicity of these venoms is due to their high content of peptide neurotoxins, which are highly active against ionic channels (Miranda *et al.*, 1970). Scorpions use this cocktail of toxins to immobilize their prey and to protect themselves. Their venoms are therefore composed of a complex mixture of polypeptides with different pharmacological activities.

Ionic channels are the principal targets of these toxic polypeptides. The most studied toxins, responsible for the high toxicity of scorpion venom to mammals and insects, modify the gating mechanism of voltage-sensitive Na^+ channels from excitable cells (Catterall, 1975, 1979, 1980, 1986, 1992, 1995; Jover *et al.*, 1980a, b; Strichartz *et al.*, 1987). Others bind with high affinity to

Methods and Tools in Biosciences and Medicine
Animal Toxins, ed. by H. Rochat and M.-F. Martin-Eauclaire

K^+ channels in various tissus, blocking their activity (Carbone *et al.*, 1982; Strong, 1990; Miller, 1995). A peptide that blocks a Cl^- channel has been purified (DeBin *et al.*, 1993), as have two new peptides, which inhibit (Zamudio *et al.*, 1997a) or increase (Zamudio *et al.*, 1997b) the [3H] ryanodine binding to the ryanodine-sensitive cardiac and skeletal sarcoplasmic reticulum Ca^{2+} channel. This review focuses on the toxins that affect Na^+ conductance.

Toxins that exclusively act on voltage-dependent sodium channels can be broadly classified according to their host specificity, as toxins active against vertebrates (mammal) or invertebrates (arthropods).

1.1 Toxins active against vertebrate voltage-sensitive Na^+ channels

Voltage-sensitive Na^+ channels are responsible for the rapid depolarization phase of the action potential of nerve, muscle, and heart cells (Catterall *et al.*, 1992). Scorpion toxins affect the opening or closing of these channels, increasing the depolarization of the membrane and the release of neurotransmitters. Studies of electrophysiological behaviors and differences in binding sites and ion-flux studies have shown that scorpion toxins active against vertebrate Na^+ channels are of two types: (i) α-type toxins slowing down the inactivation of the channel, prolonging the action potential (Naharashi *et al.*, 1974; Romey *et al.*, 1975; Catterall, 1979; Jover *et al.*, 1980a); (ii) β-type toxins, shifting the voltage-dependent activation to more negative membrane potentials and leading to repetitive firing in both muscles and nerves (Cahalan, 1975; Jover et al., 1980b; Couraud et al., 1982; Yatani et al., 1988). In both cases, the toxins exert their effect from the extracellular side of the membrane, and the Na+ influx of the cells increases.

α- and β-toxins have different modes of action, but they belong to the same family of homologous proteins. They consist of 61–70-aminoacid residues, cross-linked by four disulfide bridges in a similar pattern. The architecture of the polypeptide chain is similar, and the structural elements responsible for the various pharmacological activities have not been identified. α- and β-toxins fold similarly. They consist of a short helix and a three-stranded anti-parallel β-sheet, linked by two disulfide bridges. On one side of the molecule there is a hydrophobic surface which has been implicated in the high-affinity binding of scorpion toxins to their pharmacological targets (Fontecilla-Camps et al., 1980, 1988).

α-Toxins

α-Type toxins were the first scorpion toxins to be purified and chemically characterized from the scorpion *Androctonus australis*, a member of the *Buthinae* from North Africa (Miranda et *al.*, 1970; Martin-Eauclaire and Rochat, 1986).

They are present in the venoms of *Androctonus mauretanicus, Leiurus quinquestriatus*, and various *Buthus* subspecies from Africa and Asia (Martin-Eauclaire and Couraud, 1995). A small amount of α-type toxins are also present in American scorpion venoms, e.g. those of *Tityus serrulatus* and *Centruroïdes sculpturatus*. However, the α-toxins of these venoms have a lower affinity for their receptors than those purified from "North African" species (Diniz, 1978; Barhanin et al., 1982; Vijverberg et al., 1984; De Lima et al., 1992; Martin-Eauclaire et al., 1994a).

These toxins are principally responsible for venom toxicity in mammals. They prolong the action potential in nerve and muscle by slowing the inactivation of sodium channels (Naharashi et al., 1974; Romey et al., 1975). They specifically bind to their target in a way dependent on membrane potential (Catterall, 1979; Jover et al., 1980a). 90% of the toxicity of the *Androctonus australis* venom is due to four toxins (AaH I – IV), accounting for only 2 to 3% in weight of the crude venom obtained by electric stimulation (Martin and Rochat, 1986). The AaH II toxin itself accounts for 50% of activity in the mice (injected by subcutaneous injection) and has the highest affinity described for site 3 of the voltage-sensitive sodium channels of excitable cells. Site 3 is formed by extracellular loops of the α-subunit of the Na+ channel, between transmembrane segments S5 and S6 of domain I (Tejedor and Catterall, 1988) and domain IV (Thomsen and Catterall, 1989). It has been shown that Glu 1613 on the S3–S4 extracellular segment of domain IV of the α-subunit of the type IIA Na+ channel from rat brain (Rogers et al., 1996) is required for the α-toxin binding.

Some α-toxins cause contractile paralysis of the insect, after a time lag due to extreme prolongation of the action potential by slowing the inactivation of sodium currents. These toxins are found mostly in *Leiurus quinquestriatus* subspecies *hebraeus* and *quinquestriatus*, *Buthus tunetanus* (Eitan *et al.*, 1990; Kopeyan *et al.*, 1993; Borchani *et al.*, 1996a). They bind to similar but not identical receptor sites in rat brain and insect sodium channels. Such an α-insect toxin, Lqh α IT has been produced in insect cells and lepidopteran larvae using baculovirus (Chevanovsky *et al.*, 1995) and in *Escherichia coli* (Zilberberg *et al.*, 1997).

Comparative studies have separated the structurally α-scorpion toxins affecting the inactivation of sodium current into two groups according to their activity against mammals and insects (Gordon *et al.*, 1996).

β-*Toxins*

β-type toxins are present in the venoms of scorpions from North and South America. They consist of a single polypeptide chain of 60 – 66 residues cross-linked by four disulfide bridges (Babin *et al.*, 1974; Possani *et al.*, 1981a, b, 1982, 1985, 1991; Martin *et al.*, 1987a). Variant 3 from *Centruroides sculpturatus* Ewing has only weak activity but has an amino acid sequence very similar of the most potent β-toxins. It was the first scorpion toxin to be determined at high resolution by X-ray crystallography (Fontecilla-Camps *et al.*, 1980). β-

Toxins are those that behave like the reference toxin Css II, purified from *Centruroides suffusus suffusus* (Couraud *et al.*, 1982). They bind to site 4 of the voltage-sensitive sodium channels of rat brain synaptosomes. This binding does not depend on membrane potential, and there are no signs of allosteric effects due to toxins binding to other sites on the Na^+ channel. These toxins affect the activation process and cause repetitive firing in voltage-clamped frog myelinated nerve (Cahalan, 1975; Couraud *et al.*, 1982).

Ts γ (Possani *et al.*, 1981a; Barhanin *et al.*, 1983a, b), also defined as Ts VII (Bechis *et al.*, 1984; Céard *et al.*, 1992; Martin-Eauclaire *et al.*, 1992), is the main toxic compound purified from the venom of the Brazilian scorpion *Tityus serrulatus*. It also binds with high affinity to the insect sodium channel (Pauron *et al.*, 1985; De Lima *et al.*, 1986, 1989). This may be due to the extreme flexibility of the molecule resulting in best fit of the various receptor sites (Loret *et al.*, 1990).

Recently, using sodium channel chimeras, it was shown that the β-toxin Css IV from *C. suffusus suffusus* binds to a receptor site that requires Gly 845 in the S3–S4 loop, at the extracellular end of the S4 segment in domain II of the α-subunit (Cestèle *et al.*, 1998).

Two other toxins must be mentioned here, because they share a broader specificity for binding to sodium channel and are able to compete with radiolabeled β-type toxins bound to their receptor site on rat brain synaptosomes.

AaH IT4, purified from the venom of the Tunisian scorpion *Androctonus australis*, has a sequence with little in common with those of β-type toxins. However, AaH IT4 is recognized by anti-Css II antibodies. It competes with an α-type toxin, AaH II, a β-type toxin, Css II, and an insect toxin, AaH IT, for their respective sites on mammal and insect sodium channels and has, thus, been suggested to be an ancestral scorpion toxin (Loret *et al.*, 1991). A toxin active against crustaceans has also been isolated from the venom of *Centruroides noxius*. At very high doses (100 μM) this toxin competes with the β-type toxin Cn2 (*C. noxius* 2) for binding. Its amino acid sequence is very similar to those of β-type toxins, but the differences in sequence are concentrated into two stretches forming a continuous surface region, which may be involved in species specificity (Garcia *et al.*, 1997).

The β-toxins may also be classified according to their activity against mammals and insects as has been reported for Ts γ and AaH IT4. However, β-like toxins that act very specifically towards arthropods have also been described (Selisko *et al.*, 1996).

1.2 Toxins active against invertebrate voltage-sensitive Na^+ channels

The high toxicity of scorpion venoms to insects is mainly due to the actions of polypeptides that have various effects on the voltage-sensitive sodium channels of one or several insect families. These toxins include:

The excitatory anti-insect toxins

The first scorpion toxin to be isolated following the toxicity to an insect was AaH IT from *Androctonus australis*. It causes repetitive firing in the motor nerve of insects, resulting in immediate contractile paralysis of the prey (Zlotkin *et al.*, 1972). Binding studies with a radioiodinated derivative have shown that the toxin binds specifically only to insect neuronal membranes (locust and cockroach) independently of voltage and does not interact with the mammalian sodium channel (Gordon *et al.*, 1984, 1992; DeLima *et al.*, 1989). AaH IT has the highest specificity of all the toxin characterized as "anti-insect toxins" (De Dianous *et al.*, 1987) and one of the highest affinities found for the insect sodium channel. So, AaH IT may be used as a very specific probe for study of insect nervous system excitability and as a reference model for the design of new biopesticides. Similar toxins have been obtained from the venom of other *Buthinae* scorpions, including *Buthotus judaicus*, *Androctonus mauretanicus*, and *Leiurus quinquestriatus* (Lazarovici *et al.*, 1982; Zlotkin *et al.*, 1985). AaH IT consists of 70 amino acids. Its disulfide-bridge pattern differs from that of the α- and β-type toxins (Darbon *et al.*, 1982). This particular cross-linked pattern may decrease molecular flexibility, as shown by circular dichroism experiments, and this may be responsible for its high specificity for the insect sodium channel (Loret *et al.*, 1990). Its structure in solution has been determined by ^{1}H-nuclear magnetic resonance (NMR) and suggests that the difference in specificity between toxins that discriminate between mammalian and insect voltage-gated Na^+ channel, are due to the position of the C-terminal peptide relative to a hydrophobic surface common to all scorpion toxins (Darbon *et al.*, 1991). The high potency of AaH IT and its strict selectivity (De Dianous *et al.*, 1987) has led to various groups using its gene (Bougis *et al.*, 1989) in the design of biopesticides. Synthesis of the toxin in the hemolymph of tobacco budworms (*Heliothis virescens*) by a recombinant baculovirus increases the speed of the animal's death and, thus, limited feeding damage to the plant (Stewart *et al.*, 1991; Maeda *et al.*, 1991; McCutchen *et al.*, 1991). A recombinant-AaH IT has also been produced in *Saccharomyces cerevisiae*, using the pre-pro-signal sequence of the yeast α-mating factor precursor (Martin-Eauclaire *et al.*, 1994b).

The depressant anti-insect toxins

These toxins cause the slow progressive paralysis (described as "flaccid") of insects. This paralysis results from blocking the evoked action potentials. These toxins do not bind to rat brain neuronal membranes (Zlotkin *et al.*, 1985). They are present in the venoms of *Leiurus quinquestriatus hebraeus* (LqhIT2) and *quinquestriatus* (LqqIT2), *Buthotus judaicus* (Bj IT2), *Buthus tunetanus* (Bot IT4 and Bot IT5), and *Buthacus arenicola* (Ba IT2) (Kopeyan *et al.*, 1990; Zilberberg *et al.*, 1991; Zlotkin *et al.*, 1991, 1993; Borchani *et al.*, 1997a, b; Cestèle *et al.*, 1997a). They consist of 61-amino acid residues, and the disulfide bridges are in the same positions as those for other toxins active against Na^+ channels, except for the "contracturant insect toxins". No X-ray or NMR structures studies are available.

An anti-insect toxin, AaH IT5, more potent than AaH IT and than the depressant insect toxins against the tobacco budwarm, has been purified from *Androctonus australis* (Nakagawa *et al.*, 1997). Its amino acid sequence shows 38% similarity to the sequence of depressant insect toxins. *In vitro* competition experiments with radiolabelled "excitatory anti-insect-toxins" or depressant insect toxins to identify and characterize the receptor sites of these toxins remain to be done.

Bot IT2 from *B. tunetanus* differs from the other toxins of this group: its overall effects in current-clamp conditions are relatively similar to those of contracturant insect toxins on cockroach axon preparations (Stankiewicz *et al.*, 1996; Borchani *et al.*, 1997a). It also causes a repetitive activity similar to the contractions caused by the contracturant insect toxins. However, its sequence is closer to those of depressant anti-insect toxins (67%). Competitive binding experiments with other known sodium channel neurotoxins suggest that Bot IT2 binds to a receptor site located close to, but not identical to, the binding site of the excitatory anti-insect toxins (Cestèle *et al.*, 1997b).

Lqh2 has been produced in a bacterial expression system (Turkov *et al.*, 1997) and in baculovirus (Gershburg *et al.*, 1998).

2 Scorpion Species

Scorpions have been classified into six families: *Bothriuridae*, *Scorpionidae*, *Buthidae*, *Vejovidae*, *Chlaerilidae* and *Chactidae*. Only the venoms of scorpions *Buthidae* have been described as potentially dangerous for humans. This *Buthidae* family is divided into subfamilies on the basis of geographic and morphological criteria: *Isometrinae*, which is of minor importance; *Buthinae*, from Africa and Asia; *Centrurinae* and *Tityinae*, living in North or Central America and South America, respectively (Bucherl, 1971).

Source of venoms and toxins

α-Toxins, excitatory, and depressant anti-insect toxins have been purified mainly from Buthinae venoms. The main sources are the venoms of *Androctonus australis*, *A. mauretanicus*, and *Leiurus quinquestriatus*, available in Latoxan (Rozan, France). They can also be purified from some American scorpion venoms, such as those from *Tityus serrulatus* and *Centruroïdes sculpturatus*. However, they are minor components of such venoms and have a lower toxicity and affinity for their receptors than those purified from the Buthinae.

Most β-toxins have only been purified from the venoms of scorpions from the Centrurinae and Tityniae families. Neither α- nor β-toxins are commercially available. They can be purified as described below (see purification).

The toxins specific to insects are found in all types of venom (see the paragraphs related to each toxin in the introduction).

3 Purification of the Toxins

The available venoms are mostly obtained by the electrical stimulation of animals: they constitute complex mixtures of homologous proteins with similar physicochemical properties. Thus the purification procedure must take into account the possibility of cross-contamination between the purified toxin preparations which, if not detected, may impede the definition of the pharmacological properties.

The first step in the purification of pooled venoms is the dialysis of the water extract against water. This step makes it possible to eliminate by precipitation about 50% of the nontoxic material, e.g., mucoproteins, with no loss of toxicity (the cutoff of the Spectrapor membrane has to be 1000 to retain the soluble toxins inside the membranes). For some heavily mucous venoms (e.g., those of *Buthus* and *Leiurus*), it is necessary to dissolve the venom in an acetic acid medium (0.1 – 1 M) before starting the purification process (Martin and Rochat, 1984). Centrifugation to eliminate the insoluble material before chromatography should be conducted at high speeds (between 11000 and 15000 × g).

3.1 Chromatography

Scorpion toxins can be routinely purified from the venom of Buthidae by conventional chromatography. Several steps, including molecular filtration followed by one or several steps on anionic and cationic exchangers, are necessary to fractionate the water extract of the venom. High-performance liquid chromatography (HPLC) can also be used for analytical purposes or if small amounts of toxins are needed. Combinations of reverse-phase (C8 or C18 matrix) and ion-exchange columns are used to isolate pure toxins. HPLC is the best way to obtain small toxins (PM≅4 kDa), which are eluted well before the other larger venom proteins on C18-RP columns.

Molecular filtration: Sephadex G_{50} successfully purifies all toxins. The best buffer is ammonium acetate (Ac NH4), which is easily removed by lyophilization (two lyophilization steps are necessary). Toxins are strongly resistant to denaturation by lyophilization. However, it is not advisable to lyophilise toxins at Ac NH4 concentrations above 0.2 M. Ac NH4 is a very good solvent, and it also prevents the growth of bacteria and fungi. If necessary, the filtration can be performed in acetic acid to eliminate nonspecific interactions with the gel matrix.

Ion-exchange chromatography: Ion exchange chromatography is performed at equilibrium in Ac NH4 (0.01 – 0.2 M, pH between 6.0 and 8.5). Then, an ionic-strength gradient or a stepwise elution increases the concentration (up to 2 M) to elute all the fractions from the column.

HPLC: The crude venom is dissolved in the "starting" buffer, centrifuged at 12,000 × g, and the supernatant filtered though an HV Millipore filter (0.45 μm), before loading, to avoid clogging the columns with mucous constituents. The resolution of α- and β-toxins by reverse-phase chromatography is optimum with C8 columns using 0.15 M (12 mS) ammonium formate, pH 2.7, as solvent A and acetonitrile as solvent B (Bougis *et al.*, 1986; Martin-Eauclaire *et al.*, 1987b). With this setup, 280 nm is the only wavelength that can be used to monitor optical density, and only peptides and proteins with a substantial number of aromatic residues can be detected. Purification with conventional HPLC solvents such as 0.1% TFA in water (A) and 0.1% TFA in acetonitrile (B) gives lower quality separations, and a large proportion of acetonitrile is required to elute the toxins. However, optical density can be monitored at 215 nm, making it possible to detect every peptide fraction. The same solvents are used for venom separation on C18 columns. In these conditions, all small toxins (around 4 kDa) are detected and generally well separated, although the largest toxins (7 kDa and above) may be definitively lost or eluted in a denatured form.

Immunoaffinity chromatography: Toxins have high specific affinities for their receptor sites: K_d values in the nano to the picomolar range are not rare. If this is the case, then only 1 contamination of a toxin with another one, almost undetectable by classical biochemical analysis, may result in incorrect pharmacological conclusions being drawn.

In case of doubt, additional chromatography steps can be performed using I_gG specific for the contaminating toxin, covalently bound to protein A-Sepharose. The high antigenic polymorphism of the toxins makes possible this kind of segregation: not only between toxins with different pharmacological activities, but also between toxins of the same type but belonging to different immunological groups (Granier *et al.*, 1989).

3.2 Polyacrylamide gel electrophoresis

This can be performed to check the purity of a fraction, or to blot the molecule for immunostaining or amino acid sequence determination by Edman degradation in a gas-phase sequencer.

SDS-PAGE: Almost all the toxins are similar in size (around 4 or 7 kDa). Thus SDS gels cannot be used to demonstrate the purity of a toxin: they cannot discriminate between toxins, but only between toxins and other polypeptides. In any case, a highly cross-linked gel (20%) must be used.

Native-PAGE: The toxins are very basic proteins. Even with a buffer pH of 8.5, migration must be conducted towards the cathode. Native gels must also be highly cross-linked (15%) for adequate separation.

IEF (isoelectrofocusing) does not usually give clear enough separation to check toxin purity. This is due to the very similar and very high pHi values of scorpion toxins (easily 10 out of range of the analysis).

4 Chemical Modification of the Toxins

We will concentrate on two main chemical modifications: radioiodination for binding experiments and chemical modification of the NH_2 group of lysine residues by a photoactivable reagent, for photoaffinity or cross-linking experiments.

Iodination: Scorpion toxins are readily labeled with $Na^{125}I$, either on Tyr or His residues. AaH II and Lqq V may be radioiodinated and used as specific high-affinity markers (0.2 – 1 nM) for site 3 of mammalian sodium channels. Css II, Css IV and Css VI, Cn 2 (*Centruroides noxius* 2), and Ts VII (Ts γ) are excellent probes for site 4 (Jover *et al.*, 1980; Couraud *et al.*, 1982; Thomsen *et al.*, 1995; Barhanin *et al.*, 1982; De Lima *et al.*, 1989; Dehesa-Davila *et al.*, 1996).

Protocol 1 Labeling of the α-toxin AaH II as an example (Rochat *et al.*, 1976)

1. 5 µg of lyophilized toxin, or a solution in water of toxin (1µg/µl).
2. 30 µl of 50 mM phosphate buffer, pH 7.2.
3. 10 µl $K^{125}I$ (Amersham), corresponding to 1mCi.
4. 10 µl lactoperoxidase (0.25 µg/µl, *i.e.*, 2.5 µg) to catalyze the reaction at room temperature.
5. Start the reaction by adding 5 µl of H_2O_2 diluted 1/50,000; add another 5 µl after 1 min.
6. Stop the reaction 1 min later by diluting the mixture in the starting buffer before separation of the free iodide.

Notes:
Various methods are available for performing this separation:
Method 1:
The toxin can be precipitated by its specific IgG (1 h 30 min incubation at 37°C).
The pellet is obtained by centrifugation (3 min at 13,000 × g), washed three times with 1 ml 9% NaCl, 1 mg/ml BSA, then disolved in 40 μl pure acetic acid, which is evaporated off under an N_2 stream.
The pellet is then suspended in 50 mM NaH_2PO_4, pH 7.2, 1 mg/ml BSA.
The mixture is incubated for 15 min, and centrifugation (13,000 × g, 15 min) is used to separate the denatured IgG in the pellet from the iodinated toxin recovered in the supernatant.
Routinely, about 65% of the ^{125}I is incorporated into the molecule.
Method 2:
If specific IgGs against the toxin are not available, it is possible to eliminate free ^{125}I by HPLC (C8 or C18 column). This is the best way to separate mono-iodo-derivatives from diiodo-derivatives and native toxin. Each toxin has its own particular problems.
Method 3:
If HPLC is not possible, separation can be achieved by absorption of $^{125}I^-$ on an anionic exchanger.
(1) Dowex IX8 (500 μl) is equilibrated with the reaction buffer containing 0.1% BSA before use.
(2) The reaction medium is added to the Dowex and mixed. $^{125}I^-$ is retained by the resin, and the basic toxin is released in the supernatant.
(3) Dowex should be washed three times (total volume/1 ml) with reaction buffer containing BSA to prevent nonspecific interactions between the toxin and Dowex.
Specific radioactivities up to 2000 Ci/nmol were routinely obtained.

Protocol 2 Alternative method for scorpion toxin iodination

The method described above uses lactoperoxidase to incorporate one $^{125}I^-$ atom into a Tyr residue of the toxin.

1. Iodination can be performed with IODOGEN (1,3,4,6-tetrachloro-3α,6α,D diphenylglycouril, Sigma) as an oxidative agent. For 1 nmol of toxin, 3 nmol of IODOGEN is routinely used (may be increased up to 12 nmol).

2. IODOGEN is dissolved in CH_2Cl_2, aliquoted into various tubes (to provide ready-to-use samples), dried under a N_2 stream, and kept frozen.

3. Iodination is performed in 20 mM Tris-HCl buffer, pH 8.2.

   Notes:
   Time and temperature differ according to the protein to be labeled. For some difficult-to-label toxins incubation at 40°C for 1 h may be required.

4. After labelling, the method for eliminating free iodide is the same as that described above.

Photoaffinity labelling of the α- and β-toxins: A radioiodinated toxin can be further covalently linked after binding to its receptor using photoactivable derivatives. This chemical modification of α- and β-toxins has been used to analyze the subunit composition of Na^+ channels from rat brain, in both synaptosomes and cultured brain neurons (Sharkey *et al.*, 1984; Darbon *et al.*, 1983a, b; Jover *et al.*, 1988). Two bands are labeled with both types of toxin: a major 34-kDa component and a minor 300-kDa component (in synaptosomes) or 260-kDa component (in cultured neurons). This modification has been used to follow the downregulation and internalization of Na^+ channels in fetal neurons (Dargent *et al.*, 1994).

The toxins active against insect Na^+ channels could also be modified in this way to characterize their receptor sites. A single band of about 190 kDa is labeled for three different toxins: AaH IT, an excitatory anti-insect-toxin; Lqq IT2, a depressant anti-insect-toxin, and Ts VII (or γ), a β-toxin active on the insect nervous system (De Lima, 1988).

Protocol 3 Modification of Lqq V or Css IV by ANBS-NOS as an example

1. The neurotoxins are first iodinated as described above (see Protocol 1), then modified by the photoactivable reagent in a one-to-one molar ratio per total reactive amino group (scorpion toxin 2 x 10^{-7} M and BSA 7.6 x 10^{-6} M).

   Notes:
   The reagents used are:
   - *succinimidyl-4-azido-2 nitrophenylaminoacetate (su-ANPAA) (from Dr. Angelides, Gainesville, FL, USA),*
   - *N-succinimidyl-6 (4-azido-2'-nitrophenylamino) hexanoate (su-ANPAH)*
   - *N-5-azido-2-nitrobenzoyloxysuccinimide (ANBS-NOS) from Pierce (Rockford, IL, USA)*
2. Modification of the toxin: 100 µl of 2 x 10^{-7} M $^{125}I^-$ toxin, 100 µl of 0.4 Tris-HCl buffer, pH 9, 24 µl of ANB-NOS (1 mg/ml in acetonitrile, diluted in the dark).
3. The mixture is incubated in the dark at 20°C for 2 h
4. Photoaffinity labelling: the modified radioiodinated toxin is incubated with membranes in binding buffer under standard conditions. For the P2 synaptosomal fraction, for example:
   - one tube with
     60 µl of the modified toxin (now 10^{-7} M)
     0.6 mg of P2 in 5940 µl binding buffer, giving a final concentration of toxin of 10^{-9} M
   - to a second tube containing the same mixture, native toxin is added (to a final concentration of 10^{-6} M)
5. The tubes are incubated for 1 h at 25°C.

6. The contents of the tubes are spread out in Petri dishes (resting on a bed of ice, in the coldroom), and the membranes are irradiated for 5 min with a Phillip UV Lamp (λ max 356 nm) placed 5 cm above the samples.
7. The samples are centrifuged at 11,000 × g for 2 min.
8. The pellets are washed twice with ice-cold incubation buffer, then twice with ice-cold BSA-free incubation buffer.
9. The pellets are then analyzed by SDS-PAGE, and the gel is dried and autoradiographed.

5 Troubleshooting

Molecular filtration: Nonspecific interactions with the gel matrix, particularly with dextran gels, occur due to the high contents of aromatic and basic residues in the toxins active against the Na^+ channels. These interactions, which occur in very low ionic strength buffer, may be of considerable advantage for separating molecules of identical size and pHi (the pHi of the toxin is usually >9). Sometimes, it is worthwhile to "recycle" the toxic fractions to improve the separation (Miranda *et al.*, 1970).

HPLC: Some pure toxins dissolved in water rather than in the chromatographic buffer before loading onto a C18 column may give two fractions rather than one. This behavior is not yet understood.

A sharp peak in HPLC does not mean a "pure toxin" but only that the proteins corresponding to the peak are homogeneous in their behavior in the particular conditions used. It is often necessary to further purify this fraction by ion-exchange chromatography.

SDS-PAGE: Very often, in nonreducing conditions in the presence of SDS, the molecular weight of the toxins appears to be double that in reducing conditions.

Iodination: For iodination with lactoperoxidase, the toxin sample must be lyophilized or in water. Any product causing a decrease in the final pH of the reaction mixture should be avoided because this will also abolish the lactoperoxidase activity.

6 Conclusions

Scorpion neurotoxins constitute excellent tools for the study of ion channels, in particular the voltage-sensitive sodium channel. Scientists using this toxins should be aware of the following:

- The precise location from which the animals are collected and, thus, the origin of the venom. Toxin polymorphism is extremely high and depends on the venom source. Some toxins are never present in some venoms. For example, venom from *Androctonus australis* collected in the area around Tozeur (Tunisia) does not contain the α-toxin AaH III. However, this toxin is present in venom from animals collected in Mecheria (Tunisia) and Chellala (Algeria). The α-toxins AaH I' and AaH I", in contrast, are present only in the venom of animals collected in Chellala (Martin-Eauclaire and Rochat, 1986).
- The method of venom isolation is also important: venom obtained by electric stimulation may be extremely heterogeneous, and the purified products may not necessarily be typical of physiologic secretion.
- With these biological tools, with affinities in the nano- to picomolar range for their receptors, the pharmacological effect observed may be due to minute cross-contamination (sometimes less than 1/10 000). Thus the purification and purity criteria are of major importance.

References

Babin DR, Watt DD, Goos SM, Mlejnek RV (1974) Amino acid sequence of neurotoxin protein variants from the venom of Centruroides sculpturatus Ewing. Archiv Biochem Biophys 164: 694–706

Barhanin J, Giglio JR, Leopold P et al (1982) Tityus serrulatus venom contains two classes of toxins. J. Biol. Chem. 257: 12553–12558

Barhanin J, Pauron D, Lombet A et al (1983a) Electrophysiological characterization, solubilization and purification of the Tityus gamma toxin receptor associated with the gating of the sodium channel from rat brain. EMBO J. 2: 915–920

Barhanin J, Schmid A, Lombet A et al (1983b) Molecular size of different neurotoxin receptors on the voltage sensitive sodium channel. J. Biol. Chem. 258: 700–702

Bechis G, Sampieri F, Yuan PM et al (1984) Amino acid sequence of toxin VII, a β toxin from the venom of the scorpion Tityus serrulatus. Biochem. Biophys. Res. Comm. 122: 1146–1153

Borchani L, Stankiewicz M, Kopeyan C et al (1997a) Purification, structure, and activity of three insect toxins from Buthus occitanus tunetanus venom. Toxicon, 35: 365–382

Borchani L, Mansuelle P, Stankiewicz M et al (1997b) A new scorpion venom toxin paralytic to insects that affects Na+ channel activation. Purification, structure, antigenicity, and mode of action. Eur. J. Biochem, 24: 525–532

Bougis P, Marchot P, Rochat H (1986) Characterization of Elapidae snake venom components using optimized reverse phase high performance liquid chromatographic conditions and screening assays for α-neurotoxin and phospholipase A2 activities. Biochemistry, 25: 7235–7243

Bougis P, Rochat H, Smith LA (1989) Precursors of Androctonus australis scorpion neurotoxin structure of precursors, processing outcomes, and expression of a functional recombinant toxin II. J. Biol. Chem. 264: 19259–19265

Bucherl W (1971) Venomous animals and their venoms, vol 3, (Bücherl W, Buckley E, eds) p 41: New York: Academic Press

Cahalan MD (1975) Modification of sodium channel gating in frog myelinated nerve fibres by Centruroides sculpturatus scorpion venom. J. Physiol. 244: 511–534

Carbone E, Wanke E, Prestipino G et al (1982) Selective blockage of voltage-dependent K^+ channels by a novel scorpion toxin. Nature, 296: 90–91

Catterall WA (1975) Cooperative activation of action potential sodium ionophore by neurotoxins. Proc. Natl. Sci. USA 72: 1782–1786

Catterall WA (1979) Binding of scorpion toxin to receptor sites associated with sodium channels in frog muscle. J. Gen. Physiol. 74: 375–391

Catterall WA (1980) Neurotoxins that act on voltage-sensitive sodium channels in excitable membranes. Ann. Rev. Pharmacol. Toxicol. 20: 15–43

Catterall WA (1986) Molecular properties of voltage-sensitive sodium channels. Ann. Rev. Biochem. 55: 953–985

Catterall WA, Trainer V, Baden DG (1992) Molecular properties of the sodium channels: a receptor formultiple neurotoxins. Bull. Soc. Pathol. Exot. 85(5 pt 2), 481–485

Catterall WA (1995) Structure and function of voltage-gated ion channels. Annu. Rev. Biochem, 64: 493–531

Céard B, De Lima ME, Bougis PE, Martin-Eauclaire MF (1992) Purification of the main β-toxin from Tityus serrulatus venom using HPLC. Toxicon, 30: 105–110

Cestèle S, Kopeyan C, Oughideni R et al (1997a) Biochemical and pharmacological characterization of a depressant insect toxin from the venom of the scorpion Buthacus arenicola. Eur. J. Biochem. 243: 93–99

Cestèle S, Borchani L, El Ayeb M, Rochat H (1997b) Bot IT2: a new scorpion toxin to study receptor site on insect sodium channels. FEBS lett, 405: 77–80

Cestèle S, Qu Y, Rogers J et al (1998) Voltage sensor-trapping: enhanced activation of sodium channels by β-scorpion toxin bound to the S3-S4 loop in domain II. Neuron, 21: 919–931

Chejanovsky N, Zilberberg N, Rivkin H et al (1995) Functional expression of an alpha anti-insect scorpion neurotoxin in insect cells and lepidpterous larvae. FEBS Lett, 376: 181–184

Couraud F, Jover E, Dubois JM, Rochat H (1982) Two types of scorpion toxin receptor sites, one related to the activation, the other to the inactivation of the action potential sodium channel. Toxicon 40: 9–16

Darbon H, Zlotkin E, Kopeyan C et al (1982) Covalent structure of the insect toxin of the North African scorpion Androctonus australis Hector. Int. J. Pept. Protein Res, 20: 320–330

Darbon H, Jover E, Couraud F, Rochat H (1983a) α-scorpion neurotoxin derivatives suitable as potential markers of sodium channels. Preparation and characterization. Int. J. Peptide Protein Res. 22: 179–186

Darbon H, Jover E, Couraud F, Rochat H (1983b) Photoaffinity labelling of?α and β-scorpion toxin receptors associated with rat brain sodium channel. Biochem. Biophys Res.Commun. 115: 415–422

Darbon H, Weber C, Brawn W (1991) Covalent structure of the insect toxin of the north african scorpion Androctonus australis Hector. Biochemistry, 30: 1836–1844

Dargent B, Paillard C, Carlier E et al (1994) Neuron, 13: 683–690. Sodium channel internalization in developing neurons. Neuron, 13: 683–690

DeBin JA, Maggio JE, Strichartz GR (1993) Purification and characterization of chlorotoxin, a chloride channel ligand from the venom of the scorpion Leiurus quinquestriatus hebraeus. Am. J. Physiol. 264: 361–369

De Dianous S, Hoaro F, Rochat H (1987) Re-examination of the specificity of the scorpion Androctonus australis Hector insect toxin towards arthropods. Toxicon, 25: 411–417

Dehesa-Davila M, Ramirez AN, Zamudio FZ et al (1996) Structural and functional comparison of toxins from the venom of the scorpions Centruroides infamatus infamatus, Centruroides limpidus limpidus and Centruroides noxius. Comp. Biochem. Physiol. B. Biochem. Mol. Biol, 113: 331–339

De Lima ME, Martin MF, Diniz CR, Rochat H (1986) Tityus serrulatus bears pharmacological properties of both β-toxin and insect toxin from scorpion venoms. Biochem. Biophys. Res. Comm. 139: 296–302

De Lima ME, Couraud F, Lapied B, Pelhate M, Diniz CR, Rochat H (1988). Photoaffinity labelling of scorpion toxin receptors associated with synaptosomal Na^+ channels.Biochem Biophys. Res. Comm 151: 187–192

De Lima ME, Martin-Eauclaire MF, Hue B et al (1989) On the binding of two scorpion toxins to the central nervous system of the cockroach Periplaneta americana. Insect Biochem 19: 413–422
De Lima ME, Martin-Eauclaire MF, Chavez-Olortegui C et al (1992) Tityus serrulatus venom toxins display a complex pattern of antigenic reactivity. Toxicon 31: 223–227
Diniz CR (1978) Chemical and pharmacological aspects of Tityniae venoms. Arthropod venoms Vol 48 (Bettini S ed) pp 379-394, Berlin: Springer
Eitan M, Fowler E, Herrman R et al (1990) A scorpion venom neurotoxin paralytic to insects that affects sodium current inactivation: purification, primary structure, and mode of action. Biochemistry, 29: 5941-5947
Fontecilla-Camps JC, Almassy RJ, Suddath FL et al (1980) Three-dimensional structure of a protein from scorpion venom: a new structural class of neurotoxins. Proc. Natl. Acad. Sci. USA, 85: 7443–7447
Fontecilla-Camps JC, Habersetzer-Rochat C, Rochat F (1988) Orthorombic crystals and three-dimensional structure of the potent toxin II from the scorpion Androctonus australis Hector. Proc. Natl. Acad. Sci. USA 85: 7443–7447
Garcia C, Becerril B, Selisko B et al (1997) Isolation, characterization, and comparison of a novel crustacean toxin with a mammalian toxin from the venom of the scorpion Centruroides noxius Hoffmann. Comp. Biochem. Physiol, 116B, 3: 315–322
Gershburg E, Stockholm D, Froy O et al (1998) Baculovirus-mediated expression of a scorpion depressant toxin improves the insecticidal efficacy achieved with excitatory toxins. FEBS Lett, 422: 132–136
Gordon D, Jover E, Couraud F, Zlotkin E (1984) The binding site of the insect selective neurotoxin (AaIT) from scorpion venom locust synaptosomal membrane. Biochem. Biophys. Acta, 778: 349–358
Gordon D, Moskowitz H, Warmer C et al (1992) Localization of the receptor sites for insect-selective toxins on sodium channels by site-directed antibodies. Biochemistry, 31: 7622–7628
Gordon D, Martin-Eauclaire M-F, Cestèle S et al (1996) Scorpion toxins affecting sodium current inactivation bind to distinct homologous receptor sites on rat brain and insect sodium channels. J. Biol. Chem, 271: 8034–8045
Granier C, Novotny J, Fontecilla-Camps J-C et al (1989) The antigenic structure of a scorpion toxin. Mol. Immun, 26: 503–513
Jover E, Martin-Moutot N, Couraud F, Rochat H (1980a) Binding of scorpion toxins to rat brain synaptosomal fractions. Effects of membrane potential ions and other neurotoxins. Biochemistry 19: 463–467
Jover E, Couraud F, Rochat H (1980b) Two types of scorpion neurotoxins characterized by their binding to two separate receptor sites on rat brain synaptosomes. Biochem. Biophys. Res. Commun. 95: 1607–1614
Jover E, Massacrier A, Cau P et al (1988) The correlation between Na+ channel subunits and scorpion toxin-binding sites. A study in rat brain synaptosomes and in rat brain neurons developing in vitro. J. Biol. Chem, 263: 1542–1548
Kopeyan C, Mansuelle P, Sampieri F et al (1990) Primary structure of scorpion anti-insect toxins isolated from the venom of Leiurus quinquestriatus quinquestriatus. FEBS Lett, 261: 423–426
Kopeyan C, Mansuelle P, Martin-Eauclaire M-F et al (1993) Characterization of toxin III of the scorpion Leiurus quinquestriatus quinquestriatus: a new type of alpha-toxin highly toxic both to mammals and insects. Natural Toxins, 1: 308–312
Lazarovici P, Yanai P, Pelhate HG, Zlotkin E (1982) Insect toxic components from the venom of a chactoid scorpion, Scorpio maurus palmatus (scorpionidae). J. Biol. Chem, 257: 8397–8404
Loret E, Sampieri F, Roussel A et al (1990) Conformational flexibility of a scorpion active on mammals and insects: a circular dichroism study. Proteins, 8: 164–172
Loret E, Martin-Eauclaire M-F, Mansuelle P et al (1991) An anti-insect toxin purified from the scorpion Androctonus australis Hector also acts on the α- and β- sites of the mammalian sodium channel: sequence and circular dichroism study. Biochemistry, 30: 633–640

Maeda S, Volrath SL, Hanzlik TN et al (1991) Insecticidal effects of an insect-specific neurotoxin expressed by a recombinant baculovirus. Virology, 164: 777–780

McCutchen BF, Choudary PV, Crenshaw R et al (1991) Development of a recombinant baculovirus expressing an insect selective neurotoxin: potential for pest control. Bio/technology (NY), 9: 848–852

Martin M.-F, Rochat H (1984) Purification of thirteen toxins active on mice from the venom of tne North African scorpion Buthus occitanus tunetanus. Toxicon, 22: 279–291

Martin-Eauclaire M-F, Rochat H (1986) Large scale purification of toxins from the venom of the scorpion Androctonus australis Hector. Toxicon, 24: 1131–1139

Martin M-F, Garcia y Perez LG, El Ayeb et al (1987a) Purification and chemical and biological characterization of seven toxins from the Mexican scorpion toxin: a neural property. J. Biol. Chem, 262: 4452–4459

Martin M-F, Rochat H, Marchot P, Bougis PE (1987b) Use of high performance liquid chromatography to demonstrate quantitative variation in components of venom from the scorpion Androctonus australis Hector. Toxicon, 25: 569–573

Martin-Eauclaire M-F, Céard B, Ribeiro AM et al (1992) Molecular cloning and nucleotide sequence analysis of a cDNA encoding the main β-neurotoxin from the venom of the South American scorpion Tityus serrulatus. FEBS Lett. 302: 220–222

Martin-Eauclaire M-F, Céard B, Ribeiro AM et al (1994a) Biochemical, pharmacological, and genomic characterization of Ts IV, an a-toxin from the venom of the South American scorpion Tityus serrulatus. FEBS Lett. 342: 181–184

Martin-Eauclaire M-F, Soggard M, Ramos C et al (1994b) Production of active, insect-specific scorpion neurotoxin in yeast. Eur. J. Biochem, 223: 637–645

Martin-Eauclaire M-F, Couraud F (1995) Scorpion neurotoxins: effects and mechanisms. Handbook of neurotoxicology (Chang/Dyer, ed) pp. 683–716:New York:

Miller C (1995) The charybdotoxin family of K^+ channel-blocking peptides. Neuron, 15: 5–10.

Miranda F, Kopeyan C, Rochat C et al (1970) Purification of animal neurotoxins. Isolation and charac terization of eleven neurotoxins from the venom of the scorpion Androctonus australis Hector, Buthus occitanus tunetanus, and Leiurus quinquestriatus quinquestriatus. Eur. J. Biochem, 16: 514–523

Naharashi T (1974) Chemicals as tools in the study of excitable membranes. Physiol. Rev, 54: 813–889

Nakagawa Y, Lee YM, Lehmberg E et al (1997) Anti-insect toxin (AaIT5) from Androctonus australis. Eur. J. Biochem. 240: 496–501

Possani LD, Martin BM, Mochca-Morales J, Svendsen I (1981a) Purification and chemical characterization of the major toxins from the venom of the Brazilian scorpion Tityus serrulatus, Lutz and Mello. Carlsberg Res. Comm. 46: 195–205

Possani LD, Steinmetz WE, Dent MAR et al (1981b) Preliminary spectroscopic characterization of six toxins from latin american scorpions. Biochem. Biophys. Acta. 669: 183–192

Possani LD, Martin BM, Swendsen I (1982) The primary structure of noxius toxin a K^+ channel blocking peptide, purified from the the venom of the scorpion Centruroides noxius Hoffman. Carlsberg. Res. Comm, 47: 285–289

Possani LD, Martin BM, Svendsen I et al (1985) Scorpion toxins from Centruroides noxius and Tityus serrulatus. Biochem. J, 229: p. 739–750

Possani LD, Martin BM, Fletcher MD, Fletcher PL (1991) Discharge effect on pancreatic exocrine secretion produced by toxins purified from Tityus serrulatus scorpion venom. J. Biol. Chem, 266: 3178–3185

Pauron D, Barhanin J, Lazdunski M (1985) The voltage-dependent sodium channel of insect nervous system identified by receptor sites for tetrodotoxin and scorpion and sea anemone toxins. Biophys. Res. Commun. 131: 1226–1233

Rochat H, Tessier M, Miranda F, Lissitzky S (1976) Enzymatic radioiodination of animal toxins to very high specific radioactivity. Animal, Plant, and Microbial toxins, Vol 1 (Ohsaka A, Hayashi K, Sawai Y) pp 81–88: New York: Plenum

Rogers JC, Qu Y, Tanada TN et al (1996) Molecular determinants of high affinity binding of a-scorpio toxin and sea anemone toxin in S3-S4 extracellular loop in domain IV of the Na^+ channel a subunit. J. Biol. Chem, 271: 15950–15962

Romey G, Chicheportiche R, Lazdunski M et al (1975) Scorpion neurotoxin: a presynaptic toxin which affects both Na+ and K+ channels in axons. Biochem. Biophys. Res. Commun, 64: 115–121

Selisko B, Garcia C, Becerril B et al (1996) An insect-specific toxin from Centruroides noxius Hoffmann: cDNA, primary structure, three dimensional model, and electrostatic surface potentials in comparison with other toxin variants. Eur. J. Biochem, 242: 235–241

Sharkey RG, Beneski DA, Catterall WA (1984) Differential labelling of the α and β1 subunits of the sodium channel by photoreactive derivatives of scorpion toxins. Biochemistry 23: 6078–6086

Stankiewicz M, Grolleau B, Lapied B et al (1996) Bot IT2: a toxin paralytic to insects from the Buthus occitanus tunetanus venom modifying the activity of insect sodium channels. J. Insect Physiol. 42: 397–405

Stewart LMD, Hirst M, Ferber MG et al (1991) Construction of an improved baculovirus insecticide containing an insect-specific toxin gene. Nature, 352: 85–88

Strichartz G, Rando T, Wang GK (1987) An integrated view of the molecular toxinology of sodium channel gating in excitable cells. Ann. Rev. Neurosci. 10: 237–269

Strong PN (1990). Potassium channel toxins. Pharmacol. J, 46: 137–162

Tejedor FJ, Catterall WA (1988) Site of covalent attachment of a-scorpion toxin derivatives in domain I of the sodium channel?α-subunit. Proc. Natl. Acad. Sci. USA, 85: 8742–8746

Thomsen WJ, Martin-Eauclaire M-F, Rochat H, Catterall W (1995) Reconstitution of a β-scorpion toxin to neurotoxin receptor site 4 on purified sodium channels. J. Neurochem, 65: 1358–1364

Thomsen WJ, Catterall, WA (1989) Localization of the receptor site for α-scorpion toxins by antibody mapping: implications for sodium channel topology. Proc. Natl. Acad. Sci. USA, 86: 10161–10165

Turkov M, Rashi S, Zilberberg N et al (1997) vitro folding and functional analysis of an anti-insect selective scorpion depressant neurotoxin produced in E. coli. Protein. Exp. Purif, 10: 123–131

Vijverberg HPM, Pauron O, Lazdunski M (1984) The effects of Tityus serrulatus scorpion toxin γ on Na channels in neuroblastoma cells. Pfluges Arch. 401: 297–303

Watt DD, Simard JM (1984) Neurotoxic proteins in scorpion venom. J. Toxicol. Toxin Rev, 3: 181-

Yatani A, Kirsch GE, Possani LD, Brown AM (1988) Effects of New World scorpion toxins on single-channel and whole cell cardiac sodium currents. Am. J. Physiol, 254: H 443: H 451

Zamudio FZ, Conde R, Arévalo C et al (1997a) The mechanism of inhibition of ryanodine receptor channels by imperatoxin I, a heterodimeric protein from the scorpion Pandinus imperator. J. Biol. Chem, 272: 11886–11894

Zamudio FZ, Gurrola GB, Arévalo C et al (1997b) Primary structure and synthesis of Imperatoxin A (IpTXa), a peptide activator of Ca2+ release channels/ryanodine receptors. FEBS Lett 405: 385–389

Zilberberg N, Zlotkin E, Gurevitz M (1991) The cDNA sequences of a depressant insect selective neurotoxin from the scorpion Buthatus judaicus. Toxicon 29: 1155–1158

Zilberberg N, Froy O, Loret E et al (1997) Identification of structural elements of a scorpion α-neurotoxin important for receptor site recognition. J. Biol. Chem, 272: 14810–14816.

Zlotkin E, Miranda F, Rochat H (1972) Proteins in scorpion venoms toxic on mammals and insects. Toxicon, 10: 211–216

Zlotkin E, Kadouri D, Gordon D et al (1985) An excitatory and a depressant insect toxin from scorpion venom both affect sodium conductance and possess a common binding site. Arch.Biochem. Bioph. 240: 877–887

Zlotkin E, Eitan M, Bindokas VP et al (1991) Functional duality and structural uniqueness of depressant insect-selective neurotoxins. Biochemistry, 30: 4814–4821

Zlotkin E, Gurevitz M, Fowler E, Adams ME (1993) Depressant insect selective neurotoxins from scorpion venom: chemistry, action and gene cloning. Arch. Insect. Biochem.
Physiol. , 22: 55–73

11 Immunological Properties of Scorpion Toxins

Christiane Devaux and Mohamed El Ayeb

Contents

1 Introduction

Early studies on scorpion venom were motivated by the problems of treating envenomation in various parts of the world (Goyffon *et al.*, 1982; Freire-Maia *et al.*, 1994; Dehesa-Davila *et al.*, 1994). Over the last 30 years, there has been extensive work by our groups and others to elucidate the mechanism of envenomation and to fight against the noxious effects of the scorpion sting.

Scorpion venom contains series of small proteins which are neurotoxic when injected into laboratory animals. Two major classes of toxic molecules have been described: short neurotoxins (about 37 residues) which, bind to var-

Methods and Tools in Biosciences and Medicine
Animal Toxins, ed. by H. Rochat and M.-F. Martin-Eauclaire

ious potassium channels, and long neurotoxins (60–65 residues), which are specific for sodium channels (Rochat *et al.*, 1979; Possani, 1984; Martin-Eauclaire and Couraud, 1995; Martin-Eauclaire and Rochat, this book). Based on results obtained with mice as an animal test, it seems that only the long toxins are responsible for the noxious effects observed when people are stung by scorpions (Miranda *et al.*, 1970). These molecules, reticulated by four disulphide bridges, constitute a large family, which can be divided into two groups based on rat brain synaptosome-binding studies and electrophysiological effects. α-Neurotoxins (mainly African and Asian scorpions) and β-neurotoxins (Southern USA, Central and South America) act on different sites of the same Na^+ channels in excitable cells (Couraud at al., 1982). They have some structural features in common but differ in amino acid content and have high-level antigenic polymorphism (El Ayeb *et al.*, 1985). They not only differ from species to species, but the amount of each toxin also varies between individuals of the same species. Current treatment methods involve the administration of specific antivenom sera. Thus the large amount of variation complicates the rational design of polyvalent and efficient serotherapy. The effectiveness of this treatment is highly dependent on its rapidity because the toxic proteins diffuse throughout the body, making their subsequent capture by specific antibodies difficult. It would therefore be of value, apart from improving serotherapy, try to improve the immunity of people at risk.

The North African scorpion, *Androctonus australis hector* (Aah), is one of the most noxious in the world. Its venom contains several neurotoxins that act on voltage-dependent Na^+ channels. Toxin II (AahII) is the most potent (LD_{50} of 0.5 ng/20 g when injected intracerebroventricularly into mice) and has been extensively studied by our group biochemically, structurally, and immunologically. We used AahII as a model toxin and tested several approaches for the identification of epitopes and the design of immunogens conferring immunity in an animal model, the mouse (Granier *et al.*, 1989; Aït-Amara *et al.*, 1993; Zenouaki *et al.*, 1997; Kharrat *et al.*, 1997). In this chapter, we describe specific and sensitive radioimmunoassays and sandwich-type ELISA for characterising and quantifying, respectively, toxins or venom in biological fluids. Strategies for mapping antigenic sites on toxins will be also discussed.

2 Qualitative and Quantitative Analysis of Venoms Using Immunoassays

2.1 Radioimmunoassay (RIA)

Aim and procedure

Scorpion toxins induce immune responses in several animal species: horse, rabbit, and mouse. Polyclonal antisera specific for individual toxins can be obtained and tested for their ability to bind the corresponding radiolabelled toxins (Rochat *et al.*, 1970, 1977; Tessier *et al.*, 1978).

Purification and iodination of toxins

(see Martin-Eauclaire and Rochat, Chapter 10)

RIA involves two basic procedures. First, we assessed the binding of antisera to the corresponding toxin.

Protocol 1 Binding of antisera to the corresponding toxin (Competitive RIA)

1. [^{125}I]-labelled toxin (5×10^{-10} M) in PBS containing 0.1% BSA (PBS-B) was incubated with various dilutions of rabbit antisera for 90 min at 37°C, then overnight at 4°C.
2. Nonimmune rabbit antibody and goat anti-rabbit IgG (Jackson) were then added, and the mixture incubated for 90 min at 4°C.
3. PEG 8000 (12.5%) was added, and the mixture incubated for 90 min.
4. The immune complexes were pelleted by centrifugation at 10,000 × g.
5. Radioactivity was measured with a gamma counter (Packard Crystal II).

For protein quantification, competitive RIA was performed using radioactive and non-radioactive antigen. Each of a series of concentrations of unlabelled toxin (or analogue or sample) was incubated with rabbit immune sera for 5 h, and the [^{125}I]-labelled toxin was then added. The dilution of the rabbit serum used was that giving approximately 40% maximum binding (Fig. 1A).

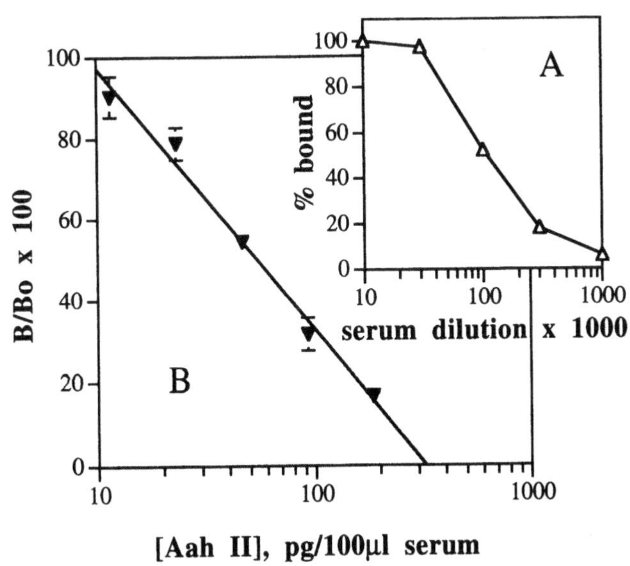

Figure 1 A typical standard curve for Aah II produced using competitive RIA data (B).

The appropriate polyclonal serum dilution was deduced from the serum titration curve (A).

Notes:
- All assays were performed in duplicate.
- The concentration of each reagent had to be determined with care to ensure accurate results. The theoretical and practical aspects of RIA are described in a laboratory manual (Chard, 1990).
- Results are expressed as B/Bo where B and Bo are the radioactivity bound to antibody in the presence (B) or absence (Bo) of unlabelled ligand. Standard curves are semilogarithmic plots of B/Bo versus the concentration of unlabelled toxin (Fig. 1B).
- Sensitivity is the minimum detection limit of an assay. It was 50 pg per 100 µl of serum for AahI, 10 pg/100 µl for AahII, and 20 pg/100 µl for AahIII (Devaux, Krifi, and El Ayeb, unpublished results).

Applications

RIA and competive RIA have been extensively used to:
- Evaluate cross-reactions between toxins. This has made it possible to establish a classification consistent with the groups defined by sequence similarities (Delori *et al.*, 1981; El Ayeb *et al.*, 1983; Martin *et al.*, 1987).
- Determine the immunoreactivity of toxin analogues obtained by chemical modification or synthesis (El Ayeb *et al.*, 1986a).
- Select and characterise monoclonal antibodies (Bahraoui *et al.*, 1988).
- Quantify individual toxins in venoms from different species and/or different geographical areas (El Ayeb *et al.*, 1985).
- Quantify individual toxins in pharmacokinetic studies (Devaux, Krifi, and El Ayeb, unpublished results).

2.2 ELISA

A sandwich-type ELISA using polyclonal antiscorpion toxin fraction (Krifi *et al.*, 1998) was developed for detecting venom in clinical and experimental envenomation by *Androctonus australis hector* (Aah).

Production of antibodies

A partially purified toxic fraction from scorpion venom was used as an immunogen for horse antivenom serum production. Antibodies from immune sera were purified by ammonium sulphate precipitation. $F(ab')_2$ were obtained by pepsin treatment and affinity chromatography purification on toxin fraction coupled to CNBr-activated Sepharose (Pharmacia). Specific $F(ab')_2$ were recovered from a second immunoaffinity column in which false-positive human serum proteins from healthy donors were coupled to CNBr-activated Sepharose. Finally a fraction (5 mg) of specific $F(ab')_2$ was labelled with peroxidase (10 mg) using glutaraldehyde-coupling reagent (Krifi *et al.*, 1998).

ELISA procedure and validation

Protocol 2 Sandwich-type ELISA

1. Microtitration plates were coated with specific $F(ab')_2$ (5 mg/ml), incubated overnight at 4°C, washed, and the remaining active sites were saturated with 5% nonfat milk by incubation for 1 h at 4°C.
2. Plates were washed.
3. Various concentrations of reference toxin fraction or serum samples were added and incubated for 90 min at 37°C.
4. The plates were washed, peroxidase conjugate (2 µg/ml) was added, and incubation was continued for 60 min at 37°C.
5. The plates were washed, and substrate solution (O-phenylenediamine) was added.
6. The plates were incubated for 5 min, and the reaction was stopped with sulphuric acid.
7. Absorbance at 492 nm was measured (Krifi *et al.*, 1998).

Notes:
When human sera were tested, assays were performed with 0.5% nonimmune horse serum in the samples and the conjugate solution to prevent nonspecific binding of heterophilic antibodies (Ho et al., 1986; Krifi et al., 1998). A reference curve was produced using known concentrations of purified toxin fraction. The curve was linear between 0.5 and 15 ng/ml. The detection limit was 0.09 ng/assay (Fig. 2A; for details see Krifi et al., 1998)

Applications

This sandwich-type ELISA is currently used to:

- Determine the toxicokinetic parameters of scorpion toxin fraction following the experimental envenomation of animals (Krifi *et al.*, unpublished data).
- Quantify the circulating toxin fraction and establish the potential correlation with clinical gradation based on symptom severity (Krifi *et al.*, 1998, and Fig. 2B).
- Evaluate the effectiveness of immunotherapy in experimental animal systems and in treatment of human envenomation.

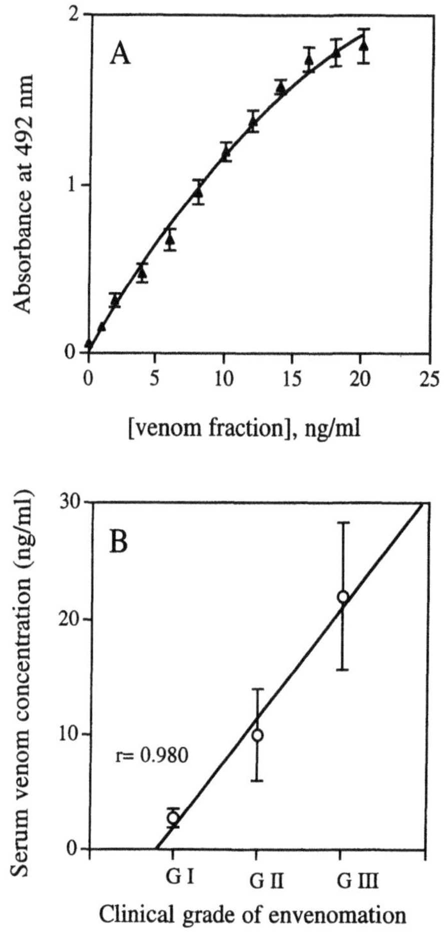

Figure 2 Sandwich-type ELISA for the determination of Aah venom antigens.

The mean reference curve for toxin fraction was determined using normal human sera (A). The correlation between venom concentration and the severity of envenomation (B) was calculated from ELISA data (from Krifi *et al.*, 1998)

3 Antigenic and Immunogenic Analysis of Toxins with Synthetic Peptides

3.1 Antigenicity and immunogenicity of synthetic peptides derived from particular domains of the Aah II molecule

Aim and procedure

Peptides mimicking accessible regions of the toxin were prepared by solid-phase synthesis to assess the antigenic properties of Aah II at the molecular level.

- Sequences were selected taking into account the known characteristics of the toxin: the sequence of AahII, sequence differences from other toxins, position of disulphide bridges, secondary structures such as α-helix or β-

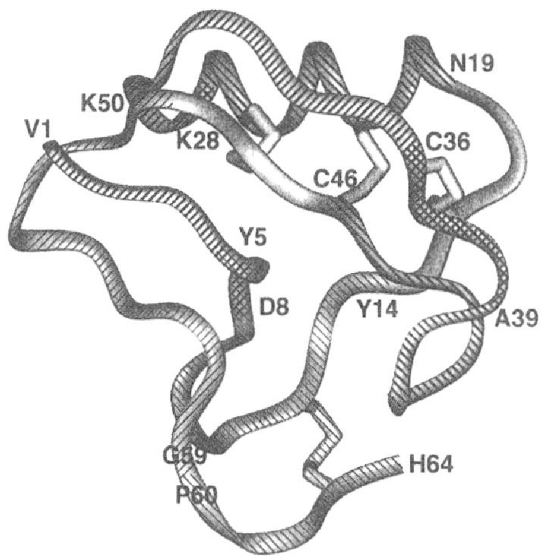

Figure 3 The three-dimensional model of the AahII molecule deduced from the crystallographic data (Fontecilla-Camps *et al.*, 1988).

Antigenic determinants were defined using peptides mimicking accessible regions (sequences 19–28, 28–39, 36–46, 50–59 and 5–14/60–64). The antigenic sites defined by the Pepscan method encompassed sequences 1–8, 4–12, 27–35, 39–45, 52–58, and 55–61. But only two peptides mimicking regions 1–8 and 50–59 were able to induce neutralising antibodies when coupled to a carrier protein.

turn, the three-dimensional model deduced by crystallography (Fontecilla-Camps *et al.*, 1981, 1988) and accessibility and segment mobility (Granier *et al.*, 1989).

- Peptides were coupled to Sepharose to further purify subpopulations of antibodies by immunoaffinity chromatography (Bahraoui *et al.*, 1986a).
- Alternatively, peptides were coupled to a protein carrier to induce antibodies with predetermined specificity (Bahraoui *et al.*, 1986b).
- Both peptide-purified antibodies and antipeptide antibodies were characterised using RIA as described above.

Results and applications

- Antibodies purified on peptides have been used to characterise the antigenicity of toxins. We found that almost the entire molecule of Aah II was antigenic (Fig. 3). However, some residues in the pharmacological site were not antigenic. These results may account for the lack of paraspecificity between the various toxin antigen groups (El Ayeb *et al.*, 1986a).
- Antibodies purified on peptides were used to check the orientation of Aah II on its Na^+ channel receptor. The antigenic region encompassing the α-helix is accessible when Aah II is bound to its receptor (El Ayeb *et al.*, 1986b).
- The ability of antibodies with predetermined specificity to recognise and neutralise the corresponding toxins was tested. One peptide (sequence 50–59) was of particular interest because antibodies raised in rabbit against the peptide linked to BSA recognised Aah II in a liquid-phase assay (RIA) and neutralise the toxin (Bahraoui *et al.*, 1986b).

All these aspects have been extensively reviewed and analysed in Granier *et al.* (1989).

3.2 Systematic analysis of overlapping peptides from the Aah II sequence (Pepscan)

Aim

Geysen *et al.* (1984, 1987) described a systematic method of searching for linear epitopes by rapidly and simultaneously synthesising large numbers of peptides (hundreds to thousands) on separate solid supports. The interaction of peptides with mono- and polyclonal antibodies was then assayed in ELISA without removing the peptides from support. To assess the effectiveness of this technique for detecting the known epitopes of a toxin and identifying new antigenic regions, we used a series of short overlapping heptapeptides covering the entire sequence of Aah II to systematically search for linear epitopes.

Procedure

Synthesis in the Pepscan format. The 58 overlapping heptapeptides covering the entire 64-residue sequence of AaH II were simultaneously synthesised using the commercial Epitope Scanning Kit (CRB, Cambridge, UK). All peptides were synthesised in duplicate, directly on polyethylene rods arranged so as to fit into 96-well microtitre plates, following each step described in the kit manual.

Notes:
- Software is provided to make it possible to draw up a schedule for each cycle of peptide synthesis. Detailed methods with useful hints are also described in a recent review by Rodda et al. (1996).
- We used the Epitope Scanning Kit from Cambridge Research Biochemicals. Several kits are now available from Chiron Mimotopes (Clayton Victoria, Australia) or their local distributors including noncleavable (solid phase) or cleavable multipin peptide synthesis kits for B-cell and T-cell epitope scanning, respectively. All reagents and solvents were of the best available quality.

Pepscan ELISA. ELISA reactions were performed on rod-bound peptides assembled on the holder. The rods were saturated by incubation for 1 h at room temperature with 1% casein in PBS containing 0.01% Tween (PBS-T). They were washed with 0.9% NaCl, 0.01% Tween (washing buffer), incubated overnight at 4°C with an appropriate dilution of serum (1:400 to 1:15,000) in PBS-T and washed again. Bound antibodies were detected by incubating blocks of rods for 1 h with either enzyme-conjugated goat anti-rabbit or rabbit anti-mouse IgG (H+L).

Notes:

- The experimental results are expressed as described by Trifilieff et al. (1991): the mean absorbance value of the 29 pins with the lowest absorbance was calculated as the noise (N) of the experiment; then the S/N ratio (the absorbance value S of each peptide divided by the noise N) was calculated. The results were plotted against the peptide number (defined as the position of a peptide's N-terminal residue in the sequence of the molecule) (Fig. 4).

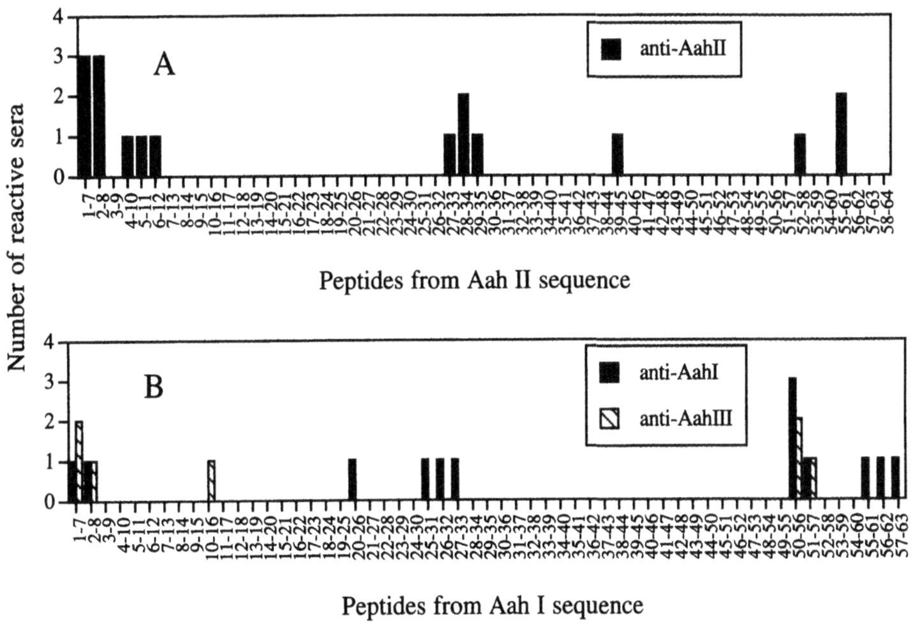


```
AahII   VKDGYIVDDVNCTYFCGR-NAYCNEECTKLKGESGYCQWASPYGNACYCYK-LPDHVRTKGP-GRCH*
AahI    KRDGYIVYPNNCVYHCVPP---CDGLCKKNGGSSGSCSFLVPSGLACWC-KDLP-NVPIKDTSRKCT-
AahIII  VRDGYIVNSKNCVYHCVPP---CDGLCKKNGAKSGSCGFLIPSGLACWCVA-LPDNVPIKDPSYKCHS
```

Figure 4 Representation of the reactivity of rabbit antitoxin sera with peptides synthesized in the Pepscan format.

Reactivity of overlapping rod-bound synthetic heptapeptides covering the entire sequence of AahII (A) or AahI (B) with five anti-AahII (A, from Devaux *et al.*, 1997) or three anti-AahI (B; black columns) and three anti-AahIII sera (B; white columns) (Devaux and Granier, unpublished). Sequences of the three main toxins of *Androctonus australis hector* scorpion venom (C).

- The best signal-to-noise ratios were obtained when casein was used as the saturating reagent rather than classical BSA.
- We used alkaline phosphatase-conjugated IgG and the paranitrophenyl phosphate substrate (Sigma) for detection. In our hands this system was the most reproducible and gave the least problems (such as excessive coloration of the pins). Absorbance at 405 nm was measured after 90 min of incubation.
- Between assays, bound antibodies were removed from the rods by sonication for 30 min at 60° C in an 0.1 M phosphate buffer, pH 7.2, containing 1% sodium dodecyl sulphate and 0.1% ß-mercaptoethanol. Rods were then thoroughly washed in water at 60°C, immersed in hot methanol for 1 min, and air-dried.
- Classical ELISA equipment (microtitre plate reader), a sonication bath (Branson 8200) for removal of antibody, and a shaking platform were necessary. Mercaptoethanol is toxic, and the disruption procedure should be carried out in a fully functional fume cupboard. Methanol is highly flammable, so sources of ignition should be avoided.
- Blocks of pins should be stored in sealed containers with a desiccant at 4°C. We have successfully reused blocks of pins, up to 50 times, over a long period of time (several years).

Results and applications

- The antigenic regions identified by this technique agree well with those previously mapped on the same toxin by other methods (Fig. 3). One difference, however, was the N-terminus, which was not clearly detected by other methods. This finding is particularly interesting because this sequence is highly conserved. Further work supports this result: rabbit antibodies raised against a synthetic peptide mimicking the sequence linked to a protein carrier bind and neutralise the toxin (Devaux *et al.*, 1993; Devaux *et al.*, 1996).
- This is a relatively simple and rapid procedure for synthesising and screening large numbers of peptides at the same time and does not require special equipment. The regeneration of pin-bound peptides allows numerous assays. When linear epitopes are identified, the contribution of each amino acid residue to antibody binding activity may be determined using replacement sets (Geysen *et al.*, 1984, 1987). This approach can also be used as an alternative to the determination of epitopes by predictive algorithms. It can be used to determine which sequence should be tested in immunisation studies when nothing is known about the structure of the protein except its amino acid sequence.
- There have been numerous studies using the Pepscan method or methods derived from it. Developments are progressing (Rodda *et al.*, 1986), and new applications and methods are emerging (Frank and Overwin, 1996; Rodda, 1996).

- Due to their complex conformation, native molecules such as toxins probably have discontinuous epitopes in which amino acids far apart in the protein sequence are brought together by protein folding. Short peptides, in contrast, have defined linear epitopes, and therefore continuous sequences. The reactivity of antibodies to peptides has sometimes been thought to be due to the presence of small quantities of denatured protein in the preparation used for immunisation (Jemmerson, 1987; Laver *et al.*, 1990). However, it has also been suggested that continuous epitopes may have conformational features similar to those of the cognate parent protein (Van Regenmortel, 1987). Peptides could then mimic part of the conformational epitope (Geysen *et al.*, 1987).
- Monoclonal antibodies directed against intact proteins are mostly specific for discontinuous epitopes. However, a small proportion react with both immunogen and peptide fragments (Pellequer *et al.*, 1991), and some monoclonal antibodies can be characterised with the Pepscan or a derived method (Marcel *et al.*, 1991; Redlich *et al.*, 1991; Gao and Esnouf, 1996). However, our antitoxin monoclonal antibodies directed against either Aah II or its nontoxic analogue did not react with rod-bound heptapeptides derived from the Aah II sequence (Yahi *et al.*, 1992; Devaux *et al.*, 1997).

3.3 Protection of peptide-immunised mice

Epitope mapping using short synthetic peptides is a well-developed technique. It is expected to be extremely useful in the development of peptide-based vaccines and peptide-based immunoassays (Lerner, 1984; Arnon and Horwitz, 1992). We have demonstrated that some synthetic peptides derived from toxins induce antibodies that recognise and neutralise native toxins when coupled to a carrier protein. Unfortunately, they did not protect immunised mice (Aït-Amara *et al.*, 1993, and Granier *et al.*, unpublished results). We took these results into account, and a synthetic nontoxic peptide [$(Abu)_8$Aah II] was designed and synthesised. It contained the entire sequence of Aah II with the eight half-cystine residues replaced by α-aminobutyric acid, and was used as an immunogen in mice and rabbits. All elicited antibodies efficiently recognised and neutralised the native toxin. Most important, the immunised mice were completely protected against challenge with several lethal doses of the corresponding native toxin, even 6 months after the immunisation protocol (Zenouaki *et al.*, 1997).

4 Conclusion and Perspectives

Several approaches were used to characterise in detail the immunological properties of the toxins. Both RIA and ELISA were used with poly- and monoclonal antitoxin antibodies, in qualitative and quantitative studies. Chemical peptide synthesis combined with immunoassays was a valuable approach for determining which antigenic sites were involved in neutralising the toxins. Almost all the accessible regions of the molecule were antigenic, but there is probably a hierarchy of antigenicity. Only a few peptides mimicking antigenic sites elicited neutralising antibodies. Peptides mimicking neutralising epitopes were useful for studying toxin neutralisation and designing vaccines, but were not always sufficient, even when coupled to a protein carrier, to induce immunity in animals. In these studies we considered only the humoral response of the immune system (B epitopes). The cellular response should also be studied by measuring the proliferating effect of peptides (T epitopes). Association of B and T epitopes, as probably occurred for the analogue $(Abu)_8$Aah II, would increase the level of protection.

Na^+ channel-specific toxins are responsible for the severe effects observed when people are stung by scorpions, which are a major public health problem. It is important to develop methods for detecting venom or toxins in biological fluids. Assays in experimental animals can be used to provide toxicokinetic data for evaluating the effectiveness of serotherapy. In cases of human envenomation, sandwich-type ELISA makes it possible to establish a correlation between the clinical severity and toxin concentration. New methods such as rapid ELISA (within 30 min) could be developed to predict the severity of envenomation and to determine specific treatment: serotherapy, appropriate drugs, or intensive care.

Acknowledgements

The authors are grateful to Prof. H. Rochat for constant support. They wish to thank Dr. C. Granier for his previous work in toxin immunology and Dr. M. N. Krifi for the development of quantitative ELISA. They are indebted to Dr. F. Sampieri for his hard work on the molecular graphics of Figure 3.

References

Aït-Amara D, Chavez-Olortegui C, Romi R et al (1993) Antibodies cross-reactive with the scorpion-toxin II from Androctonus australis Hector elicited in mice by a synthetic peptide. Natural Toxins 1: 255–262

Arnon R, Horwitz RJ (1992) Synthetic peptides as vaccines. Curr Opin Immunol 4: 449–454

Bahraoui E, El Ayeb M, Van Rietschoten J et al (1986a) Immunochemistry of scorpion alpha-toxins: study with synthetic peptides of the antigenicity of four regions of toxin II of Androctonus australis hector. Mol. Immunol. 23: 357–366

Bahraoui E, Granier C, Van Rietschoten J et al (1986b) Specificity and neutralising capacity of antibodies elicited by a synthetic peptide of scorpion toxin. J. Immun. 136: 3371–3377

Bahraoui E, Pichon J, Muller JM et al (1988) Monoclonal antibodies to scorpion toxins. Characterisation and molecular mechanisms of neutralisation. J. Immunol. 141: 214–220

Couraud F, Jover E, Dubois JM, Rochat H (1982) Two types of scorpion toxin receptor sites, one related to the activation, the other to the inactivation of the action potential sodium channel. Toxicon 20: 9–16

Chard T (1990) An introduction to radioimmunoassay and related techniques. Laboratory Techniques in Biochemistry and Molecular Biology, Vol 6 (Burdon RH and Knippenberg, eds): Amsterdam: Elsevier

Dehesa-Davila M, Possani LD (1994) Scorpionism and serotherapy in Mexico. Toxicon 32: 1015–1018

Delori P, Van Rietschoten J, Rochat H (1981) Scorpion venoms and neurotoxins: an immunological study. Toxicon 19: 392–407

Devaux C, Juin M, Mansuelle P, Granier C (1993) Fine molecular analysis of the antigenicity of the Androctonus australis Hector scorpion neurotoxin II: a new antigenic epitope disclosed by the Pepscan method. Mol Immunol. 30: 1061–1068

Devaux C, Fourquet P, Granier C (1996) A conserved sequence region of scorpion toxins rendered immunogenic induces broadly cross-reactive, neutralising antibodies. Eur. J. Biochem, 242: 727–735

Devaux C, Clot-Faybesse O, Juin M et al (1997) Monoclonal antibodies neutralising the toxin II from Androctonus Australis hector scorpion venom: usefulness of a synthetic, non-toxic analog. FEBS Lett 412: 456–460

El Ayeb M, Delori P, Rochat H (1983) Immunochemistry of scorpion α-toxins: antigenic homologies checked with radioimmunoassays (RIA). Toxicon 21: 709–716

El Ayeb M, Rochat H (1985) Polymorphism and quantitative variations of toxins in the venom of the scorpion Androctonus australis Hector. Toxicon 23: 755–760

El Ayeb M, Darbon H, Bahraoui E et al (1986a) Differential effects of defined chemical modifications on antigenic and pharmacological activities of scorpion alpha and beta toxins. Eur J Biochem 155: 289–294

El Ayeb M, Bahraoui E, Granier C, Rochat H (1986b) Use of antibodies specific to defined regions of scorpion α-toxin to study its interaction with its receptor site on the sodium channel. Biochemistry 25: 6671–6678

Fontecilla-Camps JC, Habersetzer-Rochat C, Rochat H (1988) Toxin II from the scorpion Androctonus australis hector orthorhombic crystals, three-dimensional structure, and structure function relationships. Proc. Nat. Acad. Sci. USA 85: 7443–7447

Frank R, Overwin H (1996) SPOT synthesis. Epitope Mapping Protocols (Morris GE, ed) pp 137–148: Totowa, New Jersey: Humana Press

Freire-Maia L, Campos JA, Amaral CF (1994) Approaches to the treatment of scorpion envenomation. Toxicon 32: 1009–1014

Gao B, Esnouf MP (1996) Multiple interactive residues of recognition: elucidation of discontinuous epitopes with linear peptides. J. Immunol. 157: 183–188

Geysen HM, Meloen RH, Barteling SJ (1984) Use of peptide synthesis to probe viral antigens for epitopes to a resolution of a single amino acid Proc. Natl. Acad. Sci. USA 81: 3998–4002

Geysen HM, Rodda SJ, Mason TJ et al (1987) Strategies for epitope analysis using peptide synthesis. J. Immun. Methods 102: 259–274

Goyffon M,Vachon N, Broglio N (1982) Epidemiology and clinical characteristics of scorpion envenomation in Tunisia. Toxicon 20: 337–344

Granier C, Novotny J, Fontecilla-Camps JC et al (1989) The antigenic structure of a scorpion toxin. Mol. Immunol. 26: 503–513

Ho M, Warrell MJ, Warrell DA et al (1986) A critical reappraisal of the use of enzyme-linked immunosorbent assays in the study of snake bite. Toxicon 24: 211–221

Jemmerson R (1987) Antigenicity and native structure of globular proteins: low frequency of peptide reactive antibodies. Proc. Natl. Acad. Sci. USA 84: 9180–9184

Kharrat R, Zenouaki I, BenLasfar Z et al (1997) Molecular characterisation, antigenicity and immunogenicity of anatoxic polymeric forms conferring protection against scorpion venoms. Toxicon 35: 915–930

Krifi MN, Kharrat H, Zghal K et al (1998) Development of an ELISA for the detection of venoms in sera of humans envenomed by Androctonus australis garzonii (Aag) and Buthus occitanus tunetanus (Bot). Correlation with clinical severity of envenomation in Tunisia. Toxicon 36: 887–900

Laver WG, Air GM, Webster RG, Smith-Gill SJ (1990) Epitopes on proteins: misconceptions and realities. Cell 61: 553–556

Lerner RA (1984) Antibodies of predetermined specificity in biology and medicine. Adv. Immunol. 36: 1–44

Marcel YL, Provost PR, Koa H et al (1991) The epitopes of apolipoprotein A-I define distinct structural domains including a mobile middle region. J. Biol. Chem. 266: 3644–3653

Martin M-F, Garcia y Perez LG, El Ayeb M et al (1987) Purification and chemical and biological characterisations of seven toxins from the Mexican scorpion, Centruroides suffusus suffusus. J Biol Chem. 262: 4452–4459

Martin-Eauclaire M-F, Couraud F (1995) Scorpion neurotoxins: effects and mechanisms. Handbook of Neurotoxicology (Chang LW and Dyer RS, eds) pp 683–716: New York: Marcel Dekker Inc

Miranda F, Kopeyan C, Rochat C et al (1970) Purification of animal neurotoxins. Isolation and characterization of eleven neurotoxins from the venoms of the scorpions Androctonus australis hector, Buthus occitanus tunetanus and Leiurus quinquestriatus quinquestriatus. Eur. J. Biochem. 16: 514–523

Pellequer JL, Westhof E, Van Regenmortel MHV (1991) Predicting location of continuous epitopes in proteins from their primary structures. Methods in Enzymol. 203: 176–201

Possani LD (1984) Structure of scorpion toxins. Handbook of Natural Toxins, Vol 2 (Tu A, ed) pp. 513–550: New York: Marcel Dekker

Redlich PN, Hoeprich PD, Colby CB, Grossberg SE (1991) Antibodies that neutralise human ß interferon biological activity recognise a linear epitope: analysis by synthetic peptide mapping Proc. Natl. Acad. Sci. USA 88: 4040–4044

Rochat H, Rochat C, Kupeyan C et al (1970) Scorpion neurotoxins a family of homologous proteins. FEBS Lett10: 349–351

Rochat H, Tessier M, Miranda F, Lissitzky S (1977) Radioiodination of scorpion and snake neurotoxins. Analyt. Biochem. 85: 532–548

Rochat H, Bernard P, Couraud F (1979) Scorpion toxins chemistry and mode of action. Advances in Cytopharmacology, Vol 3 (Ceccarelli and Clementi, eds) pp 325–334: New York: Raven Press

Rodda SJ (1996) T-cell epitope mapping with synthetic peptides and peripheral blood cells. Epitope Mapping Protocols (Morris GE, ed) pp 137–148: Totowa, New Jersey: Humana Press

Rodda SJ, Maeji NJ, Tribbick G (1996) Epitope mapping using multipin peptide synthesis. Epitope Mapping Protocols (Morris GE, ed) pp 137–148: Totowa, New Jersey: Humana Press

Tessier M, Delori P, Bechis G, Rochat H (1978) A sensitive radioimmunoassay of a scorpion neurotoxin. FEBS Lett 85: 163–166

Trifilieff E, Dubs MC, Van Regenmortel MHV (1991) Antigenic cross-reactivity potential of synthetic peptides immobilized on polyethylene rods. Mol Immunol, 28: 889–896

Van Regenmortel MHV (1987) Antigenic cross-reactivity between proteins and peptides: new insights and applications. Trends Biochem. Sci. 12: 237–240

Yahi N, Devaux C, Mansuelle P et al (1992) Monoclonal antibodies to toxin II from the scorpion Androctonus australis Hector: further characterisation of epitopes specificities and neutralising capacities. Toxicon 30: 723–731

Zenouaki I, Kharrat R, Sabatier JM et al (1997) vivo protection against Androctonus australis hector scorpion toxin and venom by immunization with a synthetic analog of toxin II. Vaccine 15: 187–194

12 Scorpion Toxins and Potassium Channels

Maria L. Garcia, Markus Hanner, Hans-Günther Knaus,
Robert S. Slaughter and Gregory J. Kaczorowski

Contents

1 Introduction

Potassium channels are a group of proteins that have in common the property of selectively allowing the movement of K^+ through aqueous pores in the membrane. Gating of these proteins occurs through conformational changes that are controlled by voltage and/or ligand binding. Accordingly, K^+ channels can be divided into voltage-dependent and ligand-activated channels. These channels are involved in the control of cell resting potential, and in modulation of the electrical excitability of neuronal, endocrine, and muscle cells. A large amount of information regarding the structure and function of these proteins has become available in the last few years due to two major developments: (i) the molecular cloning of complementary DNAs (cDNAs) encoding these proteins, and (ii) the discovery of high-affinity peptidyl inhibitors in the venom of different species. These peptidyl inhibitors have been useful in the development of the pharmacology of K^+ channels (Kaczorowski et al., 1996), and in determining the physiologic role that a particular K^+ channel plays in a given cell or tissue (Garcia et al., 1995). Moreover, they have also been employed as a

Methods and Tools in Biosciences and Medicine
Animal Toxins, ed. by H. Rochat and M.-F. Martin-Eauclaire

I	ChTX	Z F T N V S C T T S K E C W S V C Q R L H N T S R G K C M N K K C R C Y S
	IbTX	Z F T D V D C S V S K E C W S V C K D L F G V D R G K C M G K K C R C Y Q
	Lq_2	Z F T Q E S C T A S N Q C W S I C K R L H N T N R G K C M N K K C R C Y S
	LbTX	V F I D V S C S V S K E C W A P C K A A V G T D R G K C M G K K C K C Y ...
II	NxTX	T I I N V K C T S P K Q C S K P C K E L Y G S S A G A K C M N G K C K C Y N N
	MgTX	T I I N V K C T S P K Q C L P P C K A Q F G Q S A G A K C M N G K C K C Y P H
	C.l.l. I	I T I N V K C T S P Q Q C L R P C K D R F G Q H A G G K C I N G K C K C Y P ...
	TyKα	V F I N A K C R G S P E C L P K C K E A I G K A A G K C M N G K C K C Y P
III	$AgTX_1$	G V P I N V K C T G S P Q C L K P C K D A G M R F G K C I N G K C H C T P K
	$AgTX_2$	G V P I N V S C T G S P Q C I K P C K D A G M R F G K C M N R K C H C T P K
	$AgTX_3$	G V P I N V P C T G S P Q C I K P C K D A G M R F G K C M N R K C H C T P K
	KTX	G V E I N V K C S G S P Q C L K P C K D A G M R F G K C M N R K C H C T P K
	KTX_2	V R I P V S C K H S G Q C L K P C K D A G M R F G K C M N G K C D C T P K

Figure 1 Comparison of the amino acid sequences of charybdotoxin (ChTX), iberiotoxin (IbTX), *L. quinquestriatus* toxin 2 (Lq2), limbatustoxin (LbTX), noxiustoxin (NxTX), margatoxin (MgTX), *C. limpidus limpidus* toxin I (C.l.l. 1), tityustoxin-Kα (TyKα), agitoxin 1 ($AgTX_1$), agitoxin 2 ($AgTX_2$), agitoxin 3 ($AgTX_3$), kaliotoxin (KTX), and kaliotoxin 2 (KTX_2).

The sequences have been aligned with respect to the six cysteine residues which are in boldface type. The position of the disulfide bonds is indicated.

marker for channel purification from native tissues, thus allowing the determination of these channels' subunit composition (Garcia-Calvo et al., 1994; Giangiacomo et al., 1995; Parcej and Dolly, 1989; Rehm and Lazdunski, 1988). As a matter of fact, all auxiliary subunits of K^+ channels have been identified in this fashion. Peptidyl inhibitors of K^+ channels, because of their well-understood mechanism of action, have allowed the identification and molecular characterization of the pore region of these channels (Gross et al., 1994), and the determination of their subunit stoichiometry (MacKinnon, 1991). Detailed studies combining knowledge of the three-dimensional structure of these inhibitors with site-directed mutagenesis have provided a picture of the interactive surface of these peptides with K^+ channels, thereby allowing the derivation of the complementary shape and dimensions of the receptor in the channel pore (Aiyar et al., 1995; Goldstein et al., 1994; Stampe et al., 1994).

Peptidyl inhibitors of K^+ channels have been discovered in the venom of different species such as scorpions, snakes, bees, spiders, and sea anemone (Garcia et al., 1996). In this chapter, we will only concentrate on some aspects of those peptides derived from scorpion venoms. These peptides consist of 37–39 amino acids and display significant sequence homology between themselves. A common feature of all of these peptides is the existence of six Cys residues that form three disulfide bridges which provide the peptide with a very rigid structure. The three-dimensional structure of some of these peptides in solution has been determined by NMR techniques (Aiyar et al., 1995; Bontems et al., 1992;

Bontems et al., 1991a, b; Dauplais et al., 1995; Johnson et al., 1994; Johnson and Sugg, 1992; Krezel et al., 1995). The overall structure is very similar and contains an α-helical region that is linked by the disulfide bonds to a three-strand antiparallel β-sheet, where those residues critically involved in interaction with the channel are located. Despite the large sequence conservation between all members of this family, they can be further subdivided into different groups (Fig. 1). Members of the first group all inhibit high-conductance Ca^{2+}-activated K^+ channels, although only iberiotoxin (IbTX) and limbatustoxin are specific for this channel. All other members are targeted against voltage-gated K^+ channels. Out of all the different classes of voltage-gated K^+ channels, only members of the K_v1 family, which represent the mammalian counterparts of the *Drosophila* gene *Shaker*, are sensitive to these inhibitors. Moreover, only $K_v1.1$, 1.2, 1.3, and 1.6, but not $K_v1.4$, 1.5, and 1.7, are targets of these peptides.

In this chapter we will discuss the protocols for:

- Synthesis of the peptides by recombinant techniques
- Radiolabelling of the peptides
- Use of the peptides in receptor binding studies.

2 Materials and Chemicals

Buffers and reagents

- **Fusion protein preparation**
  - **Solution 1**
    - 50 mM NaCl
    - 10 mM Tris-HCl, pH 8.0
    - 1 mM EDTA
    - 1 mM DTT
  - **Equilibration buffer (Solution 2)**
    - 50 mM NaCl
    - 10 mM Tris-HCl, pH 8.0
    - 1 mM EDTA
    - 1 mM DTT
    - 100 μM PMSF
  - **Elution buffer (Solution 3)**
    - 350 mM NaCl
    - 10 mM Tris-HCl, pH 8.0
    - 1 mM DTT
    - 100 μM PMSF
- **Fusion protein dialysis (Solution 4)**
  - 100 mM NaCl
  - 20 mM Tris-HCl, pH 8.0
  - 0.5 mM β-mercaptoethanol

- **Mono S HR10/10**
  20 mM sodium borate, pH 9.0
  1 M NaCl in 20 mM sodium borate, pH 9.0
- **C_{18} reversed-phase HPLC**
  10 mM TFA
  Isopropyl alcohol/acetonitrile 2:1, 4 mM TFA
- **Toxin resuspension solution (Solution 5)**
  100 mM NaCl
  20 mM Tris-HCl, pH 7.4
- **GF/C filter soaking solution**
  0.5–1% polyethylenimine

Chemicals

DEAE-Sepharose	Pharmacia	17-0709-01
Factor Xa	Boehringer Mannheim	1179 896
GF/C glass fiber filters	Whatman	0987432
^{125}INa	DuPont NEN	NEZ-033A
Iodo-Gen	Pierce	28600
[^{3}H]NEM	DuPont NEN	NET-373
Polyethyleneglycol	Sigma	P-2139
Polyethylenimine	Sigma	P-3143
Trypsin	Sigma	T-8642

Equipment

- SDS-PAGE apparatus
- HPLC apparatus
- Filtration apparatus
- γ Counter
- Liquid scintillation counter

3 Biosynthesis of the Peptides

To demonstrate the identity of the peptide purified from a venom, it is necessary to accomplish its synthesis. Given the size of the peptides under consideration, 37–39 amino acids in length, solid-phase synthesis represents a feasible approach (Sugg et al., 1990). Alternatively, peptides can be produced in *Escherichia coli* as part of a fusion protein (Garcia et al., 1994; Garcia-Calvo et al., 1993; Park et al., 1991). In this way, large amounts of product can be prepared at relatively low costs. In addition, this system provides an efficient way of performing structure-activity relationship studies since toxin variants can be easily obtained by site-directed mutagenesis.

Protocol 1 Biosynthesis of the peptides

Construct the toxin gene from two synthetic oligonucleotide duplexes and insert it into an appropriate vector

We have successfully employed two types of vectors. In pCSP105, the resulting construct encodes a fusion protein of the T7 gene 9 product and the peptide separated by either a factor Xa cleavage site or by an enteropeptidase site (Garcia-Calvo et al., 1993). In pG9, there are six histidine residues between the fusion protein and the factor Xa cleavage site at the beginning of the peptide sequence (Johnson et al., 1994; Knaus et al., 1996). This may facilitate purification of the fusion protein by Ni^{2+}-affinity chromatography, although in our hands this chromatographic step does not represent a significant advantage for isolation of the fusion protein.

Transform *E. coli* BL21(DE3) cells with the corresponding plasmid

1. Select a single colony and inoculate the culture medium.
2. Grow the cells at 30–37°C with shaking.
3. When the OD at 650 nm has reached 0.6–0.7, remove a small aliquot of culture.
4. Induce the remaining with 0.5 mM IPTG and incubate for an additional 3 h at 37°C.
5. Check for induction of the fusion protein by running 12% SDS gels and stain the gels with Coomassie blue.

   Notes:
   After destaining, a major product at 57,000 should be visible in the induced culture.

6. Pellet the cells by centrifugation; wash them once with **Solution 1**.
7. Pellet the cells again and resuspend them in 10 ml of **Solution 1**.
8. Freeze the samples in liquid N_2 and store them at –70°C.

   Note:
   From now on, the protocols will refer to 1 l of starting culture.

9. Cells are thawed on ice.
10. The volume is adjusted to 50 ml with **Solution 2**, and 0.5 mg/ml lysozyme, and the suspension is kept on ice for 1 h.
11. The cell suspension is sonicated four times for 1 min each, keeping the tubes on ice between cycles.

12. The lysate is subjected to centrifugation at 27,000 × g for 15 min.
13. The supernatant is collected, and the nucleic acids are removed by precipitation with streptomycin sulfate.
14. The supernatant is loaded onto a DEAE-Sepharose fast flow column (30 ml of packed resin) that has been equilibrated with 5 vols of **Solution 2**.
15. The column is washed with 5 vols of equilibration buffer **(Solution 2)**, and the fusion protein is eluted with **Solution 3**.
16. The fractions containing the fusion protein are identified by Coomassie staining after SDS-PAGE.
17. These fractions are dialyzed overnight at 4°C against **Solution 4**.

    Notes:
    Determine the protein concentration of the sample before proceeding with the cleavage step.

There are two major procedures used to accomplish cleavage:

In the ***first approach****, factor Xa is used to release the peptide from the fusion protein. For this purpose,*

18. $CaCl_2$ is added to a final concentration of 3 mM followed by addition of 1 μg of factor Xa per milligram of fusion protein. The digestion mixture is incubated overnight at room temperature.

    Notes:
    Progress of the digestion can be easily followed by SDS-PAGE.

In the ***second approach****, trypsin is used to release the peptide from the fusion protein. In this case,*

18. $CaCl_2$ is added to a final concentration of 5 mM, and the sample is incubated with trypsin at a ratio of 5 μg enzyme per milligram of fusion protein for 30 min at room temperature.
19. After incubation with the proteolytic enzymes, the sample is loaded onto a Mono S HR10/10 column equilibrated with 20 mM Na-borate, pH 9.0.

    Notes:
    For highly positivly charged peptides, such as charybdotoxin (ChTX), margatoxin, agitoxin1 ($AgTX_1$), and agitoxin2 ($AgTX_2$), the sample is loaded directly onto the column, whereas for less charged peptides, such as IbTX, it must be dialyzed against 20 mM Na-borate, pH 9.0. The fusion protein and other proteolytic fragments are not retained by the column under these conditions.

20. When the absorbance has returned to baseline, bound toxin is eluted in the presence of a linear gradient of NaCl (0–0.5 M in 1 h) at a flow rate of 2 ml/min.

Notes:
A major absorbance peak representing the peptide of interest should be observed.

21. If peptides require N-terminal cyclization, as is the case for ChTX and IbTX, HPLC-grade acetic acid is added to the Mono S sample at a final concentration of 5%, and the mixture is incubated at either 65°C (ChTX) or 45°C (IbTX) until full cyclization is reached.

Notes:
We follow cyclization by N-terminal sequencing since the chromatographic behavior of these two species is quite similar.

The final stage in purification is achieved by reversed-phase chromatography.

22. The Mono S sample is directly injected into a semipreparative C_{18} reversed-phase HPLC column equilibrated with 10 mM TFA.
23. The peptides are eluted with a linear gradient of isopropyl alcohol/acetonitrile 2:1 (0–30%, 90 min) at a flow rate of 3 ml/min.
24. The sample is then distributed in aliquots of *ca.* 50 µg, lyophilized, and stored at –70°C.

Notes:
The composition of the purified peptide is routinely checked by N-terminal sequencing, amino acid hydrolysis, and/or mass spectroscopy.

4 Radiolabelling of the Peptides

There are two major approaches to radiolabelling the peptides. One is to take advantage of the presence of natural amino acids in the peptide, such as Tyr or His, which can be reacted in the presence of ^{125}INa to yield a high specific activity adduct. Another approach is to place amino acid residues in the sequence that can be modified so as to incorporate a radiolabeled tag at a position that is not critical for activity. In this way, those peptides that do not contain Tyr, His, or Lys residues, or those in which such residues are important for activity, can be radiolabeled after synthesis of the corresponding modified peptide.

4.1 Iodination of peptides

Iodination of peptides such as ChTX at its tyrosine residue can be achieved by two methods: Iodo-Gen and glucose oxidase/lactoperoxidase techniques (Garcia et al., 1992). The latter utilizes beads in which the enzymes have been immobilized, but because the manufacturer of these beads, Bio-Rad, has discontinued their production, we will only discuss the Iodogen methodology.

Protocol 2 Iodination of peptides (Iodo Gen)

1. A solution of Iodo-Gen is prepared in acetone.
2. An aliquot of this solution corresponding to 0.5–1 μg of the reagent is dried under argon or nitrogen at the bottom of a Reacti-vial. Once dried, the vial is rinsed with buffer to remove any aqueous contaminants.
3. A solution of ChTX containing 10–20 μg of peptide in 100 mM sodium phosphate, pH 7.3, is added followed by the addition of 2 mCi of ^{125}INa (20 μl, 2200 Ci/mmol).
4. After mixing, the vial is capped and the reaction is allowed to proceed for 15 min at room temperature.
5. One minute before the end of the reaction, the vial is open to the air, and the iodination mixture is removed with a syringe and injected into a C_{18} reversed-phase HPLC column that had been equilibrated with 10 mM TFA.
6. After extensive washing to remove any traces of unreacted iodine, retained material is eluted with a linear gradient of isopropyl alcohol/acetonitrile 2:1 (5–14%, 40 min), at a flow rate of 0.5 ml/min.
7. Changes in absorbance are monitored at 210 nm, and the peaks are collected manually.

   Notes:
   For ChTX, a major absorbance peak corresponding to unreacted peptide is first resolved. The second major absorbance peak that is well separated from unreacted toxin corresponds to monoiodotyrosine ChTX (^{125}I-ChTX), based on the specific activity of the radiolabeled peptide and on sequence analysis of the modified toxin.
8. The ^{125}I-ChTX fraction is made 0.1% (w/v) in bovine serum albumin, lyophilized, and reconstituted with **Solution 5**.
9. Aliquots of this material are frozen in liquid N_2 and stored at –70°C.

Warning: It is very important to be cautious when loading the iodination mixture onto the HPLC column because the low pH buffer could lead to the generation of iodine gas which can permeate through skin and accumulate in the thyroid gland. We always collect the loading and the washing material in 1 N NaOH to prevent the formation of iodine gas.

4.2 Labeling with [^{3}H]NEM

An alternative labelling approach that has been successfully employed with ChTX, IbTX, $AgTX_1$, and $AgTX_2$ makes use of a peptide mutant in which a Cys residue has been placed at position 19 (ChTX and IbTX; Knaus et al., 1996; Shimony et al., 1994; Sun et al., 1994) or 20 ($AgTX_1$ and $AgTX_2$; Aggarwal and

MacKinnon, 1996). The free sulfydryl is then reacted with [^{3}H]NEM to produce material with full biological activity at intermediate specific activity, ~60 Ci/mmol. These peptides can be easily made by recombinant techniques, as previously described. However, the newly introduced Cys residue is not obtained as a free sulfydryl, but instead makes a disulfide bridge either with β-mercaptoethanol or with another peptide molecule. Therefore, it is necessary to selectively reduce this disulfide bond without affecting the integrity of the constitutive disulfide bridges of the peptide.

Protocol 3 Labeling with [^{3}H]NEM

1. For IbTX-D19C, incubate the peptide with 10 mM DTT for 1 h at room temperature in 50 mM sodium phosphate, pH 7.0.
2. Apply the reaction mixture to a C_{18} reversed-phase HPLC column and elute with a linear gradient of isopropyl alcohol/acetonitrile 2:1 (10–30%, 30 min).

   Notes:
   IbTX-D19C with a free sulfydryl can be easily identified by the change in retention time as compared with unreduced material. We have found that optimal recovery of C19 reduced material is about 50%. Higher DTT concentrations or pH can lead to full disappearance of starting material, but also to the occurrence of other reduced forms of the peptide that are not of interest.
3. Reduced toxin is concentrated in a speed vac to about 20 μl, and 100 μl of 50 mM sodium phosphate, pH 7.0, is added.
4. One vial corresponding to 1 mCi of [^{3}H]NEM is opened, and the content is transferred to a glass tube.
5. The pentane is evaporated under N_2 to about 300 μl, and then 200 μl of 50 mM sodium phosphate, pH 7.0, is added.
6. After vortexing, the remaining pentane is evaporated.
7. The [^{3}H]NEM solution is added to the peptide solution, and the vial is rinsed with an additional 200 μl of 50 mM sodium phosphate, pH 7.0.
8. The reaction mixture is allowed to react for 1 h at 37°C, and then is injected into a C_{18} reversed-phase HPLC column equilibrated with 10 mM TFA.
9. Elution is achieved in the presence of a linear gradient of isopropyl alcohol/acetonitrile 2:1 (10–30%, 30 min).

   Notes:
   Two well-separated radiolabeled peaks present in equivalent amounts are obtained, and both display identical biological activity.

Handling of the peptides after column elution and storage conditions is identical to that previously described for ^{125}I-ChTX.

5 Receptor Binding Studies

Understanding of the properties of toxin interaction with a given tissue is important. This includes knowledge of the kinetics of ligand binding, pharmacology, and tissue distribution. There are different ways by which to monitor toxin interaction. In autoradiographic studies, the distribution of receptors in a defined tissue can be determined. It is also possible to determine interaction of the peptide with intact cells or tissues in which the target receptor is in its native environment. Perhaps the most common studies are those in which the ligand interaction is measured in highly enriched membrane preparations, and these will be the topic of the following discussion.

Protocol 4 Receptor binding studies

1. Prepare a highly enriched membraneous receptor preparation by purifying the membranes on a sucrose gradient.
2. Make aliquots of these membranes, freeze them in liquid N_2, and store them at −70°C.
3. Prepare a solution containing the radiolabeled peptide, with and without excess of nonradiolabeled ligand.

   Notes:
   Media composition depends on many parameters. For example, peptide interaction in most cases is known to decrease as the ionic strength of the medium is raised. At low ionic strength, affinities between 0.1–10 pM have been determined (Knaus et al., 1995; Vazquez et al., 1990). Consider that most of the peptides are highly charged, and at the low concentrations used in the assays, they will stick to glass surfaces. Therefore, use plastic tubes to handle all the solutions and include 0.1% bovine serum albumin.
4. Incubate the radiolabeled peptide with the membranes.
5. First, determine the optimal protein concentration that gives a linear and optimal signal. Then, determine the necessary incubation time to reach equilibrium.

   Notes:
   This parameter can vary significantly between different peptides. For example, with ^{125}I-MgTX or ^{125}I-ChTX it may take a couple of hours, whereas for ^{3}H-IbTX, an overnight incubation is necessary.
6. Separate free from bound ligand by filtration.

Notes:
We have found this procedure to be the most reliable way of accomplishing this task. Because of the nature of the peptides it is necessary to soak the GF/C glass fiber filters in 0.5–1% polyethylenimine for half an hour at room temperature. This procedure will significantly reduce free toxin association to the filters. It is also important to use as a quench buffer a high ionic strength medium that will cause immediate dissociation of toxin bound to low-affinity sites.

7. Determine the radioactivity associated with filters using either a γ counter for ^{125}I-labeled peptides or a liquid scintillation counter for ^{3}H-labeled peptides.

It is also possible to determine toxin association to solubilized receptor preparations. All the parameters should be identical to those described above except for the filtration step. In this case, separation of bound from free ligand is achieved by addition of γ-globulin followed by addition of the quench buffer containing 10% polyethylene glycol (~MW 8000) (Garcia-Calvo et al., 1991, 1994).

6 Troubleshooting

No induction of fusion protein:

- Check for proper orientation of the toxin gene.
- Be sure that cells do not grow above 0.7 OD, before addition of IPTG.

No cleavage of the peptide from the fusion protein:

- Use an alternative proteolytic enzyme. Trypsin appears to be the most reliable one.

The peptide does not bind to the chromatographic column:

- Be sure that columns are properly equilibrated before loading the peptide.
- Dialyze the sample before loading if the salt concentration may interfere with binding of the peptide to the resin.

Radiolabeled material elutes together with unreacted peptide:

- Improve the separation conditions by checking different buffer compositions.
- Modify the elution conditions.

Nonspecific binding of peptides is too high:

- Be sure that the filters are soaked in 0.5–1% polyethylenimine.
- Use a high ionic strength quench buffer.
- Further enrich the membrane preparation.

References

Aggarwal SK, MacKinnon R (1996) Contribution of the S4 segment to gating charge in the Shaker K^+channel. Neuron 16: 1169–1177

Aiyar J, Withka JM, Rizzi JP et al (1995) Topology of the pore-region of a K^+ channel revealed by the NMR-derived structures of scorpion toxins. Neuron 15: 1169–1181

Bontems F, Gilquin B, Roumestand C et al (1992) Analysis of side chain organization on a refined model of charybdotoxin; structural and functional implications. Biochemistry 31: 7756–7764

Bontems F, Roumestand C, Boyot P et al (1991a) Three-dimensional structure of natural charybdotoxin in aqueous solution by ^{1}H-NMR. Charybdotoxin possesses a structural motif found in other scorpion toxins. Eur. J. Biochem. 196: 19–28

Bontems F, Roumestand C, Gilquin B et al (1991b) Refined structure of charybdotoxin: common motifs in scorpion toxins and insect defensins. Science 254: 1521–1523

Dauplais M, Gilquin B, Possani LD et al (1995) Determination of the three-dimensional solution structure of noxiustoxin: analysis of structural differences with related short-chain scorpion toxins. Biochemistry 34: 16563–16573

Garcia ML, Garcia-Calvo M, Hidalgo P et al (1994) Purification and characterization of three inhibitors of voltage-dependent K^+ channels from Leiurus quinquestriatus var. hebraeus venom. Biochemistry 33: 6834–6839

Garcia ML, Garcia-Calvo M, Vazquez J, Kaczorowski GJ (1992) Charybdotoxin in study of voltage-dependent potassium channels. Methods Neurosci 8: 137–148

Garcia ML, Hanner M, Knaus H-G et al (1996) Pharmacology of potassium channels. Advances in Pharmacology (August JT, Murad F, Coyle JT, eds): San Diego: Academic Press (in press)

Garcia ML, Knaus H-G, Munujos P et al (1995) Charybdotoxin and its effects on potassium channels. Am. J. Physiol. 269: C1-C10

Garcia-Calvo M, Knaus H-G, McManus OB et al (1994) Purification and reconstitution of the high-conductance calcium-activated potassium channel from tracheal smooth muscle. J. Biol. Chem. 269: 676–682

Garcia-Calvo M, Leonard RJ, Novick J et al (1993) Purification, characterization, and biosynthesis of margatoxin, a component of Centruroides margaritatus venom that selectively inhibits voltage-dependent potassium channels. J. Biol. Chem. 268: 18866–18874

Garcia-Calvo M, Vazquez J, Smith M et al (1991) Characterization of solubilized charybdotoxin receptor from bovine aortic smooth muscle. Biochemistry 30: 11157–11164

Giangiacomo KM, Garcia-Calvo M, Knaus H-G et al (1995) Functional reconstitution of the large-conductance, calcium-activated potassium channel purified from bovine aortic smooth muscle. Biochemistry 34: 15849–15862

Goldstein SAN, Pheasant DJ, Miller C (1994) The charybdotoxin receptor of a Shaker K^+ channel: peptide and channel residues mediating molecular recognition. Neuron 12: 1377–1388

Gross A, Abramson T, MacKinnon R (1994) Transfer of the scorpion toxin receptor to an insensitive potassium channel. Neuron 13: 961–966

Johnson BA, Stevens SP, Williamson JM (1994) Determination of the three-dimensional structure of margatoxin by ^{1}H, ^{13}C, ^{15}N triple-resonance nuclear magnetic resonance spectroscopy. Biochemistry 33: 15061–15070

Johnson BA, Sugg EE (1992) Determination of the three-dimensional structure of iberiotoxin in solution by ^{1}H nuclear magnetic resonance spectroscopy. Biochemistry 31: 8151–8159

Kaczorowski GJ, Knaus H-G, Leonard RJ et al (1996) High conductance calcium-activated potassium channels: structure, pharmacology, and function. J. Biomembr. Bioenerg. 28: 255–267

Knaus H-G, Koch ROA, Eberhart A et al (1995) [^{125}I]Margatoxin, an extraordinarily high affinity ligand for voltage-gated potassium channels in mammalian brain. Biochemistry 34: 13627–13634

Knaus H-G, Schwarzer C, Koch RO A et al (1996) Distribution of high-conductance Ca^{2+}-activated K^+ channels in rat brain: targeting to axons and nerve terminals. J. Neurosci. 16: 955–963

Krezel AM, Kasibhatla C, Hidalgo P et al (1995) Solution structure of the potassium channel inhibitor agitoxin 2: caliper for probing channel geometry. Protein Sci 4: 1478–1489

MacKinnon R (1991) Determination of the subunit stoichiometry of a voltage-activated potassium channel. Nature 350: 232–235

Parcej DN, Dolly JO (1989) Dendrotoxin acceptor from bovine synaptic plasma membranes: binding properties, purification, and subunit composition of a putative constituent of certain voltage-activated K^+ channels. Biochem. J. 257: 899–903

Park CS, Hausdorff SF, Miller C (1991) Design, synthesis, and functional expression of a gene for charybdotoxin, a peptide blocker of K^+ channels. Proc. Natl. Acad. Sci. USA 88: 2046–2050

Rehm H, Lazdunski M (1988) Purification and subunit structure of a putative K^+-channel protein identified by its binding properties for dendrotoxin I. Proc. Natl. Acad. Sci. USA 85: 4919–4923

Shimony E, Sun T, Kolmakova-Partensky L, Miller C (1994) Engineering a uniquely reactive thiol into a cysteine-rich peptide. Protein Eng 7: 503–507

Stampe P, Kolmakova-Partensky L, Miller C (1994) Intimations of K^+ channel structure from a complete functional map of the molecular surface of charybdotoxin. Biochemistry 33: 443–450

Sugg EE, Garcia ML, Reuben JP et al (1990) Synthesis and structural characterization of charybdotoxin, a potent peptidyl inhibitor of the high conductance Ca^{2+}-activated K^+ channel. J. Biol. Chem. 265: 18745–18748

Sun T, Naini AA, Miller C (1994) High-level expression and functional reconstitution of Shaker K^+ channels. Biochemistry 33: 9992–9999

Vazquez J, Feigenbaum P, King VF et al (1990) Characterization of high affinity binding sites for charybdotoxin in synaptic plasma membranes from rat brain: evidence for a direct association with an inactivating, voltage-dependent, potassium channel. J. Biol. Chem. 265: 15564–15571

13 Chemical Synthesis and Characterization of Small Proteins: Example of Scorpion Toxins

Jean-Marc Sabatier

Contents

1 Introduction

In the last 10 years, numerous polypeptide toxins acting on ion channels have been isolated from diverse scorpion venoms and characterized. These toxins are useful pharmacological probes for studying ion-specific channel proteins because they interact both selectively and with high affinity with these channels and modulate their activities (Miranda *et al.*, 1970). Only small amounts of

Methods and Tools in Biosciences and Medicine
Animal Toxins, ed. by H. Rochat and M.-F. Martin-Eauclaire

natural toxins can be isolated from scorpion venoms, but large quantities are necessary for extensive study of their pharmacological and structural properties [e.g., three dimensional (3-D) structure determination]. The solid-phase peptide synthesis (SPPS) approach (Merrifield, 1986) is extremely useful for their production (Lecomte *et al.*, 1998). Using SPPS, toxin analogs or toxin-related peptides have been obtained, and this has allowed analysis of their structures and activities, and dissection of the molecular basis of toxin-ion channel interaction. Thus, chemical synthesis is proving to be one of the most powerful methods available to investigate the structure, pharmacology, and even immunology of scorpion toxins. There are two main groups of scorpion toxins, with different structural and pharmacological features: the short-chain toxins (from 29 to 39 residues, three or four disulfide bridges) which are selective blockers of calcium-activated and voltage-gated potassium channels (see book by Darbon and Sabatier, 1999); and the long-chain toxins (from 60 to 70 residues, four disulfide bridges) which are involved in voltage-gated sodium channel blockade (see review by Martin-Eauclaire & Couraud, 1994). These toxins – independent of their length and specificity – share a common structural motif, the α/β scaffold, consisting in an α-helix connected to a double-stranded β-sheet by two disulfide bridges (Bontems *et al.*, 1991). Despite this conserved scaffold, only short-chain toxins and their structural analogs have been successfully synthesized by the solid-phase method. Long-chain scorpion toxins are still difficult to obtain by the chemical synthesis route mainly because of (i) the relative heterogeneity of the high-mass products obtained after final peptide resin cleavage, a consequence of the stepwise elongation of the peptide chain during standard SPPS (Van Rietschoten *et al.*, 1986), (ii) the insolubility of many of the reduced peptide forms in most current oxidative systems (Sabatier *et al.*, 1987), and (iii) inadequate folding leading to biologically inactive oxidized species. Presumably because the low-mass toxins and their analogs can be (in most cases) obtained in good yield by chemical synthesis, and because there are a number of different types and subtypes of K^+ channels which can be either selectively blocked or have their gating properties altered by these compounds, synthetic peptide-based studies have been particularly focused on toxins which act on K^+ channels (Lecomte *et al.*, 1998).

This chapter describes some "guidelines" and basic experimental procedures acquired during many years of experience with SPPS for the successful chemical synthesis of either simple or more complex disulfide-bridged polypeptides such as scorpion toxins.

The principle of SPPS: Briefly, SPPS is based on the sequential addition (from C- to N-terminus) of α-amino group and side chain-protected amino acid derivatives to a solid support/resin (for complete peptide/SPPS reviews, see books by Stewart & Young, 1984; Atherton & Sheppard, 1989; Grant, 1992; Bodanszky, 1993; Bodanszky & Trost, 1993; Pennington & Dunn, 1994). Two protecting groups, i.e., the acid-labile t-butyloxycarbonyl (Boc) or the base-labile fluorenylmethyloxycarbonyl (Fmoc), can be used for N-α-protection depending on the synthesis strategy. After removal of the Boc or Fmoc groups by

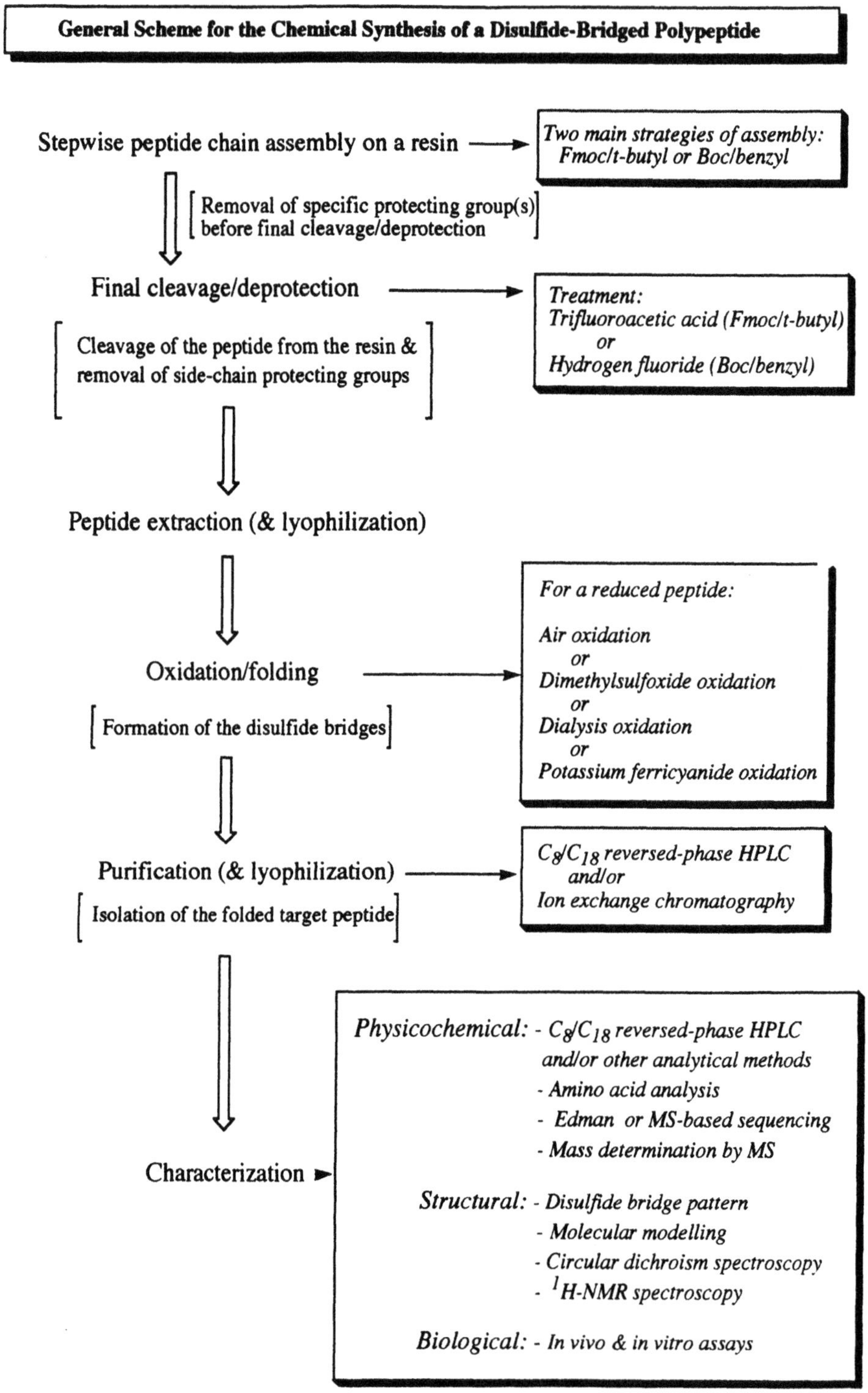
General Scheme for the Chemical Synthesis of a Disulfide-Bridged Polypeptide
Stepwise peptide chain assembly on a resin
Two main strategies of assembly: Fmoc/t-butyl or Boc/benzyl
Removal of specific protecting group(s) before final cleavage/deprotection
Final cleavage/deprotection
Treatment: Trifluoroacetic acid (Fmoc/t-butyl) or Hydrogen fluoride (Boc/benzyl)
Cleavage of the peptide from the resin & removal of side-chain protecting groups
Peptide extraction (& lyophilization)
Oxidation/folding
For a reduced peptide: Air oxidation or Dimethylsulfoxide oxidation or Dialysis oxidation or Potassium ferricyanide oxidation
Formation of the disulfide bridges
Purification (& lyophilization)
C8/C18 reversed-phase HPLC and/or Ion exchange chromatography
Isolation of the folded target peptide
Characterization
Physicochemical: - C8/C18 reversed-phase HPLC and/or other analytical methods
- Amino acid analysis
- Edman or MS-based sequencing
- Mass determination by MS
Structural: - Disulfide bridge pattern
- Molecular modelling
- Circular dichroism spectroscopy
- 1H-NMR spectroscopy
Biological: - In vivo & in vitro assays

treatment with either trifluoroacetic acid or piperidine (i.e., N-terminal deprotection), the next protected amino acid derivative is incorporated onto the growing peptide resin by activation of its α-carboxyl function in the presence of a coupling reagent such as a carbodiimide. Alternatively, preactivated amino acid derivatives (e.g., active esters) can be used directly as the acylating agent instead of adding a coupling reagent. During assembly, the C-terminal extremity of the peptide chain remains attached to the resin by a specific linker. After full assembly, the crude peptide is cleaved to yield a peptide acid or amide depending on the linker and type of resin used. The side chain-protecting groups of trifunctional residues (t-butyl for Fmoc or benzyl for Boc) are generally chosen such that they are cleaved simultaneously with cleavage of the peptide from the resin. The final cleavage of the peptide from the resin with concomitant side-chain deprotection is achieved by hydrogen fluoride or trifluoromethanesulfonic acid in the case of Boc/benzyl chemistry and trifluoroacetic acid in Fmoc/t-butyl chemistry.

2 Peptide Assembly

Fmoc and Boc strategies: Both Fmoc/t-butyl and Boc/benzyl chemistries are suitable for the SPPS of scorpion toxins. Indeed, the purity of crude synthetic products is similar for the two strategies of peptide chain assembly. Presumably, both methods can give excellent results except in particular unusual cases. However, the Fmoc-based chemistry is far more accessible to the inexperienced scientist for routine synthesis of polypeptides (Fields & Noble, 1990). In this approach, the growing peptide chain is repeatedly treated with a mild base (piperidine) for N-terminal deprotection, and trifluoroacetic acid is used only for final cleavage and side-chain deprotection of the peptide resin. In contrast, final cleavage and deprotection in Boc-based chemistry generally requires hydrogen fluoride treatment: this reagent needs user training and specialized apparatus.

Solid supports: There are many types of poly(styrene-co-divinylbenzene)-based resins available commercially which differ only by the nature of the linker. Although most are probably suitable for efficient SPPS, the most widely used for the Fmoc strategy are the 4-hydroxymethylphenyloxy resin (referred to as the HMP or Wang resin), and the 4-(2',4'-dimethoxyphenyl-Fmoc-aminomethyl)phenoxyacetamido-ethyl resin (Fmoc-amide resin), for synthesis of peptides with acid and amide C-termini, respectively. In some cases (e.g., synthesis of kaliotoxin and its analogs), the use of sterically hindered 2-chlorotrityl resin (Barlos *et al.*, 1991) is strongly recommended to prevent the side reaction of diketopiperazine formation. In the Boc/benzyl strategy, Boc-amino acyl-4-(oxymethyl)phenylacetamidomethyl, and 4-methylbenzhydrylamine resins have been successfully used for the SPPS of C-terminal acid and amide toxins and their analogs.

Side chain-protection: The choice of the appropriate side chain protecting group is crucial for trifunctional residues. Inappropriate choice of amino acid derivatives for peptide chain assembly can lead to difficulties during the synthesis or the purification step. The amino acids which need their side-chain protecting can undergo side reactions during peptide chain assembly, or during final cleavage/deprotection, or both. The side-chain protecting groups must be compatible with the chemistry selected, the type of resin, the solvent used (solubility of the derivative), and the coupling and final cleavage/deprotection procedures employed. For cysteines, the use of trityl and 4-methylbenzyl groups is highly recommended for Fmoc and Boc strategies, respectively. These groups generate the free thiol upon final cleavage/deprotection. In particular, the use of different groups (e.g., trityl or t-butylthio versus acetamidomethyl) to protect by pair the thiol of cysteines in the same synthesis is not – in my opinion – technically suitable for site-directed disulfide bridge formation, although this approach has been used with success in some cases. However, this strategy of synthesis can be helpful in case of inappropriate oxidation of a polypeptide such as a disulfide-bridged protein fragment.

Coupling procedures: Formation of the peptide-bond requires α-carboxyl activation of the protected amino acid derivative. The activating procedure must ensure high coupling efficiency and avoid side reactions. In SPPS, there are four major coupling techniques: carbodiimides, active esters, preformed symmetrical anhydrides, and phosphonium/uronium salt-based activators (HBTU, TBTU, PyBOP) which form highly reactive species [hydroxybenzotriazole (HOBt) active esters] in the presence of a tertiary base. All these coupling techniques are very efficient in SPPS; synthetic scorpion toxins and analogs have been mostly obtained using the HOBt/carbodiimide or HBTU activating procedure. For an optimized stepwise assembly of the peptide chain, a double-coupling strategy is valuable. With this strategy, an acetylation procedure for the unreacted α-amino groups (capping) can be used before N-terminal deprotection to avoid elongation of deleted peptide chains, thereby facilitating the purification step of the target peptide. Note that an excellent peptide chain assembly of a long-chain scorpion toxin (AaH toxin II) was obtained by the convergent SPPS method which consists of coupling fully protected peptides – and not protected amino acid derivatives – on a solid support (Sabatier *et al.*, 1987; Grandas *et al.*, 1989). This approach increases the homogeneity/purity of the peptide linked to the resin, and is therefore appropriate for the SPPS of high-mass peptide/protein. However, the convergent method is not suitable for routine SPPS because it is not readily accessible to a novice in the field of peptide chemistry.

Tests of coupling

Protocol 1 Ninhydrin test (qualitative; all amino acids except proline)

1. Dissolve 1 g of ninhydrin in 20 ml ethanol.
2. Dissolve 80 g of phenol in 20 ml ethanol.
3. Mix 1 ml of 1 mM potassium cyanide with 49 ml pyridine.
4. Wash a few resin beads three times with ethanol in a small glass tube.
5. Add two drops of each of the three solutions.
6. Mix and heat (120°C, 5 min).
7. The presence of free primary amino groups (over 0.5%) is detected by the blue staining of the resin beads.

Protocol 2 Chloranyl test (qualitative; for proline)

1. Prepare a solution of 2% acetaldehyde in dimethylformamide.
2. Prepare a solution of 2% chloranyl in dimethylformamide.
3. Add one drop of each solution to a sample of resin beads (1–5 mg), and stir for 5 min.
4. Green staining of the resin beads indicates the presence of secondary amines.

   Notes:
   Detecting free primary amino groups: add 200 µl of acetaldehyde solution and 50 µl of saturated chloranyl in toluene to the resin beads, and stir for a few minutes to allow staining.

Mixtures to be used for difficult Fmoc removal

Use 50% piperidine in dimethylformamide alone or in a mixture of dichloromethane/dimethylformamide/N-methylpyrrolidone (1:1:1, v/v/v) with 1% Triton X100 (3 × 1 min). Alternatively, use a mixture of 2% piperidine and 1,8-diazabicyclo(5.4.0)undec-7-ene (DBU).

Troubleshooting

If coupling is incomplete:

- Change coupling method;
- increase reaction times;
- add other solvents (N-methylpyrrolidone, dimethylsulfoxide, dimethylacetamide), or chaotropic salts (4 M KSCN in dimethylformamide) to the coupling mixture

3 Cleavage/Deprotection of the Peptide Resin

The cleavage techniques used with Fmoc/t-butyl and Boc/benzyl chemistries commonly involve acidolysis to remove both the peptide from the resin and the side-chain protecting groups from the amino acid residues. The cleavage/deprotection procedure selected according to the particular chemistry, the time of reaction, and the temperature can greatly affect the homogeneity/purity of the final product. The choice of the scavengers – dependent on the peptide sequence/side-chain protecting groups content – is particularly important because they trap the carbonium ions (benzyl/t-butyl cations) liberated during acidolysis of the side-chain protecting groups, thereby preventing them from undergoing further side reactions such as alkylation of tryptophan, methionine, tyrosine, and cysteine residues. There are numerous effective cleavage mixtures, but it is most important to choose the appropriate mixture for a given peptide and to apply the experimental protocol correctly (see *Introduction to Cleavage Techniques*, 2nd edition, The Perkin-Elmer Corporation, Foster City, CA, 1995; *Catalog and Peptide Synthesis Handbook*, Calbiochem-Novabiochem, Läufelfingen, Switzerland, 1997–98). The choice of the cleavage/deprotection method depends on the chemistry used, the peptide sequence, the type of resin, and the side-chain protection of amino acid residues (for reviews see Stewart & Young, 1984; Atherton & Sheppard, 1989; Fields & Noble, 1990).

When Boc/benzyl chemistry is used for SPPS, the most common cleavage reagent is anhydrous hydrogen fluoride (HF). However, HF-based cleavage, using either "high" or "low-high" HF technique, is not recommended for inexperienced peptide chemists because it is extremely toxic/corrosive and is therefore dangerous to manipulate. Its use also necessitates a specialized HF-resistant cleavage apparatus. An alternative is to use trifluoromethanesulfonic acid (or trimethylsilyltrifluoromethane sulfonate), which is also highly corrosive but easier to handle. Before cleavage/deprotection of the dried peptide resin, several pretreatments must be performed to remove both N-terminal Boc and specific side-chain protecting groups, e.g., dinitrophenyl (Dnp) or formyl (For) from His(Dnp)- or Trp(For)-containing peptide resin, respectively. Note that during acidic cleavage/deprotection, there should be an N to O acyl shift in Ser- and Thr-rich peptides (the acyl moiety migrates from the α-amino group to the side-chain hydroxyl group). This side reaction can be spontaneously reversed by dissolving the lyophilized peptide in 5% (w/v) ammonium carbonate for 30 min. For Ser/Thr-rich peptides containing Cys residues, reversion of the side reaction should occur during the alkaline oxidation/folding treatment.

Protocol 3 Removal of dinitrophenyl (Dnp) protection from His(Dnp)-containing N-protected peptide resin

1. Suspend the N-protected peptide resin in dimethylformamide (3 min), and filter.
2. Resuspend in dimethylformamide (10 ml per gram of peptide resin).
3. Add 20 mmol thiophenol (about 2 ml) per millimole of His(Dnp) present in the peptide resin.
4. Stir/shake the peptide resin at room temperature for 2 – 4 h until the solution turns bright yellow.
5. Filter and wash the peptide resin three times with dimethylformamide, distilled water, ethanol, and dichloromethane.

   Notes:
   A yellow color can persist due to Dnp-thiophenol noncovalently attached to the peptide resin.
6. Remove the N-terminal Boc group from the peptide resin before HF treatment (useless for TFMSA cleavage).

Protocol 4 Removal of formyl (For) protection from Trp(For)-containing peptide resin

1. Prepare a solution of 10% (v/v) piperidine in dimethylformamide. Cool the solution to 0°C in an ice-bath.
2. Place the peptide resin in a round-bottomed flask, and cool to 0°C (ice-bath).
3. Add 10 ml of solution per gram of peptide resin, and stir/shake for 2 h at 0°C.
4. Filter the peptide resin and wash three times with dimethylformamide, dichloromethane, and ethanol.
5. Dry the peptide resin under vacuum prior to cleavage/deprotection.

   Notes:
   An alternative is to remove the formyl group from the crude peptide after HF cleavage with 0.04 N NaOH, pH 11.5 (3 min), or 0.03 M hydroxylamine solution, pH 9 (2 h).

Protocol 5 Removal of the N-terminal Boc group from the peptide resin

1. Wash the N-protected peptide resin twice with dimethylformamide or N-methylpyrrolidone, and dichloromethane.
2. Add 50% (v/v) trifluoroacetic acid in dichloromethane (about 30 ml per gram peptide resin), mix for 15 min at room temperature, and filter.
3. Wash three times with dichloromethane, and methanol or ethanol.
4. Dry the peptide resin under vacuum prior to cleavage/deprotection.

   Notes:
   In most peptide synthesizers, the N-terminal Boc group is automatically removed at the end of peptide chain assembly.

Protocol 6 Standard "high" HF cleavage/deprotection of the peptide resin (Boc-based SPPS)

A special apparatus is required.

1. Use about 20 ml HF-pCresol-ethanedithiol (85:10:5) per gram of peptide resin.
2. Reaction time: 1 h; temperature: -5 to 0°C.
3. After reaction, evaporate the HF under vacuum by vigorous magnetic stirring.
4. Once the HF is removed but the reaction vessel is still frozen, add cold diethylether (or t-butylmethylether) and stir the peptide/resin/scavenger mixture with a Teflon bar.
5. Filter the diethylether solution, and wash twice with the same medium to remove scavengers from the peptide/resin mixture.
6. Extract the crude peptide from the peptide/resin mixture by adding distilled water.

   Notes:
   If not soluble, use samples to test acetic acid, formic acid, trifluoroacetic acid, NH_4OH, or denaturant such as 6 M guanidine hydrochloride. In some cases addition of acetonitrile or isopropanol may be successful.
7. Freeze the peptide solution and lyophilize.

   Notes:
   There are other scavengers which could be employed; the most commonly used are anisole, thioanisole, p-thiocresol (inappropriate for Trp-containing peptide resin), dimethylsulfide, indole, and 2-mercaptoethanol.

When the Fmoc/t-butyl strategy is used for peptide assembly, cleavage/deprotection of the peptide resin is usually performed for 90 min at room temperature (18–25°C) by trifluoroacetic acid treatment in the presence of scavengers to minimize t-butylation. Higher reaction temperatures must be avoided because most of the side reactions are temperature-dependent. Before proceeding to cleavage, the Fmoc group must be removed by piperidine treatment. The cleavage/deprotection mixture is dependent on the peptide residue content/side-chain protecting groups. For peptide resin-containing multiple pentamethylchroman (Pmc)-protected arginines, it is suitable to increase the standard cleavage reaction time by 30 min per Arg; However, the reaction time must not exceed 3 h in any case. It is possible to use a trimethylsilyl bromide-based treatment as an alternative to this cleavage/deprotection procedure (see *Introduction to Cleavage Techniques*, 2nd edition, The Perkin-Elmer Corporation, Foster City, CA, 1995).

Protocol 7 Removal of the N-terminal Fmoc group from the peptide resin

1. Wash the peptide resin twice with dimethylformamide or N-methylpyrrolidone.
2. Treat the peptide resin with 20% (v/v) piperidine in dimethylformamide or N-methylpyrrolidone for 20 min at room temperature. Use about 20 ml solution per gram of peptide resin.
3. Wash with dimethylformamide or N-methylpyrrolidone, and dichloromethane three times. Dry the peptide resin under vacuum.

   Notes:
   In automated peptide synthesizers, the N-terminal Fmoc group is usually removed at the end of peptide chain assembly.

Protocol 8 Standard trifluoroacetic acid cleavage/deprotection of the peptide resin (Fmoc-based SPPS)

The following cleavage mixtures should be freshly prepared:

- For Arg/Met-containing peptide resin (per gram):
  0.5 ml ethanedithiol,
  1 ml distilled water,
  1 ml thioanisole,
  1.5 g phenol,
  18 ml trifluoroacetic acid;
- For either Cys/Trp-containing or trityl protecting group-containing peptide resin (per gram):
  1 ml ethanedithiol,
  1 ml distilled water,
  18 ml trifluoroacetic acid;
- For other peptide resins (per gram):
  1 ml distilled water,
  19 ml trifluoroacetic acid.

Notes:
The distilled water must be omitted as a scavenger in case of N-terminal glutamic acid (pyroglutamylation);

1. Refrigerate the cleavage mixture.
2. Place the dried peptide resin in a 50-ml centrifuge tube (Falcon type, Teflon cap), and refrigerate.
3. Add the appropriate volume (see above) of the cooled cleavage mixture to the tube containing the peptide resin, and vortex for at least 90 min (< 3 h) at room temperature.
4. To separate the trifluoroacetic acid solution (which contains the free peptide) from the residual resin, vacuum-filter the acid solution through a large-pore filter directly into an excess volume (about 10-fold, 100 to 200 ml) of cold t-butylmethylether.

5. Wash the resin twice with 2 ml neat trifluoroacetic acid, and filter again into cold t-butylethylether suspension.
6. Transfer the t-butylmethylether suspension to centrifuge tubes (cap the tubes) and centrifuge for 10 min at 3000 × g.
7. Discard the supernatants by aspiration and resuspend the pellets of crude peptide in the same volume of cold t-butylmethylether.
8. Centrifuge again (10 min; 3000 × g).
9. Discard the supernatants and dissolve the pellets in distilled water or appropriate aqueous solvent(s) (see section on standard HF cleavage/deprotection), collect the fractions containing peptide, freeze the solution and lyophilize under vacuum.

Notes:
There are other possible methods (filtration and extraction) for recovering the crude peptide after cleavage/deprotection using trifluoroacetic acid. When the Boc side-chain protecting group is used (for Trp) in Trp-containing peptide, let the crude peptide sit in the aqueous solution at room temperature for 30 min prior to lyophilization (complete removal of the Boc group).

4 Disulfide Bridge Formation (Oxidation)

The *in vitro* folding process has been extensively analyzed for several proteins, including venomous animal toxins, as an experimental approach to *in vivo* protein folding (Creighton *et al.*, 1993; Creighton, 1994). The differences observed between folding *in vivo* and *in vitro*, such as the time scale of both processes, the involvement of enzymes in native half-cystine pairing in the endoplasmic reticulum of secretory cells, and subcellular interactions of the nascent chain during protein biosynthesis suggest that this "spontaneous event" is actually a highly complex phenomenon. Nevertheless, although there is no straightforward explanation of the process by which a reduced compound folds to its native structure, there is sufficient evidence from studies *in vitro* to allow some generalizations:

- The folding of reduced proteins occurs spontaneously in a given environment (e.g., pH, temperature, and ionic strength), except for some proteolytically activated proteins obtained by processing of their zymogen forms (e.g., α-chymotrypsin, insulin);
- The information leading to the stable native structure is mainly determined by the amino acid sequence of the polypeptide chain through successive short-, medium-, and long-range interatomic interactions;

- Protein folding appears to be a thermodynamically controlled process in which the rate-limiting step is theoretically the formation of the native-like species (lowest Gibbs free energy for the native protein with respect to all degrees of freedom).

In the SPPS of a multiple disulfide-bridged polypeptide, one of the most exciting and versatile steps is folding/oxidation of the reduced product. When possible, it is recommended to investigate the factors affecting the refolding process *in vitro* using the natural compound (e.g., a toxin isolated from the animal venom) previously reduced and denatured. As already stated (see the section on peptide assembly), trityl and 4-methylbenzyl side-chain groups are greatly preferred for cysteine protection in Fmoc- and Boc-based SPPS of scorpion toxins and their analogs, respectively. These acid-labile protecting groups can be removed with minimal side reactions during final cleavage/deprotection of the peptide resin, yielding the crude peptide in its reduced form. After lyophilization under vacuum of the peptide solution, spontaneous oxidation can be observed in cysteine-containing peptides. This oxidation is only partial (a few percent), and its rate is very dependent on the peptide sequence and the execution of the cleavage and lyophilization procedures. Prior to normal air oxidation of the synthetic product in alkaline buffer (*in vitro* folding of the peptide backbone), it is neither necessary nor suitable to reduce the peptide fully by ethanedithiol or β-mercaptoethanol treatment. Indeed, prereduction of the crude peptide has not been found to modify the folding/oxidation process significantly, at least for the three-disulfide-bridged toxins/analogs obtained by SPPS in our laboratory. Basically, the standard oxidation procedure could be first applied on a small scale to the crude product – without any purification of the reduced form – by directly dissolving a sample of the lyophilized peptide in the appropriate oxidation buffer, e.g., 0.2 M Tris-HCl or sodium phosphate buffer, pH 8. Trace amounts of scavengers (resulting from the peptide cleavage/deprotection) in the reaction medium probably do not interfere with the folding process. A concentrated solution of peptide can be assayed (from 10 μM to 1 mM) for oxidation as proper folding of reduced synthetic toxins in solution was found to be independent of the peptide concentration below the millimolar range. This suggests that dilution of peptide solution does not favor intramolecular half-cystine pairings, in contrast with general belief. The use of a concentrated peptide solution facilitates handling and renders easier the task of target peptide purification by preparative C_8/C_{18} reversed-phase HPLC and/or ion-exchange chromatography. The kinetics of oxidation, as well as the folding pathway, can be directly monitored by successive analyses of the reaction mixture in analytical C_8/C_{18} reversed-phase HPLC. Generally, the main peak corresponding to the hydrophobic reduced form of the peptide progressively disappears (at a variable rate), and new peaks corresponding to partially oxidized peptide intermediates are detected. With some exceptions, these unstable intermediates are generally more hydrophilic than is the reduced peptide. The content of the peptide milieu can evolve over several days depending on the peptide structure/number of half-cystines, but an equilibrium is often reached in less than 40 h at room temperature (<4 h for scorpion toxin P05-NH2; Sabatier *et al.*, 1993). At equilibrium, the oxidation process is completed, and a major hydrophilic peak will be observed which corresponds to the fully oxidized/folded target peptide. Total oxidation of the peptide can be verified by monitoring the redox potential with 5,5'-dithiobis(2-nitrobenzoic acid), i.e., Ellman's reagent. The oxidation milieu can then be filtered prior to purification since a degree of peptide aggregation is frequently observed, presumably associated with intermolecular disulfide bridge formation.

Some particular problems can arise during the folding/oxidation procedure; they are essentially: (i) insolubility of the reduced peptide in usual conditions of oxidation, e.g., neutral or basic pH values, (ii) massive precipitation/aggregation of the peptide, and (iii) formation of stable but inactive oxidized species. The way to solve these problems depends mainly on the individual peptide structure and physicochemical properties, but some chemical additives or modifications of the experimental protocol may help. For example, inclusion in the milieu of a redox mixture of 0.1 mM reduced and 1 mM oxidized glutathione accelerates oxidation by thiol-thiol interchange and reshuffling of disulfide bonds, and in some cases enhances the recovery of folded active peptide (e.g., AaH toxin II). The reduced/oxidized glutathione system has been described as acting on the stability of oxidation intermediates as follows: the reduced form stabilizes thiol groups, whereas the oxidized form stabilizes half-cystines with mixed linkages with glutathione. Thus, disulfide bonds in the intermediates are destabilized by both reduced and oxidized glutathione. Also, it has been reported that guanidine hydrochloride concentration and temperature may influence the solubility of the reduced peptide or oxidation intermediates, and affect the folding pathway (Sabatier *et al.*, 1987). Alternatively, an aqueous/organic solution containing from 5 to 20% (v/v) dimethylsulfoxide could be used as an oxidation milieu for insoluble reduced peptide to initiate half-cystine pairings, but alkaline treatment is further required to ensure completeness of the peptide disulfide bridge formation. Another method which has been developed and applied successfully to the renaturation of insoluble reduced AaH toxin II is based on a dialysis oxidation system (Sabatier *et al.*, 1987). The reduced molecules are first solubilized in 10% (v/v) acetic acid and then oxidized by air through dialysis against a series of buffers with a slow pH gradient from 2.2 to 8. This procedure is particularly convenient for oxidizing reduced polypeptide compounds totally insoluble in neutral or alkaline buffers. Other additives may help peptide oxidation, such as metal ions (e.g., trace amounts of copper), chemical oxidants (e.g., potassium ferricyanide), and natural disulfide interchange enzymes (e.g., thioredoxin, and glutaredoxin, phosphodisulfide isomerase). The basic strategies for protein folding can be found in the *Protein Guide* from Promega Corporation, Madison, WI.

Protocol 9 Air oxidation procedure

This classic method has been used for the folding of most of the polypeptides, including synthetic scorpion toxins/analogs cross-linked by two to four disulfide bridges (e.g., P05-NH2, P01, charybdotoxin, leiurotoxin I, kaliotoxin, maurotoxin, and maurocalcine).

First, use a sample to try:

1. Solubilize the crude reduced peptide in either 0.2 M Tris-HCl or 0.2 M sodium phosphate buffer, pH 8 to 8.5.

   Notes:
   The peptide concentration in the solution can be between 1 μM and 1 mM).

2. Alternatively, solubilize the peptide in an aqueous medium, and adjust the pH to 8.5 with NH_4OH.

   Notes:
   Note that the ionic strength of a buffer greatly affects its buffering capacity. In case of insolubility of the reduced peptide, add a chaotropic salt (<1 M) such as guanidine hydrochloride to the buffer until full dissolution is obtained. In case of major peptide aggregation/precipitation during oxidation, start again on a new sample by adding the chaotropic salt to the buffer prior to peptide solubilization. Be aware that high chaotropic salt concentration can provoke polymer formation.

3. Stir the peptide solution at room temperature for 48–72 h; (if possible, monitor the oxidation process by C_8/C_{18} reversed-phase HPLC).

4. At equilibrium, filter the mixture and purify the target peptide.

   Notes:
   Try another method if the air oxidation procedure fails.
   Not stirring the solution apparently does not modify the folding pathway but slightly slows the oxidation process. The oxidation reaction can be carried out in daylight or in the dark.

Protocol 10 Dimethylsulfoxide (DMSO)-based oxidation

This method should be tried as an alternative to the air oxidation procedure. It has been employed for the synthesis of the three disulfide-bridged Tityus kapa toxin and some iberiotoxin analogs.

1. Dissolve the crude reduced peptide in an aqueous/organic solution containing 10 – 20% (v/v) DMSO, e.g., DMSO:H_2O, 1:9. After 30 min, add a few drops of a buffer at alkaline pH (e.g., 0.2 M Tris-HCl, pH 8).

2. Stir the peptide mixture at room temperature for 24–48 h to complete oxidation, then filter (if necessary) and purify.

Protocol 11 Dialysis oxidation system

The dialysis oxidation method has allowed correct folding of the four disulfide-bridged scorpion toxin II from AaH and three disulfide-bridged sea anemone toxins.

1. Dissolve the reduced peptide in a solution of 10% (v/v) acetic acid.

   Notes:
   The optimal peptide concentration is about 10 μM.

2. Load the peptide solution into membrane tubing with appropriate cutoff (3500 Da if possible; cutoff < peptide mass).
3. Dialyze twice – with agitation – against 800 ml alkaline buffer at room temperature in the dark for 48 h.
4. Recover the peptide solution from the membrane tubing, filter (if necessary), and purify.

   Notes:
   The alkaline buffer can be:
   - 1 mM reduced and 1 mM oxidized glutathione,
   - 20 mM guanidine hydrochloride,
   - 0.1 M sodium chloride,
   - 0.1 M sodium phosphate, pH 8.

   The molarity of the guanidine hydrochloride in the buffer can be increased depending on the solubility of the peptide intermediates.

Protocol 12 Potassium ferricyanide [$K_3Fe(CN)_6$]-based oxidation

This alternative procedure has not been reported for scorpion toxin folding.

1. Dissolve the reduced peptide in distilled H_2O, and adjust to pH 7 with 1 M NH_4OH.

   Notes:
   The peptide concentration in the solution should be between 1 μM and 1 mM.

2. Prepare a solution of 200 mg potassium ferricyanide in 100 ml distilled H_2O.
3. Add the potassium ferricyanide solution dropwise to the peptide solution until a yellow color is maintained for at least 1 h.
4. Decrease the pH to 4.5 with 10% (v/v) acetic acid, and load the solution onto a cation-exchange chromatography column.
5. Wash the column with distilled H_2O (about half of the loaded volume) to remove salts.
6. Apply a gradient from 0 to 50% (v/v) acetic acid, collect and analyze fractions (e.g., by C_8/C_{18} reversed-phase HPLC).

   Notes:
   The fractions containing the target peptide should be pooled and lyophilized prior to purification.

Protocol 13 Ellman test (determination of the peptide thiol content)

1. Dissolve 40 mg of 5,5' dithiobis (2-nitrobenzoic acid) (DTNB; Ellman's reagent) in 10 ml of 0.1 M sodium phosphate buffer, pH 8.
2. Collect a sample from the reaction mixture and dilute with DTNB-free buffer to give 3 ml of solution containing a few nmol of peptide.
3. Add 100 µl of the prepared solution of DTNB to the peptide solution, and wait for 15 min.
4. Measure the optical density (OD) at 410 nm using a 1-cm path length cell.
5. Prepare a blank (reference solution) by adding 100 µl of the solution of DTNB to 3 ml buffer, and measure the OD at 410 nm.
6. Determine the concentration of thiol groups using the following formula: [thiol] = [OD(sample) – OD(reference)] / 13650.

5 Purification of Crude Peptide

The purification step is carried out after final cleavage/deprotection treatment for cysteine-free peptides, but after the oxidation process for disulfide-bridged compounds, such as scorpion toxins and their analogs. During the oxidation/folding procedure of SPPS, the reduced forms of these peptides can be suitably oxidized in alkaline media to give "native-like" disulfide bridge patterns using either diluted or concentrated peptide solution (e.g., 1 mM peptide concentration). Therefore, it is more convenient for purification purposes to oxidize reduced peptides in highly concentrated peptide solutions. This permits (i) the handling of small volumes of peptide solutions, which is appropriate for direct purification of large amounts of crude oxidized peptides by either preparative C_8/C_{18} reversed-phase HPLC, or ion-exchange chromatography, or both; and (ii) to eliminate an unnecessary intermediate step, peptide lyophilization required if volumes of solutions are large, but which could lead to peptide-specific solubility problems due to large amounts of salts. Short-chain scorpion toxins/analogs (<40 residues) can generally be obtained with high purity (>98% homogeneity) by a one-step purification procedure using C_8/C_{18} reversed-phase HPLC, but rarely by ion-exchange chromatography. Ion-exchange chromatography should therefore only be used if necessary as a complementary method to improve final peptide purity. However, high-mass peptides or highly acidic/basic peptides which contain numerous similarly charged groups are often best separated by ion-exchange chromatography. In silica-based reversed-phase HPLC, the appropriate type of column packings (C_{18}, C_8, $C_{4,\ \mathrm{or\ diphenyl}}$) depends on the peptide's hydrophobicity/size; as most of the synthetic scorpion toxins are hydrophilic, C_{18} packing has been almost always been employed for their purification in conjunction with the water/acetonitrile/trifluoroacetic acid gradient elution system. A wavelength of 230 nm (210–230 nm) is usual for monitoring, but 280 nm (250–280 nm) can be used for the elution of peptides containing aromatic residue(s), i.e., Trp, Tyr, and Phe (for peptide/protein purification techniques, see books by Harris & Angal, 1989; Mant & Hodges, 1991; Scopes, 1993; Doonan, 1996; Katz, 1996).

Solubility of polypeptides: recipes

Determination of the net peptide charge: The peptides which possess an overall positive or negative charge are basic or acidic, respectively. Neutral peptides are globally uncharged. A positive charge is assigned to the free N-terminal amino group, and to each Arg, Lys, and His residue contained in the peptide sequence. A negative charge is assigned to the free C-terminal carboxyl group, and to each Asp and Glu residue of the peptide. Add the charges to determine the tendency or physicochemical properties of a peptide, i.e., whether it is basic or acidic.

Basic peptides: They dissolve in H_2O and/or acidic conditions, e.g., 10% (v/v) acetic acid. The best solvents are H_2O, formic acid, acetic acid, and trifluoroacetic acid. The acids can be used either diluted or pure (not recommended; dilute thereafter). However, be aware that peptides are often highly sensitive to prolonged acidic treatments even at low temperatures (<4°C); this sensitivity is sequence-dependent and leads to side reactions and associated by-product formation. Met- and Trp-containing peptides are particularly acid-sensitive. Acid conditions with elevated temperature may hydrolyze peptide bonds (Asp-Pro bond is particularly acid-labile). Always refrigerate peptide-containing acidic solutions to prevent or minimize side reactions.

Acid peptides: They generally dissolve in H_2O, and sometimes in acetic acid. They are mostly soluble in alkaline conditions (e.g., dilute ammonia, ammonium bicarbonate buffer, etc.). Never try extreme basic conditions, because peptide integrity could be affected.

Neutral peptides: These peptides are usually the most difficult to solubilize. They may require pure acids (e.g., neat trifluoroacetic acid, glacial acetic acid, or formic acid), organic solvents (e.g., acetonitrile, methanol, propanol, or isopropanol), or chaotropic salts (e.g., guanidine hydrochloride or urea).

Notes:

Peptide solubility may vary greatly with the volume (or concentration) of a solution; dilution may help to increase the amount of dissolved peptide. For peptides which dissolve slowly, sonicate for a few minutes.

Storage of polypeptides

When lyophilized, peptides should be kept refrigerated at 4°C or below, possibly under a nitrogen atmosphere. When in solution, it is recommended to prepare and store frozen at –20°C separate samples of the same peptide to avoid repetitive freezing/thawing which could lead to some loss of biological activity. In some cases, glycerol can be added to the solution (5 – 20% glycerol, v/v) to increase peptide stability. During prolonged storage/handling (or in acidic conditions), Met-containing peptides may oxidize to form methionine sulfoxide derivative(s). This side-reaction can be reversed by peptide reduction with N-methyl-mercaptoacetamide (MMA).

Protocol of methionine sulfoxide reduction by MMA: Briefly, dissolve the peptide in 10% (v/v) acetic acid (1 – 10 mg peptide/ml); add excess MMA (10 equiv.) and incubate the solution under nitrogen at 37°C for 24–36 h; monitor by C_8/C_{18} reversed-phase HPLC (the normal methionine peptide is more hydrophobic – and therefore has greater retention time – than its methionine sulfoxide counterpart).

6 Physicochemical and Pharmacological Characterization

Basically, SPPS of a polypeptide can be considered as successful only after appropriate physicochemical/pharmacological characterization of the synthetic product. There are several ways to conduct such a characterization, and the choice depends on the nature and biological activity of the compound. Furthermore, there are a number of intermediate levels of peptide characterization from standard/basic analyses to highly specialized peptide-specific determinations (see books by Copeland, 1993; Walker, 1994; 1996; Kamp *et al.*, 1997). In all chemical synthesis of peptides, the first steps of characterization – apart from assessing the homogeneity/purity of the synthetic product by analytical chromatography-based methods – are unquestionably physicochemical, involving both amino acid analysis after acid/alkaline peptide hydrolysis and mass determination. Additionally, N-terminal Edman sequencing of the peptide may be useful (Smith, 1996). Both mass spectrometry and Edman sequencing are appropriate because their high sensitivity allows detection of even small amounts of contaminants in the peptide sample. For synthetic disulfide-bridged scorpion toxins/analogs, half-cystine pairings should be established (Gray, 1993) to investigate structure-activity relationships (Sabatier *et al.*, 1996; Drakopoulou *et al.*, 1998). The disulfide bridge pattern can be determined by enzymatic (or chemical) cleavage using single or mixtures of enzymes. The enzymes are selected for their specificity and compatibility. Ideally, they must be chosen so as to cleave between all half-cystine positions within the peptide sequence to allow a complete and unambiguous attribution of each half-cystine pairing. After proteolytic cleavage, the resulting peptide fragments can be separated by C_8/C_{18} reversed-phase HPLC, and identified by amino acid analysis, Edman sequencing, and/or mass spectrometry. Alternatively, recent advances in mass spectrometry (notably with the introduction of the matrix-assisted laser desorption ionization-time of flight technique or MALDI-TOF) (Winston & Fitzgerald, 1997) allow the mass spectrometer to be used directly in its post-source decay (PSD) fragmentation mode to produce peptide fragment ions from which the half-cystine pairings can be established by MALDI-PSD ion-mass analysis. By either of the two methods, i.e., enzyme-based cleavage and mass spectrometry fragmentation, the knowledge of the disulfide bridge pattern and resulting structural constraints is useful for computer-assisted molecular modeling of the peptide (Sternberg, 1996; Creighton, 1997). These conformational investigations can be completed with experimental data obtained using circular dichroism (analysis of secondary structures in solution) (Menez *et al.*, 1980; Sabatier *et al.*, 1994) and/or ^{1}H-NMR spectroscopy (3-D structure determination in solution). Pharmacological characterization can involve a variety of approaches and technologies which are obviously dependent on the type and specificity of the analyzed product. In the case of scorpion toxins and their analogs, both *in vivo* and *in vitro* assays are suitable and complementary; they mostly address: (i) the peptide-induced lethality in animal models (i.e. determination of the 50% lethal dose), (ii) the ability of the synthetic product to compete with the binding of radiolabeled ion channel-specific toxins to rat brain synaptosomes, and (iii) the efficacy of blockage of currents associated with various types/subtypes of ion channels produced in *Xenopus* oocytes. Of course, when the SPPS simply concerns the exact reproduction of a natural toxin which has been extensively characterized for its physicochemical and biological properties, the synthetic toxin should be totally indistinguishable from its natural counterpart.

Protocol 14: Peptide hydrolysis for amino acid content determination

1. Introduce a peptide sample (from 1 to 10 nmol) into an appropriate tube, add about 600 µl of a solution containing 6 M HCl and 1% (w/v) phenol in either H_2O, acetic or propionic acid, and seal under high vacuum.
2. Stop the hydrolysis reaction after either:
   - 20 h at 110°C (for a low degradation of Ser, Thr, and Tyr),
   - 72 h at 110°C (for complete hydrolysis of more acid-resistant peptide bonds between aliphatic residues, e.g., Val-Val; however, some residue-specific degradation occurs in these conditions depending on the vacuum, e.g., Ser, Thr, and Tyr).

   Notes:
   A good compromise between both condition is to hydrolyze at 120°C for 20 h. Alternatively, an efficient peptide hydrolysis is possible by a one-hour acid treatment at 150°C.
3. Dry the hydrolysate by heating under vacuum, and dissolve in a buffer for amino acid analysis.
4. Analyze the hydrolysate content using an amino acid analyzer.

   Notes:
   During acid hydrolysis, Asn and Gln are transformed in their acid counterparts Asp and Glu, respectively. The Trp is totally degraded during acid – but not alkaline – hydrolysis.

References

Atherton E, Sheppard RC (1989) Solid Phase Peptide Synthesis: A Practical Approach, IRL Press, Oxford, England

Barlos K, Chatzi O, Gatos D, Stavropoulos G (1991) 2-Chlorotrityl chloride resin. Studies on anchoring of Fmoc-amino acids and peptide cleavage. Int. J. Peptide Protein Res **37**: 513–520

Bodanszky M (1993) Peptide Chemistry: A Practical Textbook, 2nd edn. New York: Springer

Bodanszky M, Trost B (1993) Principles of Peptide Synthesis, 2nd edn. New York: Springer

Bontems F, Roumestand C, Gilquin B et. al., (1991) Refined structure of charybdotoxin: common motifs in scorpion toxins and insect defensins. Science **254**: 1521–1523

Copeland RA (1993) Methods for Protein Analysis: A Practical Guide to Laboratory Protocols. London: Chapman and Hall

Creighton TE (1994) The energetic ups and downs of protein folding. Nature Struct. Biol. **1**: 135–138

Creighton TE (1997) Protein Structure: A Practical Approach, 2nd edn. Oxford: IRL Press

Creighton TE, Bagley CJ, Cooper L et al. (1993) On the biosynthesis of bovine pancreatic trypsin inhibitor (BPTI). Structure, processing, folding, and disulphide bond formation of the precursor in vitro and in microsomes. J. Mol. Biol. **232**: 1176–1196

Darbon H, Sabatier J-M (1999) Animal toxins and potassium channels. Perspectives in Drug Discovery and Design Vol. 15/16. Dordrecht: Kluwer Academic Publishers

Doonan S (1996) Protein Purification Protocols. Totowa, New Jersey: Humana Press

Drakopoulou E, Vizzavona J, Neyton J et al. (1998) Consequence of the removal of evolutionary conserved disulfide bridges on the structure and function of charybdotoxin and evidence that particular cysteine spacings govern specific disulfide bond formation. Biochemistry **37**: 1292–1301

Fields GB, Noble RL (1990) Solid phase peptide synthesis utilizing 9-fluorenylmethoxycarbonyl amino acids. Int. Peptide Protein Res. **35**: 161–214

Grandas A, Albericio F, Josa J et al. (1989) Convergent solid phase peptide synthesis. VII. Good yields in the coupling of protected segments on a solid support. Tetrahedron **45**: 4637–4648

Grant GA (1992) Synthetic Peptides: A User's Guide. New York: WH Freeman

Gray WR (1993) Disulfide structures of highly bridged peptides: A new strategy for analysis. Protein Science **2**: 1732–1748

Harris EL, Angal S (1989) Protein Purification Methods: A Practical Approach. Oxford: IRL Press

Kamp RM, Choli-Papadopoulou T, Wittmann-Liebold B (1997) Protein Structure Analysis: Preparation, Characterization, and Microsequencing. New York: Springer

Katz ED (1996) High-Performance Liquid Chromatography: Principles and Methods in Biotechnology. Chichester: Wiley

Lecomte C, Sabatier J-M, Van Rietschoten J, Rochat H (1998) Synthetic peptides as tools to investigate the structure and pharmacology of potassium channel-acting short-chain scorpion toxins. Biochimie 80: 151–154

Mant CT, Hodges RS (1991) High-Performance Liquid Chromatography of Peptides and Proteins: Separation, Analysis, and Conformation. Boca Raton, Florida: CRC Press

Martin-Eauclaire M-F, Couraud F (1994) Scorpion neurotoxins: effects and mechanisms. Handbook of Neurotoxicology (Chang & Dyer eds) pp 683–716: New York: Hong-Kong publishers

Menez A, Bouet F, Guschlbauer W, Fromageot P (1980) Refolding of reduced short neurotoxins: circular dichroism analysis. Biochemistry **19**: 4166–4172

Merrifield RB (1986) The solid-phase peptide synthesis. Science **232**: 341–347

Miranda F, Kopeyan C, Rochat H, Rochat C, Lissitzky S (1970) Purification of animals neurotoxins. Eur. J. Biochem. **16**: 514–523

Pennington MW, Dunn BM (1994) Methods in Molecular Biology, Vol. 35: Peptide Synthesis Protocols. Totowa, New Jersey: Humana Press

Sabatier J-M, Darbon H, Fourquet P, Rochat H, Van Rietschoten J (1987) Reduction and reoxidation of the neurotoxin II from the scorpion Androctonus australis Hector. Int. J. Peptide Protein Res. **30**: 125–134

Sabatier J-M, Fremont V, Mabrouk K et al. (1994) Leiurotoxin I, a scorpion toxin specific for Ca^{2+}-activated K^+ channels: Structure-activity analysis using synthetic analogs. Int. J. Peptide Protein Res. **43**: 486–495

Sabatier J-M, Lecomte C, Mabrouk K et al. (1996) Synthesis and characterization of leiurotoxin I analogs lacking one disulfide bridge: evidence that disulfide pairing 3–21 is not required for full toxin activity. Biochemistry **35**: 10641–10647

Sabatier J-M, Tessier-Rochat M, Granier C et al. (1987) Convergent solid phase peptide synthesis VI: synthesis by the Fmoc procedure with a modified protocol of two protected segments, sequence 5–17 and 18–31 of the neurotoxin II of the scorpion Androctonus australis Hector. Tetrahedron **43**: 5973–5980

Sabatier J-M, Zerrouk H, Darbon H et al. (1993) P05, a new leiurotoxin I-like scorpion toxin: synthesis and structure-activity relationships of the α-amidated analog, a ligand of Ca^{2+}-activated K^+ channels with increased affinity. Biochemistry **32**: 2763–2770

Scopes RK (1993) Protein Purification: Principles and Practice, 3rd edn. New York: Springer

Smith BJ (1996) Methods in Molecular Biology, Vol. 64: Protein Sequencing Protocols. Totowa, New Jersey: Humana Press

Sternberg MJE (1996) Protein Structure Prediction: A Practical Approach. Oxford: IRL Press

Stewart J, Young J (1984) Solid Phase Peptide Synthesis. Rockford, Illinois: Pierce Chemical Company

Van Rietschoten J, Sabatier J-M, Tessier M et al. (1986) Convergent solid phase peptide synthesis: recent progresses. Protides of the Biological Fluids (Peeters ed) pp 23–26: Pergamon Press

Walker JM (1994) Methods in Molecular Biology, Vol. 32: Basic Protein and Peptide Protocols. Totowa, New Jersey: Humana Press

Walker JM (1996) The Protein Protocols Handbook. Totowa, New Jersey: Humana Press

Winston RL, Fitzgerald MC (1997) Mass spectrometry as a readout of protein structure and function. Mass Spectrom. Rev. **16**: 165–179

14 Characterization of Animal Toxins Binding to Ionic Channels from Electrophysiological Recordings

Maurice Gola and Marcel Crest

Contents

1 Introduction

Physiologically useful high-affinity peptidic toxins acting on voltage-gated and some ligand-gated ionic channels have been isolated from various animal species, scorpions, snakes, spiders, mollusks.These toxins act in the picomolar to nanomolar concentration range by occluding the outer mouth of the conduction pore of the channel or by altering the normal channel gating. Electrophysiological techniques able to detect ionic fluxes have proved highly successful in identifying new toxins and characterizing the toxin-channel interaction. The way by which the electrophysiological data are analyzed depends on whether the experiment is conducted on a large number of channels (macroscopic current) or at the unitary level (single channel). The rationale of these methods is described here with illustrations from K^+ channel-blocking toxins.

Methods and Tools in Biosciences and Medicine
Animal Toxins, ed. by H. Rochat and M.-F. Martin-Eauclaire

2 Macroscopic Whole-Cell Current

2.1 Toxin-channel affinity

The first step in determining the characteristics of the toxin-channel interaction concerns the measure of the affinity of the toxin for the channel under investigation and the stoichiometry of the reaction. This is done by simply constructing **dose-response curves.**

Most if not all toxins acting selectively on well-characterized Na^+, K^+, and Ca^{2+} channels bind reversibly to a single class of sites in a one-to-one manner. Like any reversible bimolecular reaction, forward (binding) and backward (unbinding) rate reactions characterize the toxin binding

$$T + \text{Channel} \underset{k_-}{\overset{k_+}{\leftrightarrow}} \text{Toxin-bound channel} \quad (1)$$

where T is the toxin, k_+ (in $M^{-1}\ s^{-1}$) and k_- (in s^{-1}) are the forward and backward rate reactions, respectively. The rate reactions define the association constant $K_a = k_+ / k_-$ (in M^{-1}), and the dissociation constant $K_d = k_- / k_+$ (in M). Since K_d is expressed in concentration units, it gives a direct evaluation of the toxin efficiency. Although the affinity must be rigorously evaluated by K_a, it is commonly accepted to speak in terms of affinity when presenting K_d data. The lower the K_d, the higher the affinity.

The forward flux in (1), Φon (toxin binding), is proportional to the concentration of the toxin, the "concentration" of free channels, and the forward rate reaction

$$\Phi on = k_+[C]\,[T]$$

where the brackets denote concentrations. C stands for the free channels. Similarly, the backward flux Φoff (toxin unbinding) is given by

$$\Phi off = k_-[TC]$$

where TC is the toxin-channel complex.

At equilibrium, Φon = Φoff, which leads to

$$[C]\,[T] / [TC] = k_- / k_+ = K_d$$

Let C_0 the number of channels in the absence of toxin and $y = C / C_0$, the fractional number of free channels in the presence of the toxin. Under whole-cell voltage-clamp conditions, $[C_0]$, $[C]$, and $[TC]$ will correspond to the control current C_0, the current in the presence of toxin C, and the blocked current TC.

$$C + TC = C_0$$

The number of toxin molecules in the bathing saline, even at very low concentration, is far in excess, and therefore [T] does not change during the binding reaction. The above relations lead to

$$y = [C] / [C_0] = [C] / ([C] + [TC]) = 1 / (1 + [TC] / [C])$$

$$\mathbf{y = 1 / (1 + [T] / K_d)} \quad (2)$$

This is a hyperbolic function of the toxin concentration, the shape of which depends on K_d. It shows that when $[T] = K_d$, $y = 0.5$; K_d is the toxin concentration that produces a half-maximal effect. Note that in the case of current blocking toxins, y represents the remaining current and not the blocked current, and that y decreases with increasing [T]. It is appropriate to plot y (the relative current in the presence of toxin) as a function of the decimal logarithm of [T]. The resulting Z-shaped curve is symmetrical relative to the point $[T] = K_d$ and $y = 0.5$ (Fig. 1B). The straight line tangent to the center of the curve intersects the [T] axis at 7.4 K_d. Any departure from these characteristics reveals the presence of either more than one binding site (steep dose-response curve, e.g.,,

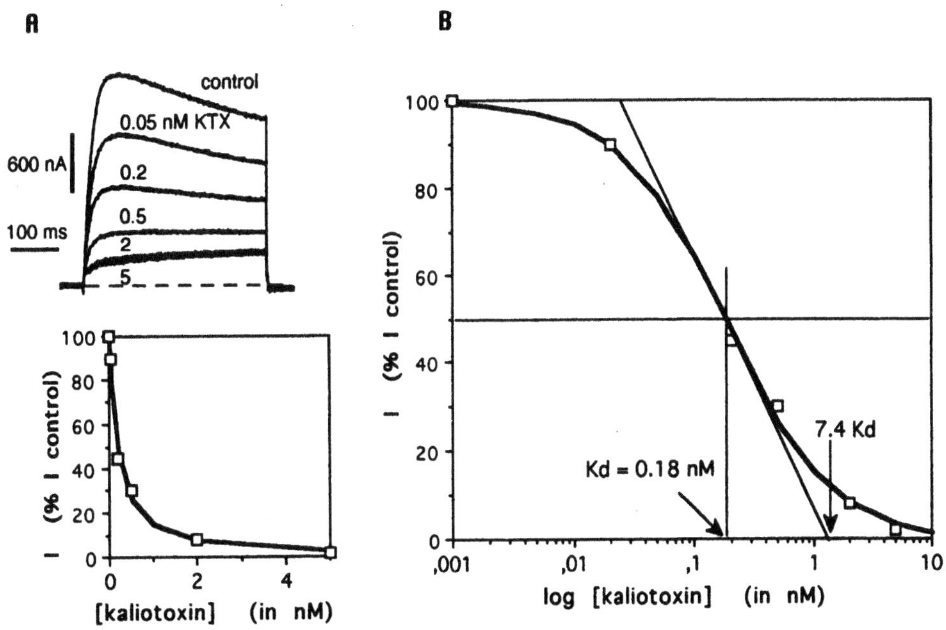

Figure 1 Whole-cell Kv1.3 current block (A) induced by increasing concentrations of the current-blocking toxin kaliotoxin (KTX) and corresponding dose-response curve.

Kv1.3 current (I) expressed in percentage of control value. KTX resistant current subtracted. B, semilogarithmic plot of relative I versus [T]. The continuous curve was drawn according to relation (1) with $K_d = 0.18$ nM.

charybdotoxin on cloned hSlo channels, Gribkoff *et al.*, 1996) or a heterogeneous channel population or cooperative effects which might influence the binding of more toxins to more channels.

The dose-response curve y = f log[T] is constructed from cumulative (step increase in toxin concentration) and noncumulative (various toxin concentrations separated by wash periods) experiments. The cumulative method is obviously faster than the noncumulative one and appropriate when the toxin washing-out is slow, as it generally occurs with high-affinity toxins. In either case, the toxin concentration is increased according to a geometric progression such as 1, 2, 5, 10, 20, 50, 100... which leads to the almost equally spaced logarithmic series 0, 0.3, 0.7, 1, 1.3, 1.7, 2...

2.2 Kinetics

In the nonstationary state, the toxin-binding rate d[TC]/dt is given by the speed with which the binding occurs minus the speed of toxin unbinding

$$d[TC]/dt = \Phi on - \Phi off = k_+ [T][C] - k_- [TC]$$

which, according to the above definitions, leads to the first-order differential equation

$$dy/dt + y (k_+ [T] + k_-) = k_-^{\,*} \qquad (3)$$

As above, y stands for the fractional number of free channels, i.e., the relative whole-cell current in the presence of the toxin. Because [T] remains unchanged during the binding reaction, (3) is also valid for the following reversible reaction

$$C \underset{k_-}{\overset{k_+ [T]}{\leftrightarrow}} TC$$

which clearly emphasizes that the on-reaction rate is proportional to the toxin concentration.

At equilibrium, dy/dt = 0, hence

$$y = k_- / (k_+ [T] + k_-)$$
$$y = 1/ (1 + [T]/Kd)$$

$$\begin{aligned} ^* \; dy/dt &= (d[C]/dt/[C_0]) = \{k_-[TC] - k_+[T][C]\}/[C_0] \\ &= \{k_-([C_0] - [C]) - k_+[T][C]\}/[C_0] \\ dy/dt &= k_-(1 - y) - k_+y[T] \\ dy/dt &+ y(k_+[T] + k_-) = k_- \end{aligned}$$

The on- and off-rates are obtained by solving the first-order differential equation (3) whose general solution is

$$y = A \exp(-t/\tau) + B \; ** \qquad (4)$$

where A and B depend on the initial conditions. The time constant τ (in s) of the response is

$$\tau = 1/(k_+ [T] + k_-) \qquad (5)$$

Relations (4) and (5) show that, when the toxin acts in a one-to-one way on a single class of channel sites, the on- and off-effects do proceed monoexponentially with the time constants τ_{on} and τ_{off}. It also shows that the on-reaction proceeds at a rate $k_{on} = 1/\tau_{on} = (k_+ [T] + k_-)$ (in s^{-1}) which increases linearly with the toxin concentration. Conversely, recovery from the toxin effect (now [T] = 0) has a rate $k_{off} = 1/\tau_{off} = k_-$ which does not depend on the concentration of the previously present toxin.

In order to ascertain that the toxin acts according to a bimolecular reaction on a single class of channel sites, **these predictions must be assessed, and the resulting rate constants must be consistent with the shape and location of the dose-response curve.**

Display of kinetics data

On-reaction. Assume that T is a channel blocker. Before adding the toxin (t = 0), $y_{t=0}$ in (4) is 1 and $\exp(-t/\tau_{on}) = 1$; A + B =1.

In the presence of T and at equilibrium (steady state = ss), $\exp(-t/\tau_{on}) = 0$ and $y_{ss} = B$, with $y_{ss} = 1/(1 + [T]/K_d)$.

The on-reaction is therefore described by

$$y = (1 - y_{ss}) \exp(-t/\tau_{on}) + y_{ss}$$
$$(y - y_{ss})/(1 - y_{ss}) = \mathbf{(I - Iss)/(Ic - Iss) = exp(-t/\tau_{on})} \qquad (6)$$

where I, Ic (c = control) and Iss stand for the ionic current at time t (I), before adding the toxin (Ic), and the toxin-resistant current (Iss).

Off-reaction. Similarly, solving (4) for the off-reaction leads to

$$\mathbf{(Ic - I)/(Ic - Iss) = exp(-t/\tau_{off})} \qquad (7)$$

Semilogarithmic plots of (6) and (7) give straight lines with the respective slopes $-1/\tau_{on}$ and $-1/\tau_{off}$.

** $y = A\exp(-t/\tau)+B$

$dy/dt = -1/\tau A \exp(-t/\tau) = -1/\tau(y - B)$

$dy/dt + y/\tau = B/\tau$, which by identifying with (3) gives

$\tau = 1/(k_+[T] + k_-)$ and $B = k_-/(k_+[T] + k_-)$

2.3 Example, block of the shakerlike Kv1.3 channel by kaliotoxin (KTX)

The following data are representative of the effects (here blockage) of most natural toxins acting on ionic channels. They were obtained from Kv1.3 currents expressed in *Xenopus* oocytes injected with 0.2 ng Kv1.3 cRNA. Figure 2AB shows the progressive current block induced by adding 1 nM KTX, and the slow current recovery upon removing the toxin. At 1 nM, KTX blocks 85% of the current, which according to (2) corresponds to a dissociation constant K_d = 0.18 nM. The semilogarithmic plots in Figure 2C and 2D are drawn according to (6) and (7). Note, however, that these relations are based on natural (neperian) logarithms (Log or $\log_e$), whereas the kinetic data are usually displayed on

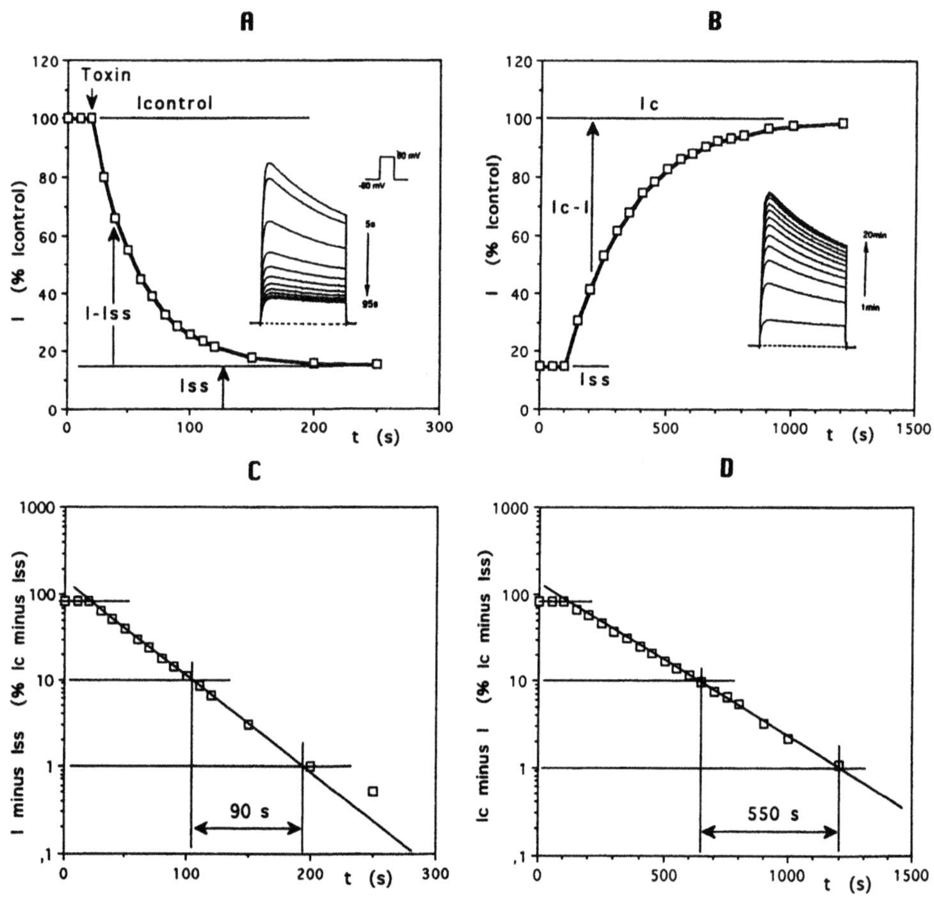

Figure 2 Kinetic of Kv1.3 current block by kaliotoxin.

A and B, time course of current block (A) and recovery from block (B). Kaliotoxin added at 1 nM. Insets: Superimposed currents in response to a test pulse applied every 5 s (A) or 1 mn (B). Ic, I control; Iss, I steady-state. C and D, semilogarithmic plots of data from A (C) and B (D).

decimal logarithmic scales (log or $\log_{10}$) as in Figure 2CD. In such a plot, the time required for a 10-fold change in the y variable (90 s in Fig. 2C; 550 s in Fig. 2D) gives the $\log_{10}$-based time constant of the current change. The $\log_e$-based time constants in (6) and (7) are obtained by dividing the above values by 2.3 ($\log_e 10 = 2.3$), which leads to $\tau_{on} = 39$ s and $\tau_{off} = 239$ s.

$1/\tau_{off} = k_- = 4.2\ 10^{-3}\ s^{-1}$
$1/\tau_{on} = k_+[T] + k_- = 25.6\ 10^{-3}\ s^{-1}$
and $k_+[T] = 21.4\ 10^{-3}\ s^{-1}$
since $[T] = 10^{-9}$ M, $k_+ = 21.4\ 10^6\ M^{-1}\ s^{-1}$
$K_d = 4.2\ 10^{-3}\ s^{-1} / 21.4\ 10^6\ M^{-1}\ s^{-1} = 0.19$ nM, a figure in good accordance with the dose-response curve shown in Figure 1B.

2.4 Choice of significant experimental parameters

When the above experiments give results consistent with a one-to-one toxin-channel interaction, the information is redundant, since three independently determined parameters τ_{on}, τ_{off}, and K_d are obtained, whereas the interaction is fully characterized by only two of these parameters. The choice of the experimental parameters used to calculate the rate constants therefore depends on the confidence attached to these parameters. We hereafter consider some practical limitations which may affect the rate constant measurement.

The off-time constant τ_{off} is independent of [T], and therefore each run as in Figure 2 gives an additional evaluation of τ_{off}. However, high-affinity toxins have very small unbinding rate constants. For instance, the unbinding of kaliotoxin 2 from Kv1.3 channels has a rate constant $k_- = 1.4\ 10^{-3}\ s^{-1}$ (Laraba-Djebari *et al.*, 1994), which corresponds to a recovery time constant $\tau_{off} = 12$ mn. 90% recovery requires a wash period of 12 $\log_e 10 = 28$ mn! It is clear that run-down processes often induced by whole-cell patch-clamp electrodes may hinder reliable information on k_-.

Reliable τ_{on} data may be limited by the speed of toxin binding. The toxin BgK from the sea anemone *Bunodosoma granulifera* blocks Kv1.1 channels with the following characteristics:

$k_+ = 15\ 10^6\ M^{-1}\ s^{-1}$, $k_- = 27\ 10^{-3}\ s^{-1}$

(Cotton *et al.*, 1997). With [T] > 30 nM, the binding occurs with $\tau_{on} < 2$ s. This requires the use of a fast concentration clamp device. In addition, the binding rate increases with [T], limiting the range of useful [T].

Finally, K_d seems the easiest data to be precisely determined in a large [T] range. This conclusion is correct provided that the actual 100% blockage level is precisely known, which requires **[T] > 100 K_d** and the possibility to subtract the toxin-insensitive currents.

3 Microscopic Single-Channel Current

At the macroscopic level, both blocked and free channels are simultaneously present in a ratio dependent on the toxin concentration. Reversible macroscopic reaction implies that the channel, when observed at the unitary level, alternates continuously from the free and blocked states in a statistically determined way.

3.1 Opening probability

Schema (1) still applies at the unitary level, but the rate constants must now be interpreted in a probabilistic way: k_+ measures the probability that the toxin binds to the channel, and k_- is a measure of the unbinding probability. Therefore, at the unitary level, blocked and unblocked states alternate in a statistically determined manner. The task of sorting out the significant rate parameters (i.e., the rate constants) from unitary channel recording has been thoroughly considered by Colquhoun and Hawkes (1983). Instead of evaluating the fraction of whole-cell current sensitive to the toxin, we must now consider the fraction of time the channel stays in the free or blocked states. These quantities must be measured over a period long enough to contain statistically significant events.

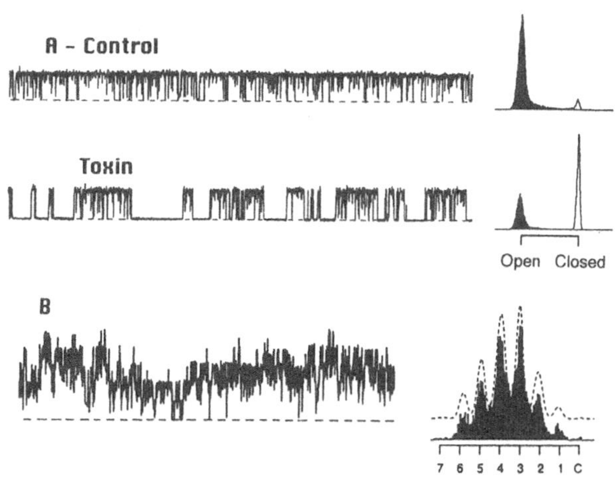

Figure 3 Quantifying the potency of a channel-blocking toxin from unitary channel recordings.

A, single-channel. In the amplitude histograms attached to the recordings, the filled peaks correspond to the open state. Their relative area gives the opening probability Po = 0.91 (control) and 0.48 (toxin), which indicates that the toxin concentration was about K_d
B, multichannel patch. This patch contains at least seven channels. The amplitude histogram was constructed from a 250-s recording period. The superimposed dashed histogram is the theoretical distribution (binomial law) predicted for seven identical and independent channels characterized by Po = 0.51 (C, closed state; 1 – 7, number of simultaneously opened channels).

Schema (1) considered that the channel has only two states, open (toxin-free) and blocked. At equilibrium, the probability that the channel is opened is

$$P_o = k_-/(k_+[T] + k_-) = 1/(1 = [T]/K_d)$$

which is similar to (2).

Steady-state Po can be obtained from single-channel recordings by summing the duration of a large series of opening events and dividing the result by the duration of observation. It is more simply deduced from amplitude histograms of the same recording period. The amplitude histogram has two peaks that may be fitted with Gaussian curves. The relative area under the peak corresponding to the open state (and not its amplitude) gives Po (Fig. 3A).

When several channels are present, P_o can still be evaluated from amplitude histograms provided that the N channels present are identical and behave independently (Fig. 3B). This is checked by comparing the N+1 peaks in the experimental amplitude histograms with the distribution predicted by the binomial law for N identical channels characterized by the individual opening probability P_o (see Colquhoun and Hawkes, 1983).

3.2 Open and blocked lifetime distributions

Amplitude histograms give Po from which K_d can be calculated. Information on the kinetics of the toxin-channel interaction requires knowledge of the lifetimes the channel spends in either state. The stochastic interpretation by Colquhoun and Hawkes leads to the following rule: the lifetime in either state of (1) is exponentially distributed with time constant = 1/transition rate that leads away from the state. The distribution of open times is a single-exponential with $\tau_O = 1/k_+ [T]$. Similarly, the single-exponential distribution of blocked times has a time constant $\tau_B = 1/k_-$.

Significant lifetime distributions are constructed from hundreds of events. When the blocked times last for minutes (e.g., charybdotoxin on KCa channels, Giangiacomo *et al.*, 1992) each run (at given [T]) requires long-lasting (hours in this case) period recordings. However, if it is proved that events obey schema (1), the rate constants are simply derived from the mean lifetime of either state: the mean lifetime in the open state is $1/k_+ [T]$, and the mean lifetime in the blocked state is $1/k_-$.

4 Voltage-Dependent Channels

Still, now the channels have been considered permanently opened before adding the blocker. This situation is not realistic since voltage-dependent gatings are reversible and therefore the channels must also spend some time in the closed state according to

$$\text{Closed} \overset{\alpha(V)}{\underset{\beta(V)}{\leftrightarrow}} \text{Open}$$

where α(V) and β(V) are the V-dependent rate parameters. When the blocker binds with similar affinity to the channel, whatever its state, the corresponding reaction is

$$\text{Closed-blocked} \overset{T}{\leftrightarrow} \text{Closed} \overset{V}{\leftrightarrow} \text{Open} \overset{T}{\leftrightarrow} \text{Open-blocked}$$

which is strictly equivalent to (1). A more complex situation arises when the toxin affinity depends on the channel state. This situation is not speculative. The charybdotoxin binding site in KCa channels is located relatively deep in the pore and is more directly accessible when the channel is open (review by Miller, 1995). Deriving the four T-related rate constant from unitary or whole-cell recordings is not straightforward (see Anderson *et al.*, 1988).

Finally, the blocker may act only once the channel opens

$$\text{Closed} \overset{V}{\leftrightarrow} \text{Open} \overset{T}{\leftrightarrow} \text{Blocked} \qquad (8)$$

This commonly occurs with a number of nonpeptidic channel blockers (e.g., quaternary ammoniums on potassium channels) (see Hille, 1992). At equilibrium, equalizing the on- and off-fluxes in (8), as done for (1), and introducing $K_v = \alpha(V)/\beta(V)$, the opening probability in the presence of the blocker is

$$P_o = 1/(1 + 1/K_v + [T]/K_d) \qquad (9)$$

and the control P_o (with [T] = 0) is $P_{o,c} = 1/(1 + 1/K_v)$

Semilogarithmic plots of (9) still result in Z-shaped symmetrical curves identical to that shown in Figure 1B. For a given K_d, the location of the curve on the x-axis is shifted, however, to increasing [T] by an amount which increases at low K_v, i.e., at low $P_{o,c}$. The apparent K_d, given by the toxin concentration that half-reduces $P_{o,c}$, is $Kd/P_{o,c}$.

This seems to indicate that the toxin binding rate constants are P_o- and therefore V-dependent, which was not initially assumed in (8). Voltage-dependent binding exists (e.g.,, charybdotoxin on KCa channels, Anderson *et al.*, 1988; MacKinnon and Miller, 1988) and this dependence relies mainly on the electrostatic nature of the binding and not on the channel state. Therefore, what is the significance of the change in apparent K_d with P_o? It would mean that the toxin is able to shift the V-dependent channel gating. Except in a few

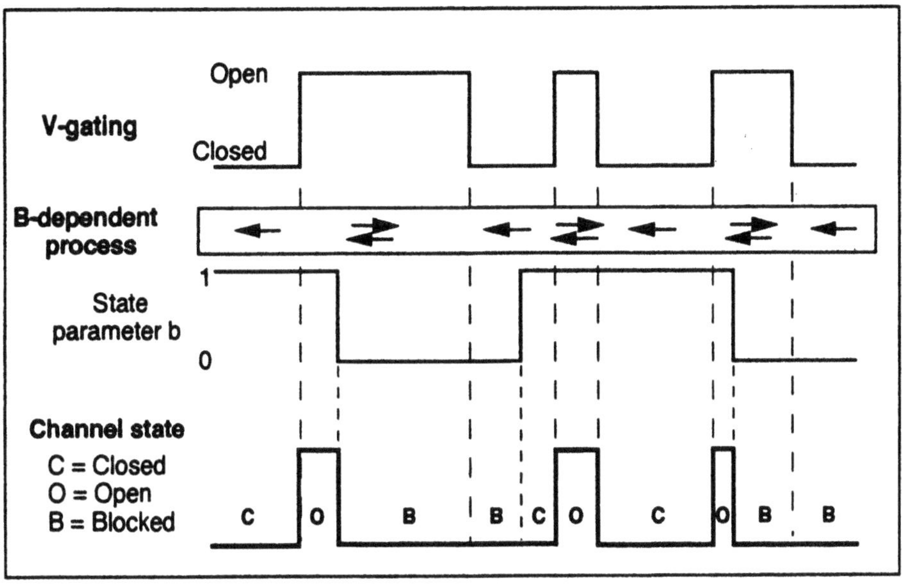

Figure 4 Rationale for simulating the effects of blocking toxins (B-dependent process) acting on the open state of V-gated ionic channels.

specific cases (e.g.,, iberiotoxin on KCa channels, Galvez *et al.*, 1990; hanatoxin on Kv2.1 channels, Swartz and MacKinnon, 1995), this conclusion is not correct. Finally, it appears that scheme (8) is not realistic simply because it incorporates two processes, channel gating and toxin binding, which are physically unrelated and occur independently of each other. Therefore, in order to perform computer simulation of single-channel events and to compare the reconstructed and experimental behaviors, we considered the two processes separately by introducing a V-dependent step and a blocker (B)-dependent state parameter b; b = 0 when the blocker is bound to the channel, else b = 1

$$\text{V-dependent step: Closed} \underset{\beta(V)}{\overset{\alpha(V)}{\leftrightarrow}} \text{Open}$$

$$\text{B-dependent variable: b} \underset{k_-}{\overset{k_+[T]}{\leftrightarrow}} (1-b)$$

where the rate constants k_+ and k_- depend on the channel state. Results are expressed as the product by b (0 or 1) of the unitary current (Figure 4).

5 Conclusion

The search for toxins acting on ionic channels was mainly aimed at evaluating the contribution of the corresponding ionic currents to the cell physiology. The recent discovery of numerous toxins highly specific of well-characterized channels has considerably enlarged their usefulness. The new challenges concern the intimate mechanisms that govern the toxin-channel interaction, the mapping of the channel structure, and ultimately the design of new therapeutic agents derived from natural and synthetic toxins.

In synergy with alternative methods, electrophysiological techniques play a crucial and specific role in these researches. They give a dynamical view of the toxin-channel interaction, particularly when observed at the single-channel level. These dynamical data are of high significance for designing new molecules still required to discriminate among the numerous channel isoforms.

References

Anderson CS, MacKinnon R, Smith C, Miller C (1988) Charybdotoxin block of single Ca^{2+}-activated K^+ channels. Effects of channel gating, voltage, and ionic strength. J. Gen. Physiol, 91: 317–33

Colquhoun D, Hawkes AG (1983) The principles of the stochastic interpretation of ion-channel mechanisms. Single-Channel Recording (Sakmann B, Neher E, eds) pp 135–175: New York: Plenum Press

Cotton J, Crest M, Bouet F et al (1997) BgK from sea anemone Bunodosoma granulifera, an inhibitor for Kv1 channels: revision of the amino acid sequence, disulfide bridge assignement, chemical synthesis and biological activity. E. J. Biochem, 244: 192–202

Galvez A, Gimenez-Gallego G, Reuben JP et al (1990) Purification and characterization of a unique, potent, peptidyl probe for the high conductance calcium-activated potassium channel from venom of the scorpion Buthus tamulus. J. Biol. Chem, 265: 11083–90

Giangiacomo K, Garcia ML, Mc Manus OB (1992) Mechanism of iberiotoxin block of the large conductance calcium-activated potassium channel from bovine aortic smooth muscle. Biochem, 31: 6719–6727

Gribkoff VK, Lum-Ragan JT, Boissard CG et al (1996) Mol Pharmacol 50: 206–217

Hille B (1992) Mechanism of block. Ionic channels of excitable membranes, 2nd edition (Hille B, ed) pp 390–422: Sunderland, Massachussetts: Sinauer Associates

Laraba-Djebari F, Legros C, Crest M et al (1994) The kaliotoxin family enlarged: purification, characterization, and precursor nucleotide sequence of KTX2 from Androctonus australis venom. J. Biol. Chem, 52: 32835–32843

MacKinnon R, Miller C (1988) Mechanism of charybdotoxin block of the high-conductance, Ca^{2+}-activated K^+ channel. J. Gen. Physiol, 91: 335–49

Miller C (1995) The charybdotoxin family of K^+ channel-blocking peptides. Neuron, 15: 5–10

Swartz KJ, MacKinnon R (1995) An inhibitor of the Kv2.1 potassium channel isolated from the venom of a Chilean tarantula. Neuron, 15: 941–949

15 "Three-Fingered" Toxins from Hydrophid and Elapid Snakes: Artificial Procedures to Overproduce Wild-Type and Mutated Curaremimetic Toxins

Jean-Claude Boulain, Frédéric Ducancel, Gilles Mourier, Pascal Drevet and André Ménez

Contents

Methods and Tools in Biosciences and Medicine
Animal Toxins, ed. by H. Rochat and M.-F. Martin-Eauclaire

1 Introduction

Toxic proteins from animal venoms usually act on molecular targets that are critically involved in the function of a physiological system of a prey. With the ultimate view of designing new pharmacological tools as well as original drugs acting on these targets, animal toxins are the subject of extensive molecular analyses. In particular, the topographies by which toxins bind to their targets and hence affect the function of the associated physiological system clearly need to be identified.

Such an identification can be achieved by a mutational approach, provided an appropriate artificial system is available to overproduce the toxin. During the past decade, we have developed a number of approaches to overproduce toxins. In this respect, bacterial systems looked most attractive because animal toxins usually (though not always) undergo no posttranslational modification. Then, biohazards associated with the production of recombinant animal toxins were examined by a national control committee and, a fully active toxin was decided not to be directly produced in a microorganism. At least one *in vitro* activation step was considered to be necessary to generate the fully active toxin. Production of a fused toxin in *Escherichia coli* received the agreement of the committee. Since animal toxins usually possess several disulfide bonds, we decided to exploit the presence of the fusion partner to tentatively address the toxin hybrid to the periplasm of the bacteria where disulfide bonds can be formed, and hence where the native structure of the toxin was anticipated to be directly acquired.

Protein A from *Staphylococcus aureus* (Nilsson et al., 1985a, b) and its derived fragments, called ZZ, which bind to IgG (Löwenadler et al., 1987; Nilsson et al., 1987) were selected as preferential fusion partners and, thus the first synthesis of a recombinant snake toxin was described 10 years ago (Ducancel et al., 1989; Boyot et al., 1990). Using similar systems we sucessfully produced various fused recombinant proteins, including a curaremimetic toxin and its mutants (Hervé et al., 1992; Pillet et al., 1993; Zinn-Justin et al., 1994; Trémeau et al., 1995; Drevet et al., 1997); a snake phospholipase A_2 (Hogdson et al., 1993); dendrotoxin, which acts on potassium channels (Danse et al., 1994); a scorpion toxin (Bouhaouala-Zahar et al., 1996); a protein from spider venom (Gasparini et al., 1994); and other proteins (Ezan et al., 1994). It was clearly established that the resulting recombinant mature toxin has biological properties (Boyot et al. 1990) and a three-dimensional structure (Arnoux et al., 1994) that are identical to those of the wild-type toxin. In addition to these recombinant approaches, the chemical synthesis of various animal toxins rich in disulfides also proved to be of great interest to produce artificially disulfide-rich toxins, in large quantities (Vita et al., 1995; Mourier et al., 1996).

This chapter will be focused on the experimental procedures that allow the production of short-chain curaremimetic toxins (Fig. 1). Two toxins will be principally considered. These are (i) erabutoxin, from the sea-snake *Laticauda*

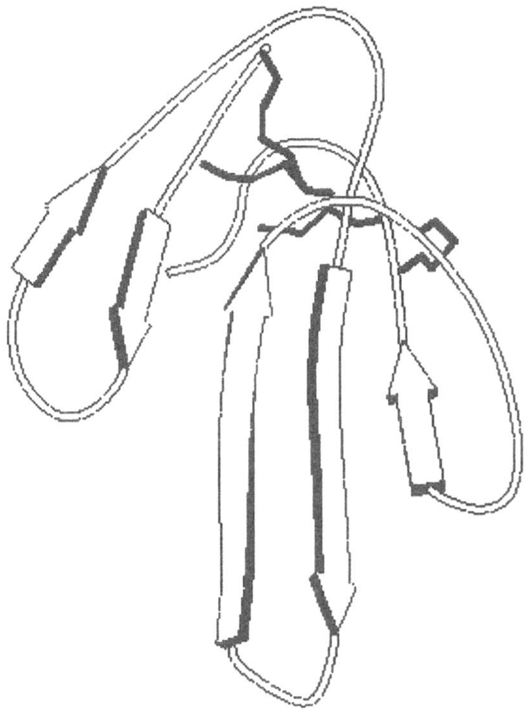

Figure 1 Schematic spatial representation of a typical "three-fingered" toxin.

The arrows represent the ß-sheets and, the disulfides are shown in bold.

semifasciata (Sato and Tamiya, 1971; Tamiya, 1973), whose complementary DNA (cDNA) (Tamiya et al., 1985) was previously described and (ii) toxin α from the land snake *Naja nigricollis* (Eaker and Porath, 1967) for which a synthetic cDNA oligonucleotide was synthesized (Drevet et al., 1997). It must be stressed, however, that the procedures that are described here can be applied to most if not all "three-fingered" toxins and most probably to a large number of other proteins rich in disulfide bonds which need to be overproduced as a wild-type or mutated protein or to be labelled with stable isotopes.

2 Materials and Chemicals

Strains and plasmids

The pET3a derivative plasmid was used as described by Studier et al. (1990) and, according to the supplier's (Novagen) recommendations, using the *E. coli* strain BL21(DE3)LysS as host cell. The pEZZ gene fusion vector was from Pharmacia and used to transform the HB101 strain. DNA experiments were carried out according to Sambrook et al. (1989).

Buffers and reagents

- **TSB medium (Solution 1)**
  17 g tryptone
  5 g yeast extract
  3 g soytone
  5g Nacl
  2.5 g K_2HPO_4 , pH 7.4
- **Lysis buffer (Solution 2)**
  Tris 30 mM, pH8
  EDTA 5 mM
  Sucrose 20%
  Lysozyme 0.1 mg/ml
- **PBT (Solution 3)**
  0.1 M phosphate buffer pH8
  0.1% Tween20
- **Renaturation buffer (Solution 4)**
  0.1 M phosphate buffer
  5 mM EDTA
  appropriate ratio of reduced over oxidized glutathione
- **Renaturation buffer (Solution 5)**
  0.1 M sodium phosphate buffer, pH 7.8, containing
  reduced glutathione 4 mM and
  oxidized glutathione 2 mM.
- **TST buffer (Solution 6)**
  50 mM Tris-buffer, pH 7.6
  150 mM NaCl
  0.05% Tween 20
- **Solvant A (Solution 7)**
  0.1% TFA in H_20
- **Solvant B (Solution 8)**
  60% acetonitrile
  0.1% TFA in H_20

Chemicals

Recombinant technology:

Acetic acid	Carlo Erba	401391
Acetonitrile	Prolabo	20 060.296
Ammonium acetate	Prolabo	21 200.297
Ampicillin	Sigma	A 9518
Chloramphenicol	Sigma	C 0378
CNBr	Sigma	C 6388
EDTA	Merck	108418
Glucose	Prolabo	24 379.294

K_2HPO_4	Prolabo	26 931.263
KH_2PO_4	Prolabo	26 926.298
Lysozyme	Sigma	L 6876
NaCl	Prolabo	27 800.291
Oxidized glutathione	Sigma	G 4376
Reduced glutathione	Sigma	G 4251
Protamine sulfate	Sigma	P 4380
Soytone	Difco	436-01
Sucrose	Prolabo	27 480.294
TFA	Prolabo	20 751.181
Trizma base	Sigma	T 1503
Tryptone	Difco	123-17-3
Yeast extract	Difco	127-17-9

Chemical synthesis:

- **Fmoc-protected amino acids** [Novabiochem, Bachem] were used with the following side-chain protections: t-butyl ester (OtBu) [Glu,Asp], t-butyl ether (tBu) [Ser, Thr, Tyr], trityl (Trt) [Cys, His, Asn, Gln], 2,2,5,7,8-pentamethyl-chromane-6-sulfonyl (Pmc) [Arg], t-butyloxycarbonyl (Boc) [Trp].
- **Solvents**: N-methyl pyrrolidone (NMP), dichloromethane (DCM), methanol, piperidine, trifluoroacetic acid were from SDS (France); phenol, ethane-dithiol, thioanisole were from Aldrich.
- **Coupling reagent**: dicyhexylcarbodiimide (DCCI) and hydroxybenzotriazole (HOBT) [Novabiochem].
- **Rink amide resin** (Novabiochem).

Equipment

Recombinant technology:

- Incubator
- Sterilizer
- DNA gel electrophoresis apparatus
- Water bath
- Thermal cycler
- Ultraviolet (UV) monitor

Chemical Synthesis:

- Automatic peptide synthesizer
- Glassware and a good fume hood
- HPLC
- Lyophilizer

3 Cytoplasmic Production of a Land Snake "Three-Fingered" Toxin

The T7 expression system has been extensively used to produce various recombinant proteins in the cytoplasm of *E. coli*. The T7 RNA polymerase/T7 promotor pair is selective and quite active, since the final product may represent more than 50% of total cell proteins. Based on this system, several pET vectors have been constructed by Studier et al. (1990). The cDNA sequence encoding the toxin was fused to a nucleotide sequence coding for two synthetic domains of the protein A (Löwenadler et al., 1987; Nilsson et al., 1987), which was previously introduced into a pET vector. This system not only received the agreement of a control committee but proved to be appropriate to stabilize foreign proteins in *E. coli* (Nygren et al., 1994; Murby et al., 1994).

3.1 Construction of a cytoplasmic expression vector

In a first step, the sequence of the ZZ protein, a synthetic double domain of protein A of *Staphylococcus aureus*, without its signal sequence has been cloned into the pET3a vector downstream of the T7 promotor. For the vector to be of general use, we introduced a polylinker with the *Kpn* I, *Sma* I, *Xma* I, *Bcl* I, *Spe* I, *Nco* I, and *Bam HI* sites at the end of the ZZ sequence. The resulting plasmid was named pCP (Fig. 2).

To insert the toxin gene into the pCP vector, two appropriate restriction sites (*Kpn* I and *Bam HI* for example) were engineered at each end of the gene. Evidently, for the fusion to occur appropriately, the toxin coding sequence had to be in frame with the ZZ coding sequence. In addition, a methionine codon was introduced immediately upstream of the first codon of the mature toxin. Since the sequence of the mature toxin has no methionine residue, we anticipated that a subsequent treatment of the resulting chimaeric protein with CNBr should generate the mature toxin. The modification was obtained in a one-step procedure using polymerase chain reaction (PCR) and specifically designed primers. Instead of a methionine it is also possible to insert an amino acid sequence that is specifically recognized by an enzyme, for example, bovine enterokinase (LaVallie et al., 1993).

The cDNA encoding erabutoxin a from *Laticauda semifasciata* and a synthetic gene encoding toxin α from *Naja nigricollis* were inserted in the pCP vector. The resulting vectors were used to transform a BL21(DE3)LysS strain. Selection was made on LB agar plate supplemented with ampicillin 200 mg/l and chloramphenicol 30 mg/l.

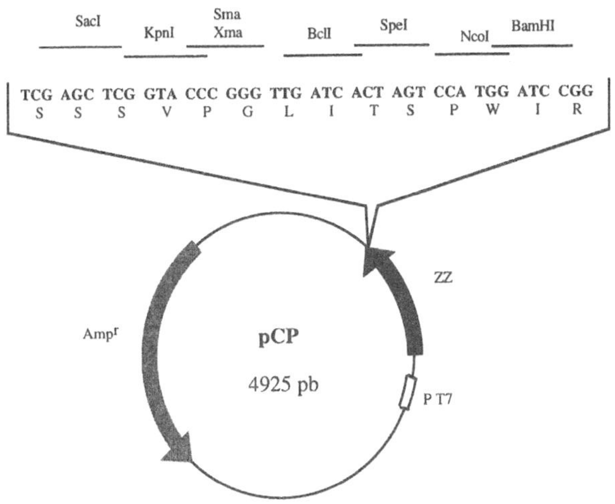

Figure 2 pCP showing detailed multiple cloning sites downstream of the ZZ region, under control of the T7 promoter.

3.2 Production of a recombinant toxin in the bacterial cytoplasm

In general, freshly transformed cells should be used. The transformation procedure was performed according to the supplier recommendations (Novagen). We proceeded as follows:

Protocol 1 Production of a recombinant toxin in the bacterial cytoplasm

1. An overnight culture of BL21(DE3)LysS transformed cells was diluted 100 times in TSB medium (DIFCO) **(Solution 1)** supplemented with 200 mg/l of ampicillin, 30 mg/l of chloramphenicol, and 5 g/l of glucose, and incubated at 37°C.
2. When OD_{600} nm reached a value of 0.6, IPTG (0.5 mM final) was added.
3. After 3 h of culture (OD_{600} = 1.7 to 3) the cells were harvested by centrifugation and resuspended in 1/10 volume of lysis buffer **(Solution 2)** and submitted to three cycles of freezing and thawing.
4. DNA and cell fragments were precipitated by addition of protamine sulfate at 2 mg/ml as a final concentration, and discarded by centrifugation for 15 mn at 12000 × g.
5. Then, PMSF 0.5 mM and Na_2SO_3, 0.3 M, were added to the supernatant.
6. The level of production was estimated from the major band of hybrid observed in SDS-polyacrylamide gel electrophoresis (PAGE).

   Notes:
   If no band corresponding to the hybrid or to the ZZ domain alone appears, the hybrid may have been precipitated due to an inadequate treatment with protamine sulfate. It is possible that a proportion of the synthesized protein forms inclusion bodies.

3.3 *In vitro* refolding and purification of a recombinant toxin from the bacterial cytoplasm

Protocol 2 *In vitro* refolding and purification of a recombinant toxin from the bacterial cytoplasm

1. IgG Sepharose equilibrated in PBT **(Solution 3)** was incubated with the cellular extract for 3 h at 4°C, poured in a column and washed with PBT until OD_{280} of the eluted buffer reached a constant value.
2. The gel was then washed with two volumes of renaturation buffer **(Solution 4)** and incubated for 15 h at room temperature.

   Notes:
   The ratio of reduced over oxidized gluthatione had to be adjusted for each protein.
3. Washing was performed with 5 mM ammonium acetate, pH 5, and the hybrid protein was eluted with 0.5 M ammonium acetate, pH 3.4.
4. The eluted fraction was immediately frozen and lyophilized.
5. The lyophilized hybrid was dissolved in HCl 0.1 N in the presence of a 500-fold molar excess of CNBr for 24 h.
6. The resulting recombinant toxin was separated from ZZ moiety and other by-products by chromatography on a reverse-phase HPLC column (Vydac C_4, 5 mm 10 x 250 mm). The column was equilibrated in 0.1% TFA, and elution was performed using a $TFA/CH_3CN/H_2O$ gradient system.
7. The recombinant toxin was further purified by chromatography on a Mono S (Pharmacia) HPLC column equilibrated in ammonium acetate 0.01 M, pH 4.5, and eluted by a gradient of 0.01 – 1.5 M amonium acetate at the same pH.

Using this procedure, we obtained 2–10 mg of purified toxin per liter of culture, irrespective of the type of curaremimetic toxin or whether the DNA source was a cDNA or a synthetic gene. Purity of the recombinant toxin was checked by SDS-PAGE and isoelectric focusing.

4 Labelling of a "Three-Fingered" Toxin with Stable Isotopes

Having in hand an expression vector that allowed the production of large amounts of toxins, we decided to exploit it to uniformly label a toxin with ^{15}N. The simplest way was to use a minimal medium with $(^{15}NH_4)_2SO_4$ as the sole source of nitrogen.

Protocol 3 Labelling of a "three-fingered" toxin with stable isotopes

1. In this large-scale production, the BL21(DE3)LyS tranformed cells were precultured in LB medium containing 200 mg/L ampicillin and 30 mg/L chloramphenicol for 6 h.
2. 10 ml of this preculture were then inoculated in 1 L of M9 medium using $(^{15}NH_4)_2SO_4$ as a source of nitrogen. We supplemented the medium with ampicillin 200 mg/L and 1 ml of ($MnCl_2$, 5 g/l; $FeCl_3$, 6 g/l; $ZnCl_2$, 5 g/l; $CoCl_2$, 0.1 g/l; $CaCl_2$, 1 g/l; $CuSO_4$, 0.4 g/l; H_3BO_3, 0.1 g/l). This second preculture was left overnight at 37°C and used to innoculate 25 l of the same medium in a 30-l fermenter (Biolaffite, St. Germain en Laye, France).
3. Induction was carried out as described previously, and the cells were harvested by centrifugation after 3 h of culture.
4. The cell pellet was resuspended in 500 ml of lysis buffer **(Solution 2)** and submitted to three cycles of freezing and thawing.
5. Cellular debris was discarded by centrifugation for 30 min at 12,000 × g, then 0.5 mM PMSF and 0.3 M Na_2SO_3 were added to the supernatant before the protein was subjected to purification and renaturation as described above (see Protocol 2).

We obtained 44 mg of ZZ toxin per liter of culture. Thus, 55 mg of highly purified ^{15}N-labelled toxin were obtained (Drevet et al., 1997). The protein was indistinguishable from the wild-type and unlabelled toxin. It is now currently used in NMR experiments aiming at understanding the dynamical properties of the toxin.

5 Periplasmic Production of a Sea Snake Toxin

Production of a fully folded and active snake toxin can be directly achieved by producing the fused toxin into the periplasm of *E. coli*. This procedure was particularly appropriate using protein A as the fused moiety. The vector used was pRIT5 (Ducancel et al., 1989). It is a secretion vector designed for translocation of fusion proteins in the periplasmic space of *E. coli* (Nilsson et al., 1985a, b). It was then substituted by pEZZ vector (Löwenadler et al., 1987; Nilsson et al., 1987), which contains the protein A sequence signal and two synthetic "Z" domains which are derived from the IgG binding domain of protein A. This vector permits direct secretion of the synthesized ZZ-toxin hybrid into the growth medium of *E. coli* (Pillet et al., 1993; Trémeau et al., 1995). The resulting hybrid toxin was capable of binding to its receptor though with a much lower affinity, as compared with the wild-type and unfused toxin.

5.1 Construction of an efficient expression/secretion vector

Insertion of the cDNA encoding erabutoxin a (Tamiya et al., 1985) into the pEZZ fusion vector implies several sequence modifications. These include

- introduction of a methionine codon at position –1 of the toxin sequence, which constitutes a potentially cleavable site of the fusion protein by cyanogen bromide
- deletion of the signal sequence of the toxin
- fusion in frame of the toxin coding sequence to ZZ coding sequence

These modifications can be obtained with classical DNA manipulations as described by Ducancel et al. (1989) or in one step experiment using PCR technology and appropriate primers. The pEZZ-erabutoxin vector was then used to transform HB 101 *E. coli* strain.

5.2 Production and purification of erabutoxin a from the bacterial growth medium

Protocol 4 Production and purification of erabutoxin a from the bacterial growth medium

1. Bacteria were grown at 37°C in a 5-liter fermenter with TSB (tryptic soy broth) medium supplemented with 5 g/l of glucose and 200 mg/l of ampicillin. The pH was maintained at 7.2 with 5 M NaOH addition, and oxygen concentration was adjusted continuously to 80% saturation.
2. Feeding of bacteria was achieved by adding glucose 40% as a carbon source, and the medium was adjusted to give a growth rate of 0.3 $OD_{600\ nm}$/h.
3. The culture was stopped at the early stationary phase (25–30 $OD_{600\ nm}$).
4. Cells were discarded by centrifugation, and the culture supernatant was filtered through an 0.2 µm filter.
5. It was then concentrated 20-fold by ultrafiltration on a membrane with a cutoff mass of 10 kDa.
6. To prevent proteolytic degradation of the hybrid, the concentrated solution was rapidly applied on a 10-ml IgG-Sepharose column. The gel was washed first using TST buffer **(Solution 6)** until no protein was detected in the flow-through.
7. It was then washed with 10 ml of 5 mM ammonium acetate pH 5 to remove salts and detergent.

8. The bound proteins were eluted with 0.5 M acetic acid, pH 3.4, and immediately lyophilized.
9. The hybrid was dissolved in HCl 0.1 N in the presence of a 500-fold molar excess of CNBr for 24 h.
10. The resulting recombinant toxin was separated from the ZZ moiety and other products by chromatography on a reverse-phase HPLC column and further purified by chromatography on a mono S HPLC column as described in **(Protocol 2)**.

According to this procedure, we currently obtained 0.5 mg of purified toxin per liter of culture. The resulting recombinant erabutoxin a was indistinguishable from the toxin purified from venom in its biological properties (Boyot et al., 1990; Hervé et al., 1992; Pillet et al., 1993) and three-dimensional structure (Arnoux et al., 1994). Note that the production yield varies with the type of mutation that is introduced in the sequence of erabutoxin a.

6 Chemical Synthesis of a Land Snake Toxin

Due both to their relatively large size (60–75 amino acids) and their high density of disulfides, the three-fingered toxins have been considered to be difficult to synthesize chemically. However, during recent years many efforts have been made by chemists to solve this problem, and synthesis of several disulfide-rich small proteins was successfully achieved (Kuroda et al., 1992; Nishio et al., 1993; Byrnes et al., 1995). This is also the case for a curaremimetic toxin, toxin α from *Naja nigricollis*, a 61 residue-containing protein with four disulfides (Mourier et al., 1996). The method has now so much progressed that one can anticipate that various toxins of comparable sizes or even slightly larger, like phospholipases A_2, will also be successfully synthesized by chemical means.

The assembly of the toxin peptide chain was carried out by the general stepwise solid-phase method of Merrifield (Merrifield et al., 1963; Stewart and Young, 1984), using the base-labile 9-fluorenyl methoxycarbonyl (Fmoc) amino protecting group (Atherton and Sheppard, 1989) and dicyclohexyl carbodiimide (DCCI)/hydroxybenzotriazole (HOBT) as coupling reagents, and N-methyl pyrrolidone (NMP) as a solvent.

6.1 Synthesis

The synthesis was performed using an automatic peptide synthesizer (ABI 433, Perkin-Elmer), with a standard 0.1-mmol-scale program and 0.05-mmol resin bearing the rink amide linker.

Protocol 5: Synthesis of a land snake toxin

1. The C-terminal asparagine was introduced by coupling the side-chain carboxyl group of N-α-Fmoc-L-Aspartic acid-t-butyl ester to the resin.

   Notes:
   Upon treatment with TFA, the peptide resin could be cleaved at any time, thus generating an Asn residue which, interestingly, was free of racemization and of dipeptide formation (Breipohl et al., 1990; Albericio et al., 1990).

2. The coupling was performed using 1 mmol of Fmoc-amino acid (a 20-fold molar excess), 1 mmol of HOBT in NMP, and 1 mmol of DCCI in NMP. Each coupling run was for 38 min.
3. Deprotection of the Fmoc group resulted from two treatments (3 min and 11 min) with 20% piperidine in NMP, followed by several washings with NMP.
4. At the end of the synthesis, the peptide resin (250 mg) was treated with TFA (10 ml), crystalline phenol (0.75 g), 1,2-ethanedithiol (0.25 ml), thioanisole (0.5 ml), and 0.5 ml distilled water.

   Notes:
   The peptide was cleaved from the resin and the protecting groups were removed from amino acid side chains.

5. After 2 h incubation, the mixture was filtered in cold tert-butyl methyloxyde and centrifuged.
6. The precipitate was washed 3 times with tert-butyl methyloxyde and centrifuged.
7. The pellet was dissolved in 10% acetic acid and lyophilized.
8. The protein was purified by reverse-phase HPLC using a Vydac C18 column (250 × 10 mm) equilibrated in solvent A **(Solution 7)** on which was applied a gradient of 0 – 35% of solvent B **(Solution 8)** in 6 min, 35 – 50% of B **(Solution 8)** in 29 min, and 50 – 100% of B **(Solution 8)** in 10 min. The flow rate was 3 ml/min, and the detection was followed at 280 and 214 nm.

   Notes:
   Yield: the major peak corresponds to 53% of the crude mixture.

6.2 Disulfide bond formation

The presence of a mixture of reduced/oxidized glutathione is of great help for formation of toxin disulfide bonds. With toxin α, we proceeded as it follows:

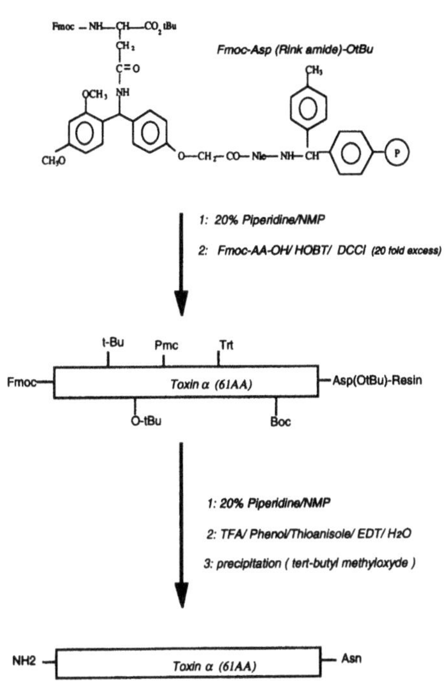

Figure 3 Solid-phase synthesis of toxin α from *Naja nigricollis*.

Protocol 6: Disulfide bond formation

1. The reduced protein was subjected to an oxidative reaction in 0.1 M sodium phosphate buffer (pH 7.8), in the presence of a ratio of 4 mM/2 mM of reduced (GSH) to oxidized glutathione (GSSG) **(Solution 5)** during 12 h at room temperature in the dark.

   Notes:
   45% of the crude mixture of deprotected toxin was thus appropriately oxidized.

2. Typically, the deprotected and reduced synthetic protein (1 mg) was dissolved in 200 ml of 0.1% TFA and immediately diluted in 10 ml oxidizing buffer, which consisted in phosphate buffer 0.1 M, pH 7.8, 4 mM reduced glutathione, and 2 mM oxidized glutathione **(Solution 5)**.

3. After 12 h, the pH was lowered to 3 by adding 30% TFA, and the mixture was loaded on a Vydac semipreparative column (250 x 10 mm) equilibrated with 0.1% TFA in water.

4. The gradient previously used to purify the reduced toxin **(Solution 8)** was then applied to the column.

5. The concentration of eluted toxin α was estimated from its absorbance at 278 nm, its ε_M being equal to 9000.

6.3 Physicochemical controls

The following controls were performed:

- analysis of amino acid composition
- determination of the N-terminal amino acid sequence
- though disulfide pairings are sometimes difficult to determine, effort should be made to check that the expected native disulfide bonds of a synthetic toxin have indeed formed
- identification of the mass using mass spectrometry
- identification of secondary structure by circular dichroism
- 1-D NMR spectrometry

7 How to Choose the Appropriate Synthesis Method?

A **periplasmic expression system** appears to be particularly adequate if one wishes to directly generate a fully active recombinant toxin. In our case, the fused toxin could be easily recovered in the external medium and only a single *in vitro* cleavage step remained to be performed to get the final product. This solution is particularly appropriate if many mutants of the toxin have to be produced (Pillet et al., 1993). However, the amount of generated toxin is unlikely to be large.

A **cytoplasmic expression system** can be interesting when the periplasmic system does not work well. For example, some mutants of a curaremimetic toxin could not be produced with a periplasmic system. We suspected that the fused mutated toxin did not fold quickly enough in the bacteria, making it highly susceptible to proteolytic degradation. By switching to a cytoplasmic expression system, the considered mutant was produced and adequately folded *in vitro* (Trémeau et al., 1995; Ducancel et al., 1996). A cytoplasmic expression system also offers the advantage of allowing the production of large amounts of toxin, which can be useful to proceed to a structural analysis of the toxin and/or to its labelling with stable isotopes (Drevet et al., 1997). However, one must emphasize that the presence of several disulfide bonds in the toxin implies that an adequate and efficient oxidation system is designed and that the disulfides of the resulting recombinant toxin be carefully checked to ensure they are identical to those of the wild-type toxin.

The **chemical approach** is highly recommended when a high quantity of toxin is required only once, for example, with the view to performing an NMR study. It may also be of interest if introduction of nonnatural (D-amino acids or else) is needed. However, this approach is clearly too expensive to be used in a systematic mutational approach, unless the toxin is of small size (Dauplais et al., 1997).

8 Troubleshooting

No hybrid protein is recovered from the bacterial cytoplasm:

- The extraction procedure may not be efficient. Depending on the toxin, the hybrid may be soluble or may form inclusion bodies. The extraction procedure has to be consequently adapted.
- Proteolysis may occur *in vivo*, particularly when the produced protein is soluble in the cytoplasm. Different fusion systems should be tested to obtain a stable hybrid.
- The recombinant plasmid is not stable. This may occur when the produced protein is highly toxic for the bacteria. Use freshly transformed cells and a totally blocked promotor when it is not induced.

No hybrid protein is recovered from the bacterial medium:

- Export and/or folding of the hybrid does not occur appropriately. Check whether the hybrid has been synthesized, and localize it in the cell fractions. If no protein is produced, control the vector sequence. If the expression vector is correct, test its stability; otherwise, construct a cytoplasmic expression vector and develop an *in vitro* procedure for the folding.
- Only the ZZ domain is recovered. This may indicate that the toxin has been degraded by proteolysis. Add various protease inhibitors and grow cells at temperatures below 37°C.

Acknowledgements

This work was supported by the Commissariat à l'Energie Atomique (CEA). Biohazards associated with the experiments described in this publication have been examined by the French National Control Committee.

References

Recombinant technology

Arnoux B, Ménez R, Drevet P et al (1994) Three-dimensional crystal structure of recombinant erabutoxin a at 2.0 Å resolution. FEBS Lett 342: 12–14

Bouhaouala-Zahar B, Ducancel F, Zenouaki I et al (1996) A recombinant insect-specific α-toxin of Buthus occitanus tunetanus scorpion confers protection against homologous mammal toxins. Eur. J. Biochem, 238: 653–660

Boyot P, Pillet L, Ducancel F et al (1990) A recombinant snake neurotoxin generated by chemical cleavage of a hybrid protein recovers full biological properties. FEBS Lett, 266: 87–90

Danse JM, Rowan EG, Gasparini S et al (1994) On the site by which α-dendrotoxin binds to voltage-dependent potassium channels: site-directed mutagenesis reveals that the lysine triplet 28–30 is not essential for binding. FEBS Lett, 356: 153–158

Dauplais, M, Lecoq, A, Song, J et al (1997) On the convergent evolution of animal toxins. Conservation of a diad of functional residues in potassium channel-blocking toxins with unrelated structures. J. Biol. Chem, 272: 4302–4309.

Drevet P, Lemaire C, Gasparini S et al (1997) High-level production and isotope labelling of snake neurotoxins, disulfide-rich proteins. Protein Expression and Purification, 10 (293–300)

Ducancel F, Boulain JC, Trémeau O, Ménez A (1989) Direct expression in E. coli of a functionally active protein A-snake toxin fusion protein. Protein Eng, 3: 139–143

Ducancel F, Mérienne K, Fromen-Romano C et al (1996) Mimicry between receptors and antibodies. Identification of snake toxin determinants recognized by the acetylcholine receptor and an acetylcholine receptor-mimicking monoclonal antibody. J. Biol. Chem, 271: 31345–31353

Eaker DL, Porath J (1967) The amino acid sequence of a neurotoxin from Naja nigricollis venom. Japan J. of Microbiol, 11: 353–355

Ezan E, Ducancel F, Gillet D et al (1994) Recombinant technology in the preparation of immunogen and enzymatic tracer. Application to the development of an enzyme immunoassay for rat prolactin. J. Immun. Meth, 169: 205–211

Gasparini S, Kiyatkin N, Drevet P et al (1994) The low molecular weight protein which co-purifies with α-latrotoxin is structurally related to crustacean hyperglycemic hormones. J. Biol. Chem, 269: 19803–19809

Hervé M, Pillet L, Humbert P et al (1992) Role and environment of the conserved Lys-27 of snake curaremimetic toxins as probed by chemical modifications, site-directed mutagenesis and photolabelling experiments, Eur. J. Biochem, 208: 125–131

Hodgson D, Gasparini S, Drevet P et al (1993) Production of recombinant notechis 11'2L, an enzymatically active mutant of a phospholipase A_2 from Notechis scutatus scutatus venom, as directly generated by cleavage of a fusion protein produced in E. coli. Eur. J. Biochem, 212: 441–446

LaVallie ER, DiBlasio EA, Kovacic S et al (1993) A thioredoxin gene fusion expression system that circumvents inclusion body formation in the E. coli cytoplasm. Bio/Technology, 11: 187–193

Löwenadler B, Jansson B, Paleus S et al (1987) A gene fusion system for generating antibodies against short peptides. Gene, 58: 87–97

Murby M, Nguyen TN, Binz H et al (1994) Production and recovery of recombinant proteins of low solubility. Separations for Biotechnology 3 (Pyle DL, ed): Bath: Bookcraft

Nilsson B, Abrahmsén L, Ulhén M (1985a) Immobilization and purification of enzymes with staphylococcal protein A gene fusion vectors. EMBO 4: 1075–1080

Nilsson B, Holmgren E, Josephson S et al (1985b) Efficient secretion and purification of human insulin-like growth factor I with a gene fusion vector in Staphylococci. Nucleic Acids Res, 13: 1151–1162

Nilsson B, Moks T, Jansson B et al (1987) A synthetic IgG-binding domain based on staphylococcal protein A. Protein Eng, 1: 107–113

Nygren PA, Stahl S, Uhlén M (1994) Engineering proteins to facilitate bioprocessing. TIBTECH, 12: 184–188

Pillet L, Trémeau O, Ducancel F et al (1993) Genetic engineering of snake toxins: role of invariant residues in the structural and functional properties of a curaremimetic toxin, as probed by site-directed mutagenesis. J. Biol. Chem, 268: 909–916

Sambrook J, Fritsch EF, Maniatis T (1989) Molecular cloning. Cold Spring Harbor, New York: Cold Spring Harbor Laboratory Press

Sato A, Tamiya N (1971) The amino sequences of erabutoxins, neurotoxic proteins of sea-snake (Laticauda semifasciata venom). Biochem. J, 122: 453–461

Studier FW, Rosenberg AH, Dunn JJ, Dubendorff JW (1990) Use of T7 polymerase to direct expression of cloned genes. Methods Enzymol 185: 60–89

Tamiya N (1973) Erabutoxins a, b, and c in sea snake Laticauda semifasciata venom. Toxicon, 11: 95–97

Tamiya T, Lamouroux A, Julien J.F et al (1985) Cloning and sequence analysis of the cDNA encoding a snake neurotoxin precursor. Biochimie, 67: 185–189

Trémeau O, Lemaire C, Drevet P et al (1995) Genetic engineering of snake toxins: the functional site of erabutoxin a, as delineated by site-directed mutagenesis, includes variant residues. J. Biol. Chem, 270: 9362–9369

Vita C, Roumestand C, Toma F, Ménez A (1995) Scorpion toxins as natural scaffolds for protein engineering. Proc. Natl. Acad. Sci. USA, 92: 6404–6408

Zinn-Justin S, Pillet L, Ducancel F et al (1994) Engineering of protein epitopes: a single deletion in a snake toxin generates full binding capacity to a previously unrecognized antibody. Protein Eng, 7: 917–923

Chemical synthesis

Albericio F, Van Abel R, Barany G (1990) Solid-phase synthesis of peptides with C-terminal asparagine or glutamine. Int. J. Pept. Prot. Res, 35: 284–286

Atherton E, Sheppard RC (1989) Solid Phase Peptide Synthesis, a Practical Approach. Oxford: IRL Press at Oxford University Press

Breipohl G, Knolle J, Stüber W (1990) Facile SPS of peptides having C-terminal Asn and Gln. Int. J. Pept. Prot. Res, 35: 281-283

Byrnes ME, Mahnir VM, Kem WR, Pennington MW (1995) Synthesis and characterization of dendrotoxin: a potent potassium channel inhibitor. Protein Peptides Lett 1: 239–245

Kuroda H, Chen YN, Watanabe TX et al (1992) Solution synthesis of calsiseptine, an L-Type specific calcium channel blocker. Pept. Res, 5 (5): 265–268

Merrifield RB (1963) The synthesis of a tetrapeptide, J. Am. Chem. Soc, 85: 2149–2154

Mourier G, Zinn-Justin S, Trémeau O, Ménez A (1996) Chemical synthesis and in vitro folding of the curaremimetic toxin α from Naja nigricollis: a three-finger fold protein. Proceedings of the 24th European Peptides Symposium, Edinburgh pp. 659–660

Nishio H, Kumagaye Y, Kubo S et al (1993) Synthesis of ω-agatoxin IV A and its related peptides. Biochem Biophys. Res. Comm, 1447–1453

Stewart JM, Young JD (1984) Solid Phase Peptide Synthesis, 2nd edn. Pierce Chemical Company

16 *Elapidae* Toxins: The Fasciculins, and their Interaction with Acetylcholinesterase

Pascale Marchot and Pierre E. Bougis

Contents

Methods and Tools in Biosciences and Medicine
Animal Toxins, ed. by H. Rochat and M.-F. Martin-Eauclaire

1 Introduction

Fasciculins, isolated from the venom of mambas (genus, *Dendroaspis*), are named after the long-lasting muscle fasciculations they produce in mice (Rodriguez-Ithurralde et al., 1983). They are potent and selective inhibitors of acetylcholinesterase (AChE[1]: EC 3.1.1.7) (Karlsson et al., 1984; Harvey et al., 1984). These 7-kDa proteins, with their three extended loops emerging from a dense core that contains four disulphide bridges, belong to the family of three-fingered snake toxins that includes the selective nicotinic receptor blockers, α-neurotoxins (Lee and Chang, 1966; Changeux et al., 1970; Endo and Tamiya, 1991) and κ-neurotoxins (Chiappinelli, 1983; Oswald et al., 1991), the subtype-specific muscarinic receptor agonists (Adem et al., 1988; Segalas et al., 1995), the L-type calcium channel blockers, calciseptine and FS2 (de Weille et al., 1991; Albrand et al., 1995), the GPIIb-IIIa antagonist and platelet aggregation inhibitor, RGD-containing dendroaspin (or mambin) (McDowell et al., 1992; Sutcliffe et al., 1994), and the cell membrane lytic cardiotoxins (or cytotoxins) (Bougis et al., 1981; Dufton and Hider, 1991). Thus, despite a highly conserved structural motif, the toxins in this family have diverse targets and display highly selective individual modes of action.

Four fasciculins are known that differ only by one to three residues: Fas1 and Fas2 from the venom of *Dendroaspis angusticeps* (Eastern green mamba) (Rodriguez-Ithurralde et al., 1983), ToxC from the venom of *D. polylepis polylepis* (black mamba) (Joubert and Taljaard, 1978), and Fas3, which was isolated from a particular batch of *D. viridis* (Western green mamba) venom and found to have the same primary structure as ToxC (Marchot et al., 1993) (Fig. 1; Tab. I). No fasciculin has been found in other *D. viridis* venoms (Marchot et al., 1986, 1993), or in that of *D. jamesonii* (Jameson's mamba). Fasciculins inhibit AChE from mammals, electric fish, and some snake venoms with K_i values in the pico- to nanomolar range; in contrast, avian, insect, and some other

```
           1        10        20        30        40        50        60
           |        |         |         |         |         |         |
Fas2       TMCYSHTTTSRAILTNCGENSCYRKSRRHPPKMVLGRGCGCPPGDDNLEVKCCTSPDKCNY
Fas1       TMCYSHTTTSRAILTNCGENSCYRKSRRHPPKMVLGRGCGCPPGDDYLEVKCCTSPDKCNY
Fas3/ToxC  TICYSHTTTSRAILKDCGENSCYRKSRRHPPKMVLGRGCGCPPGDDYLEVKCCTSPDKCNY
                  [----LoopI----]    [-----LoopII-----] [--LoopIII-]
```

Figure 1 Amino acid sequences of the wild-type fasciculins. Substitutions that differentiate fasciculins Fas1 and Fas3 (or ToxC) from Fas2 are underlined.

Modified from Marchot *et al.* (1993).

Table 1 Natural fasciculins: useful parameters

Fasciculin	Fas1	Fas2	Fas3	ToxC
Venom* of origin	Da	Da	Dv IP	Dpp
% (w/w) in the venom	~1.3	~2.7	nd	0.9
Amino acids:				
Ala (A)	1	1	1	1
Arg (R)	5	5	5	5
Asp (D)	3	3	4	4
Asn (N)	3	4	2	2
1/2-Cys (C)	8	8	8	8
Glu (E)	2	2	2	2
Gln (Q)	0	0	0	0
Gly (G)	5	5	5	5
His (H)	2	2	2	2
Ile (I)	1	1	2	2
Leu (L)	3	3	3	3
Lys (K)	4	4	5	5
Met (M)	2	2	1	1
Phe (F)	0	0	0	0
Pro (P)	5	5	5	5
Ser (S)	5	5	5	5
Thr (T)	6	6	5	5
Trp (W)	0	0	0	0
Tyr (Y)	4	3	4	4
Val (V)	2	2	2	2
Total amino acids	61	61	61	61
Monoisotopic mass	6794	6745	6804	6804
Molar absorptivity (ε_{276nm}, M^{-1} cm^{-1})	6300	4900	6300	6300
Abs. coeff. at 1 mg ml^{-1} (A_1)	0.93	0.73	0.93	0.93
Theoretical +/- charges at pH 7.0[a]	+9/-5	+9/-5	+10/-6	+10/-6
Theoretical net charge at pH 7.0[a]	+4	+4	+4	+4
Theoretical +/- charges at pH < 6.5[b]	+11/-5	+11/-5	+12/-6	+12/-6
Theoretical net charge at pH < 6.5[b]	+6	+6	+6	+6
Disulfide bridges (all fasciculins): Cys3–Cys22, Cys17–Cys39, Cys41–Cys52, Cys53–Cys59				

* Da, *Dendroaspis angusticeps*; Dv IP, *D. viridis* venom (batch 1963 n°3 Institut Pasteur); Dpp, *D. polylepis polylepis*; [a]His residues assumed to be unprotonated; [b]His residues assumed to be protonated; amino acid differences are indicated in bold; nd, not determined.

Table 2 Inhibition of Cholinesterases by the Fasciculins (substrate: ATCh for AChEs, BuTCh for BuChEs)[‡]

Enzyme and source	Fasciculin	Temp. (°C)	K_i, K_d, IC_{50} (pM)[‡] k_i, k_{on} (10^8 M^{-1} min^{-1}) k_{-i}, k_{off} (10^{-3} min^{-1})		Reference
Mammalian AChE					
Human, erythrocyte, membrane-bound G2	Fas2	22	800		Harvey et al. (1984)
"	Fas2	37	110 / 220*		Harvey et al. (1984)
"	Fas2	nr	16,000		Cerveñansky et al. (1991)
"	Fas2	23	11 (K_i)	$k_i = 16.2$; $k_{-i} = 17.4$	Eastman et al. (1995)
cerebrospinal fluid	Fas2	37	200		Cerveñansky et al. (1991)
Rat, brain, membrane-bound G4	Fas2	37	35		Harvey et al. (1984)
"	Fas1	25	14 (K_d)		Marchot et al. (1993)
"	Fas2	25	25 (K_d)	$k_{on} = 5.4$; $k_{off} < 6$	Marchot et al. (1993)
"	Fas3	25	0.4 (K_d)		Marchot et al. (1993)
muscle	Fas2	37	120		Harvey et al. (1984)
stomach	Fas2	37	11		Cerveñansky et al. (1991)
intestine	Fas2	37	160		Cerveñansky et al. (1991)
striatum membrane-bound G4	Fas2	37	100		Cerveñansky et al. (1991)
Pig, striatum, membrane-bound G4	Fas2	37	200		Cerveñansky et al. (1991)
Mouse, diaphragm	Fas2	37	52,000		Cerveñansky et al. (1991)
brain	Fas2	37	1600		Cerveñansky et al. (1991)
brain, recomb. memb.-bound G2	Fas2	37	2.3 (K_i)	$k_i = 7$; $k_{-i} = 4.4$	Radic et al. (1994)
"	Fas2	23	$k_i = 1.9$		Radic et al. (1994)
"	Fas2	nr	4.8 (K_i)		Radic et al. (1995)
"	Fas2	nr	18 (K_i)	$k_i = 1.9$; $k_{-i} = 3.5$	Radic et al. (1995)
	Fas2	nr	0.09/62* (K_i)	$k_i = 490/0.71$*; $k_{-i} = 4.4$	Radic et al. (1997)
brain, recombinant soluble G1	Fas1	25	2.5 (K_i)	$k_i = 7.9$; $k_{-i} = 2.0$	Bougis, unpublished
"	Fas2	25	4.0 (K_i)	$k_i = 5.4$; $k_{-i} = 2.2$	Bougis, unpublished
"	Fas3	25	0.3 (K_i)	$k_i = 2.7$; $k_{-i} = 0.1$	Bougis, unpublished

Enzyme and source	Fasciculin	Temp. (°C)	K_i, K_d, IC_{50} (pM)[‡] k_i, k_{on} (10^8 M^{-1} min^{-1}) k_{-i}, k_{off} (10^{-3} min^{-1})	Reference
Rabbit, heart atria	Fas2	37	680	Cerveñansky et al. (1991)
jejunum	Fas2	37	56	Cerveñansky et al. (1991)
Guinea pig, brain	Fas2	37	100	Cerveñansky et al. (1991)
erythrocytes	Fas2	37	65	Cerveñansky et al. (1991)
Bovine, brain, solubilized	Fas1	37	22	Puu & Koch (1990)
" in liposomes	Fas1	37	16	Puu & Koch (1990)
Electric fish AChE				
Gymnote (*Electrophorus electricus*) electroplax				
crude extract	Fas2	22	300	Harvey et al. (1984)
affinity purified	Fas2	37	40 (K_i, Dixon plot)	Duran et al. (1994)
affinity purified and solubilized G4	Fas2	25	200	Remy et al. (1995)
Ray (*Torpedo marmorata*) electroplax				
membrane-bound G2	Fas2	25	140	Cousin et al. (1996)
Reptilian AChE				
Gartner snake (*Thamnophis sirtalis*), brain	Fas2	37	200	Cerveñansky et al. (1991)
" erythrocytes	Fas2	37	140	Cerveñansky et al. (1991)
Cobra venom (mostly soluble G1)				
(nonspecified)	Fas2	22–37	no inhibition	Harvey et al. (1984)
(*Naja naja siamensis*)	nr	nr	no or low inhibition at 10^{-5} M	Karlsson et al. (1990)
(*Naja haje*)	Fas2	4–20[§]	>>10^{-6} M (IC_{50})	Frobert et al. (1997)
(*Naja kouthia*)	Fas2	4–20[§]	>>10^{-6} M (IC_{50})	Frobert et al. (1997)
"	Fas2	37	310,000 (K_i)	Duran et al. (1996)
(*Naja naja*)	Fas2	4–20[§]	>>10^{-6} M (IC_{50})	Frobert et al. (1997)
(*Naja nivea*)	Fas2	4–20[§]	>>10^{-6} M (IC_{50})	Frobert et al. (1997)
(*Ophiophagus hannah*)	Fas2	4–20[§]	50 (IC_{50})	Frobert et al. (1997)
(*Haemacatus haemacatus*)	Fas2	4–20[§]	>>10^{-6} M (IC_{50})	Frobert et al. (1997)

Enzyme and source	Fasciculin	Temp. (°C)	K_i, K_d, IC_{50} (pM)[‡] k_i, k_{on} (10^8 M^{-1} min^{-1}) k_{-i}, k_{off} (10^{-3} min^{-1})	Reference
Krait (*Bungarus multicinctus*)	ToxC	37	25,000 (IC_{50})	Lin et al. (1987)
"	Fas2	4–20[§]	8,500 (IC_{50})	Frobert et al. (1997)
(*Bungarus fasciatus*)	Fas2	25	50,700	Cousin et al. (1996)
" purified G1	Fas2	4–20[§]	12,000 /15,000 / 16,000[@] (IC_{50})	Frobert et al. (1997)
(*Bungarus caeruleus*)	Fas2	4–20[§]	13,000 (IC_{50})	Frobert et al. (1997)
Avian AChE				
Chick, biventer cervicis	Fas2	22–37	no inhibition	Harvey et al. (1984)
brain	Fas2	37	no inhibition	Harvey et al. (1984)
whole brain extract	Fas2	37	20,000	Duran et al. (1996)
Insect AChE				
House fly (*Musca domestica*)				
head synaptosomal fraction	Fas3	25	no inhibition at 10^{-8} M	Marchot (1986)
head	nr	nr	no or low inhibition at 10^{-5} M	Karlsson et al. (1990)
Fruit fly (*Drosophila melanogaster*)	nr	nr	no inhibition	Cerveñansky et al. (1991)
Pseudo-cholinesterases (BuChE or "ChE")				
Human plasma, total	Fas2	37	1.3 x 10^{-6} M	Cerveñansky et al. (1991)
" (G1)	Fas2	37	0.5 x 10^{-6} M	Cerveñansky et al. (1991)
" (G4)	Fas2	37	0.7 x 10^{-6} M	Cerveñansky et al. (1991)
Human serum (affinity purified)	Fas2	37	BzTh: 0.7 x 10^{-6} M	Duran et al. (1996)
"	Fas2	37	BTCh: ~10^{-6} M (IC_{50})	Duran et al. (1996)
Horse, serum	Fas2	37	3.0 / 0.02 x 10^{-6} M*	Harvey et al. (1984)
purified	Fas2	37	BzCh: ~10^{-6} M (IC_{50})	Duran et al. (1996)
"	Fas2	37	BTCh: ~10^{-6} M (IC_{50})	Duran et al. (1996)
Mouse, recombinant	Fas2	37	210 / 290 / 200 x 10^{-6} M[#] (K_i)	Radic et al. (1994)
Rat, liver	nr	nr	no or low inhibition at 10^{-5} M	Karlsson et al. (1990)
Hen, plasma	nr	nr	no or low inhibition at 10^{-5} M	Karlsson et al. (1990)

snake venom AChEs are relatively resistant, and high micromolar concentrations are required to inhibit mammalian butyrylcholinesterases (BuChE) (Tab. 2). Dissociation constants of Fas1 and Fas3 are 2-fold and 60-fold lower, respectively, than that of Fas2 for synaptosomal rat brain AChE (Marchot et al., 1993).

The crystal structures of Fas1 (1.9 Å; Le Du et al., 1992; PDB entry 1FAS) and Fas2 (2.0 Å; Le Du et al., 1996; 1FSC) reveal a large difference in the conformation of loop I, possibly reflecting two extreme conformations of the fasciculin molecule in solution. Coanalysis of these structures with that of *Torpedo californica* AChE (TcAChE, 2.8 Å; Sussman et al., 1991; 1ACE) and data from AChE mutagenesis (Radic et al., 1994) led to suggestions about the nature of the fasciculin structural determinants responsible for binding to and inhibition of AChE.

AChE, an α/β hydrolase, terminates the action of the neurotransmitter, acetylcholine, at cholinergic synapses in the central and peripheral nervous systems (Massoulié et al., 1993; Taylor and Radic, 1994); the active center triad of AChE is located nearly centrosymmetric to the subunit at the bottom of a deep and narrow gorge (Sussman et al., 1991). Several lines of evidence indicate that fasciculin binds to a peripheral site on AChE, a region distinct from the catalytic center and located at the rim of the active site gorge. This site shares a common region with the binding site of peripheral site cationic inhibitors and with the site at which the substrate binds, when present in excess (Karlsson et al., 1984; Marchot et al., 1993). Fasciculin appears not to occlude access of small molecules, such as substrate or active center inhibitors, to the catalytic site of the enzyme in solution (Marchot et al., 1993; Radic et al., 1995). Rather, it influences AChE catalysis in an allosteric fashion, although a steric gating effect may also restrict the rate of entry into the gorge of substrates whose catalysis is rate-limited or near-limited by diffusion (Eastman et al., 1995; Radic et al., 1995; van den Born et al., 1995; Rosenberry et al., 1996). Similarly, an electrostatic gating effect may restrict the rate for entry of charged sustrates (Radic et al., 1997). Site-directed mutagenesis and chemical modification experiments with Fas2 also give insight into complex formation (Tab. 3).

The recently solved structures of complexes formed between Fas2 and mouse AChE (mAChE) (3.2 Å, Bourne et al., 1995; 1MAH) and Fas2 and TcAChE (3.0 Å, Harel et al., 1995; 1FSS) revealed that three domains of Fas2 anchor it to the enzyme. A large contact area was demonstrated, consistent with the low dissociation constants of the complexes; the Fas2 and AChE residues participating in the binding interface were unambiguously established, and major hydrophobic interactions were identified. The structural analyses suggested, however, that Fas2 totally occludes substrate entry into the catalytic site, a departure from the kinetic evidence which suggests limited substrate accessibility.

In summary, the availability of the three-dimensional structure of TcAChE has prompted renewed interest in the highly selective interaction between fasciculin and its binding site on AChE. However, neither the structural studies

Table 3 Summary of site-directed mutagenesis and chemical modifications of fasciculin Fas2

Fas2 residues	Surface buried in the complex (Å²)[1]	[% of interface area][1]	Mutagenesis data	(K_i/K_i' ratio)[2,3]	Modification data	[residual Fas2 activity][4–7]
Thr1					α-NH_3^+	[0.72]
Tyr4	21	[1.8]			not reactive[4]	
His6	12	[1.0]				
Thr7	19	[1.7]				
Thr8	92	[8.0]	T8A-T9A	(18[2], 0.8[3])		
Thr9	103	[9.0]				
Ser10	22	[1.9]				
Arg11	105	[9.1]	R11Q	(6.3[2], 0.25[3])		[0.27]
Ala12	32	[2.8]				
Glu19					γ-COOH	[~1]
Tyr23					not reactive[4]	
Arg24	17	[1.5]	R24T	(0.08)[2]	not reactive	
Lys25			K25L	(1.08)[2]	ε-NH_3^+	[0.35]
Arg27	82	[7.1]	R27W	(0.012)[2]		[0.15]
Arg28			R28D	(1.1)[2]		[0.87]
His29	22	[1.9]	H29D	(73[2], 0.55[3])		
Pro30	102	[8.9]	ΔP30	(0.0052)[2]		
Pro31	77	[7.6]	P31R	(0.0016)[2]		
Lys32	51	[4.4]	K32G	(0.32)[2]	ε-NH_3^+	[0.28]
Met33	112	[9.8]	M33A	(0.13)[2]		
Val34	51	[4.4]	V34A-L35	(0.93)[2]		
Leu35	53	[4.6]				
Arg37	32	[2.8]			not reactive	
Asp45			D45K	(0.7)[2]	β-COOH	[~1]
Asp46					β-COOH	[~1]
Asn47	40	[3.5]				
Leu48	15	[1.3]				
Glu49					γ-COOH	[~1]
Lys51			K51S	(1.08)[2]	ε-NH_3^+	[0.43]
Asp57					β-COOH	[~1]
Lys58					ε-NH_3^+	[0.40]
Tyr61	88	[7.7]			α-COOH	[~1]
					^{125}I-Tyr	[1][4]

[1]Bourne et al., 1995; [2]Marchot et al., 1997; [3]Marchot et al., 1998; [4]Marchot et al., 1993; [5]Cerveñansky et al, 1994; [6]1995; [7]1996.

nor the mutagenesis data explain the residual catalytic activity or the phosphorylation of the active site serine of the complexed enzyme in solution. The divergence between the kinetic and structural data has been explained in two ways: (i) a second portal, distinct from the active site gorge, may allow entry of substrate and catalytic site inhibitors in the fasciculin-AChE complex; and (ii) a conformational change in the enzyme, not obvious in the crystal structure, may open a gap between the gorge wall and the bound fasciculin (Bourne et al., 1995, and references therein).

The purpose of this chapter is to provide the reader not with an extensive review of fasciculins but, instead, with specific information that can be used on a daily basis.

2 Materials and Chemicals

Electric rays and eels

Torpedo californica: electroplate tissue (live/frozen)	Aquatic Research Consultants, San Pedro, CA
Torpedo marmorata, live	Oceanographic centers, Arcachon and Roscoff, France
Electrophorus electricus, live	World-wide Scientific Animals, Apopka, FL

Commercial AChE

membrane-bound AChE	eel electric organ	Sigma C 2888
	bovine erythrocytes	Sigma C 5021 Biozyme ACHE2
	human erythrocytes	Sigma C 5400
amphiphilic AChE	human erythrocytes	Sigma C 0663
recombinant AChE	human	Sigma C 1682
beaded agarose-linked AChE	electric eel	Sigma C 2511

Mamba venoms

D. angusticeps, D. jamesoni, D. polylepis p., D. viridis	Latoxan, Rosans, France
D. angusticeps, D. jamesoni, D. polylepis p.	Sigma

Fasciculins

Fas1	Sigma	F3918
Fas2	Latoxan, Rosans, France	
	Alomone Lab., Jerusalem, Israel	
	Alexis, Grünberg, Germany	
	Sigma	F4293

Miscellaneous

ATCh, DTNB, AChE inhibitors	Sigma
Chromatography paper (3M); CM-52 resin	Whatman
CM-Sepharose CL-6B and Sephadex G-50 resins	Pharmacia
Dowex 1x8 resin	BioRad
Goat anti-rabbit serum	Jackson ImmunoResearch
^{125}I-Na (100 mCi ml^{-1})	Amersham, NEN
Lactoperoxidase; protease-free BSA, PEG 8000	Sigma
Spectrapore dialysis tubing	Fischer, PolyLabo
Ultrafiltration concentrators	Amicon, Filtron

Software

Enzfitter	Elsevier Biosoft, Cambridge, UK
SigmaPlot	SPSS Inc., Chicago, IL
GraphPad PRISM	GraphPad soft. Inc., CA

Books

- *Snake Toxins* (Harvey AL, ed) 1991, Pergamon Press, NY
- *Multidisciplinary Approaches to Cholinesterase* Functions (Shafferman A and Velan B, eds) 1992, Plenum Publishing Corp., NY
- *Enzymes of the Cholinesterase Family* (Quinn DM, Balasubramanian AS, Doctor BP, and Taylor P, eds) 1995, Plenum Publishing Corp., NY
- *Structure and Function of Cholinesterases and Related Proteins* (Doctor BP, Taylor P, Quinn DM, Rotundo RL, and Gentry MK, eds) Plenum Publishing Corp., NY (1998)
- *Goodman and Gilman's Pharmacological Basis of Therapeutics* (Hardman JG and Limbird LE, eds) 1995 (9th Ed) McGraw-Hill, NY, cf. *Anticholinesterase Agents and Agents Acting at the Neuromuscular Junctions and Ganglia*, by P Taylor, chapters 8 and 9, pp. 161–198

The ESTHER database

The ESTHER (for ESTerases, α/β Hydrolase Enzymes, and Relatives) database contains a wealth of information on enzymes of the α/β hydrolase fold family: cholinesterases, carboxylesterases, and lipases, as well as adhesion molecules and hormone precursors with the same fold, but devoid of catalytic activity (Cousin et al., 1997, 1998). The site has sequence alignments and a BLAST comparison tool for homology searches, well-developed sections about kinetics and inhibitors of cholinesterases with special attention devoted to the fasciculin-acetylcholinesterase interaction, a gene structure review, a compilation of mutations with access to three-dimensional images, links to other servers (including protein data banks), a mailing list, news, and more. This database is frequently updated and can be found at http://www.ensam.inra.fr/cholinesterase

3 Purification and Analysis of Fasciculins

3.1 Purification by conventional chromatography and HPLC

Elapidae snake venoms, including mamba venoms, contain many 5–10 kDa polypeptides which coelute as a large, major peak upon fractionation of the venom by gel filtration. Because of their diverse net charges, however, a combination of gel filtration and ion-exchange chromatography has proven effective, irrespective of order.

The fasciculins were first purified and sequenced well before their anticholinesterase activity was known. Thus, ToxC (first named BS_1P_1, then CS_1P) was initially purified *from D. polylepis polylepis* venom by a succession of gradient chromatography on Amberlite CG-50, gel filtration through Sephadex G-50, and gradient elution on phosphocellulose Whatman P-11, all steps being performed in NH_4HCO_3 (Strydom, 1976). Fas2 (first named F_{VII}, then Ta_1) was first purified from lyophilized *D. angusticeps* venom (Viljoen and Botes, 1973).

Rodriguez-Ithurralde and colleagues (1983) purified Fas1 and Fas2 by gel-filtration of *D. angusticeps* venom (0.5 g) through a Sephadex G-50 (3.2 × 75 cm) column equilibrated and eluted with 0.1 M NH_4Ac, pH 6.9 (flow rate, 21 ml h^{-1}); the major and active fraction, fraction III, was subjected to ion-exchange chromatography on a Bio-Rex 70 (2.0 x 28 cm) column equilibrated in 0.2 M NH_4Ac, pH 7.3, with the following sequence of steps: washing with 0.1 M NH_4Ac, pH 7.3 (50 ml); application of the active fraction from gel filtration and elution with 0.1 M NH_4Ac (230 ml); elution with a concave gradient extending from 0.1 M to 1.25 M NH_4Ac, pH 7.3 (flow rate, 40 ml h^{-1}); peaks 3 and 4 contained Fas1 and Fas2, respectively.

Karlsson and colleagues purified Fas1 and Fas2 by gel filtration of *D. angusticeps* venom (1 g in 4.8 ml 0.1 M NH_4Ac) on a Sephadex G-50 (2.0 x 147 cm) column equilibrated and eluted with 0.1 M nonbuffered NH_4Ac (flow rate, 17 ml h^{-1}); four fractions were obtained of which the main one (fraction III) had anticholinesterase activity. A 20-ml sample of this fraction was subjected to ion-exchange chromatography on a Bio-Rex 70 (2 x 31 cm) column equilibrated with 0.20 M NH_4Ac, pH 7.3, and washed with 200 ml of 0.03 M NH_4Ac prior to sample loading; elution used 120 ml of 0.03 M NH_4Ac, and then a 2-l concave gradient from 0.03 to 1.25 M NH_4Ac (flow rate, 30.8 ml h^{-1}); Fas1 and Fas2 eluted between 600 and 700 ml as peaks 5 and 6, respectively; some of the other peaks contain dendrotoxins (Cerveñansky et al., 1991).

Marchot and colleagues purified Fas1 and Fas2 by gel filtration of *D. angusticeps* venom (5 g) on four Sephadex G-50 (5 x 100 cm) columns connected in series (0.1 M NH_4Ac, pH 6.8; flow rate, 120 ml h^{-1}) followed by two successive cation-exchange chromatographic steps of the active fractions on CM-Sepharose CL-6B in NH_4Ac, pH 6.8, in gradient (2 x 1 l of 0.1 to 1 M NH_4Ac; 2.5 x 22 cm column; flow rate 25 ml h^{-1}) and stepwise conditions (0.15 M NH_4Ac; 1 x 200 cm column; flow rate 20 ml h^{-1}), respectively (cf. Le Du et al., 1989). Fas3 was purified by gel filtration of *D. viridis* venom (5.3 g of batch 163 n°3 from Institut Pasteur, Paris) using the same four Sephadex G-50 columns as above

(0.1 M NH_4Ac, pH 8.5; flow rate, 60 ml h^{-1}); two successive cation-exchange chromatographic steps on CM52 in NH_4Ac, pH 7.2, in gradient (0.5 to 2 M NH_4Ac; 2.5 x 45 cm column; flow rate, 60 ml h^{-1}) and stepwise conditions (0.2 M NH_4Ac; 1.5 x 33 cm column; flow rate, 20 ml h^{-1}), respectively; reverse-phase HPLC on a semipreparative ultrasphere-octyl (10 x 250 mm) column with a linear gradient of acetonitrile in NH_4-formate, pH 2.7 (flow rate, 5 ml min^{-1}); and ion-exchange HPLC on an IEX-535-K carboxyspherogel (6 x 150 mm) column under isocratic conditions (NH_4Ac, pH 5.0, 15% acetonitrile; flow rate, 1 ml min^{-1}) (cf. Marchot et al., 1993). In both cases, the venom (10% w/v) was first dialysed (4–10 h, 4°C) against distilled/deionised water with Spectrapore3 (cutoff limit, 3500 Da) and centrifuged (10,000 × g, 30–60 min, 4°C) to sediment insoluble material; the supernatant was adjusted to 0.1 M NH_4Ac, pH 6.8 (Fas1 and Fas2) or pH 8.5 (Fas3), before loading on the column. Purity of the final products was assessed by amino acid analysis, SDS-polyacrylamide gel electrophoresis (PAGE) in reducing and nonreducing conditions, and analytical reverse-phase HPLC on octodecyl (4.6 x 25 cm) columns (see below).

Protocol 1 Purification of fasciculins by HPLC

1. We routinely purify and analyse fasciculins on an analytical scale using reverse-phase HPLC on a 4.6 x 250 mm RP-C18 column (Merck Art. 50995, Darmstadt, Germany) maintained at 25°C.
2. Elution involves a 120-min linear gradient from 20 to 40% acetonitrile (B) in 0.1% trifluoroacetic acid (A), with a flow rate of 1 ml min^{-1}.
3. Routine analysis can be performed with a steeper gradient of 5 – 100% in 45 min.

Notes:

- On CM-Sepharose CL-6B, Fas2 elutes *before* Fas1.
- Two all-HPLC methods for purification of anticholinesterase toxins from *D. angusticeps* venom were reported as Liquid Chromatography Application Notes 353 and 355 by former LKB-Produkter, Bromma, Sweden (Pharmacia Biotech, Uppsala, Sweden). However, no evidence for homogeneity of the final products was provided.

3.2 Lyophilisation and handling of fasciculins

Purified fasciculins can be lyophilised from water or highly volatile salt solutions, NH_4Ac, NH_4-formate, or TFA, with no loss in activity. However, the salt concentration of the sample to be freeze-dried should not exceed 0.1–0.2 M. Lyophilisation from acetic acid may inactivate fasciculin.

Like most three-finger snake toxins, purified fasciculins are largely resistant, but not totally, to proteolysis and temperature changes. Therefore, care must be taken in storing and handling fasciculin to avoid loss in activity; below are a few rules:

- Lyophilised, pure fasciculins can be stored for years in tightly closed containers at 4°C.
- Concentrated stock solutions in water or buffer can be stored at –20°C; although fasciculin concentrated solutions can be freeze-thawed several times with no loss in activity, it is advisable to aliquot concentrated stock solutions so as to avoid too many freeze-thaw cycles.
- Dilute solutions should contain 0.1 mg ml^{-1} BSA to prevent fasciculin from sticking to the vessel walls.

3.3 Electrophoresis of fasciculins with a PhastSystem

Because fasciculins are small molecules, the use of a highly reticulated PhastGel high density (Pharmacia Biotech, Uppsala, Sweden) for SDS-PAGE and native-gel electrophoresis ensures optimal resolution. Fasciculins can be stained by Coomassie blue (Heukeshoven and Dernick, 1988) or silver (PhastSystem development technique file n° 210).

Native-PAGE

Only this method (PhastSystem file n° 300) resolves Fas1 from Fas2 and does so only partially. Samples must be loaded at the anode, since the cationic fasciculins migrate toward the cathode at pH 8.5.

SDS-PAGE in nondenaturing conditions

This method (PhastSystem file n° 111), which employs SDS but no reducing agent, is used prior to Western blots with an anti-fasciculin serum.

SDS-PAGE in denaturing conditions

This method (PhastSystem file n° 111) employs SDS, and β-mercaptoethanol as reducing agent. Although fasciculins are monomeric entities of about 7 kDa, bands of about 14 kDa are often observed. The high basicity of the fasciculin molecule might cause association with more SDS than "regular" proteins. Cationic short α-neurotoxins from *Elapidae* venoms (Marchot personal observation) and scorpion toxins (Martin-Eauclaire and Rochat, this book) behave identically.

Isoelectric focusing

Since the high pI of fasciculins (>9.5) generally makes them migrate out of the ampholyte range, this method (PhastSystem file n° 100) is of limited usefulness. However, it can be of value for examining contamination of a fasciculin preparation by proteins or peptides of lower pI values.

Notes:
- Since all three fasciculins and other venom toxins that coelute with fasciculins upon gel filtration have similar molecular mass (about 7 kDa), the yield of a single band upon SDS-PAGE in reducing conditions is not a criterion of purity. Methods that take into account the charge of the protein must be used in parallel.
- See Marchot et al. (1996) for electrophoresis of fasciculin-AChE complexes.

4 The AChE Inhibition Assay

The most widely used AChE assay is that of Ellman et al. (1961) by which activity is assayed as ΔA_{412nm} min^{-1} with ATCh as substrate and DTNB as chromophore; the reaction product is thiocholine (ε_{412nm} = 13,600 M^{-1} cm^{-1}).

Notes:
- An alternative method is that of Augustinsson et al. (1978) who assayed AChE activity as ΔA_{324nm} min^{-1} with ATCh as substrate and 4,4'-dithiopyridine as chromophore; the reaction product is 4-thiopyridine (ε_{324nm} = 19,800 $M^{-1}cm^{-1}$). The Augustinsson assay can have two advantages over the Ellman assay: lower enzyme activities can be detected (higher extinction coefficient), and trace amounts of haemoglobin (which absorbs at 410 nm) do not interfere.

4.1 Preparation of AChE

Membrane-bound AChE

Tetrameric (G4) AChE from rat brain crude synaptosomal fraction (P_2) can be prepared by a combination of methods described by Gray and Whittaker (1962), Israel et al. (1970), and Stadler et al. (1979):

Protocol 2 Preparation of membrane-bound tetrameric AChE

1. Brains (without encephalon) of freshly sacrificed rats are homogenised in a Potter (800 rpm, 4°C) as a 10% (w/v) suspension in 5 mM K_2HPO_4, pH 7.4, 0.32 M sucrose.
2. The homogenate is centrifuged (1000 x g, 10 min, 4°C).
3. Supernatant (S_1) is collected and stored, and the pellet is rehomogenised and centrifuged as previously.
4. Supernatant (S_2) is added to supernatant S_1 and centrifuged (10,000 x g, 30 min, 4°C).
5. Pellet (P_2) is resuspended in 25 mM Hepes, pH 7.2, 10 mM glucose, 140 mM choline, 5.4 mM KCl, 1.8 mM $CaCl_2$, 0.8 mM $MgCl_2$ (pH to be adjusted with 1 M Tris-base).

Dimeric (G2) AChE from *Torpedo californica* or *marmorata* electric organ can be prepared according to methods of Reed et al. (1976) and Morel et al. (1977) as described in Bougis et al. (1986):

Protocol 3 Preparation of membrane-bound dimeric AChE

1. Fresh or thawed electroplate fragments (30 g) are minced and homogenised with an Ultra-Turrax apparatus (20 s, 4°C) in 1.2 mM $NaPO_4$, pH 6.7, 280 mM NaCl, 3 mM KCl, 3.4 mM $CaCl_2$, 1.8 mM $MgCl_2$, 300 mM urea, 5.5 mM glucose, 250 mM sucrose, 0.1 mM PMSF, 3 mM NaN_3 (190 ml).
2. After equilibration with O_2, $NaHCO_3$ is added to bring the pH to 7.0–7.2.
3. The homogenate is centrifuged (600 x g, 10 min, 4°C) and the supernatant discarded; the pellet is taken up in a minimum volume of the above buffer and sonicated (10 s, 4°C).
4. The opalescent suspension obtained is filtered through a fine 200-µm mesh steel screen to eliminate connective tissue.
5. The volume is brought to 90 ml with buffer and the preparation centrifuged (100,000 x g, 90 min, 4°C).
6. The pellet is homogenised in 5 mM $NaPO_4$, pH 7.4, 877 mM sucrose, 0.1 mM PMSF, 3 mM NaN_3 (40 ml), and centrifuged (5000 x g, 20 min, 4°C).
7. The supernatant is then diluted with an equal volume of 5 mM $NaPO_4$, pH 7.4, and centrifuged (100,000 x g, 90 min, 4°C).
8. The final pellet is taken up in 5 mM $NaPO_4$, pH 7.4, 250 mM sucrose, 0.1 mM PMSF, 3 mM NaN_3 to a protein concentration of 6–8 mg ml^{-1}.

Notes:
- Total protein content can be determined by a protein assay.
- AChE active sites can be titrated with DFP, MEPQ, or fasciculin (cf. Marchot et al., 1993).
- Flash-frozen P_2 or electroplate membrane fragments can be stored aliquoted in liquid nitrogen for months with no loss in AChE activity.

Soluble AChE

The PIPLC-released, low-salt soluble AChE (G2) from *Torpedo marmorata* or *californica* electric organ can be prepared according to Futerman et al. (1985) and Sussman et al. (1988), and the high-salt soluble (A8–12) and trypsin-released low-salt soluble (G4) AChE from *Electrophorus electricus* electric organ, according to Massoulié and Bon (1976) and Grassi et al. (1988), respectively. These purification protocols involve affinity resins and will not be described herein.

4.2 Monitoring AChE activity with a spectrophotometer

Prepare:

- **buffer**
  0.1 M $NaPO_4$, pH 7.0–8.0
  BSA 0.1 mg ml^{-1}
  store at 4°C, equilibrate at experimental temperature before use
- **10 mM DTNB**
  weight 40 mg, add 15 mg $NaHCO_3$, and dissolve in 10 ml of buffer without BSA,
  store at –20°C; keep on ice during the experiment
- **50 mM ATCh**
  weight 99 mg, and dissolve in 10 ml buffer without BSA,
  store at –20°C; keep on ice during the experiment
- **AChE**
  dilute stock solution in buffer and keep on ice

Protocol 4 Spectrophotometric assay

1. In a disposable microcuvette add the volume of buffer required for a final volume of 1 ml
2. AChE, to a final catalytic site concentration of 5–10 pM (solubilised AChE) or to 0.1 mg ml^{-1} (synaptosomal rat brain membrane AChE)
2. 30 µl DTNB, to a final concentration of 0.3 mM
4. 10 µl ATCh, to a final concentration of 0.5 mM
   Vortex, then record ΔA_{412nm} every min for 5 min

Notes:

- As the rate of ATCh hydrolysis increases with temperature (usual values are 20–22°C, 25°C, or 37°C) and pH (usual values are 7.0, 7.5, or 8.0), identical conditions must be used for data to be compared.
- To monitor spontaneous ATCh hydrolysis, always run a control experiment (blank) replacing AChE with buffer.

4.3 Monitoring AChE activity with a microtitration plate reader

This method is particularly valuable when screening fractions from AChE or fasciculin chromatography. Final concentrations (AChE, DTNB, ATCh) are the same as above but for a final volume of 200 µl per well. This should give a pathlength value of 0.5 cm, i.e., the half-value for a cuvette. Samples are duplicated or quadruplicated, and blanks included to monitor spontaneous ATCh hydrolysis. Shake the plate, then record ΔA_{412nm} every 20 s for 1–5 min (405-, 412-, or 414-nm filters are usually available on plate readers). It is worth using a multichannel pipette for convenient and fast filling of the plate.

4.4 Data processing

If the substrate, at the concentrations employed, does not interact with the peripheral site, the scheme for fasciculin association with AChE is:

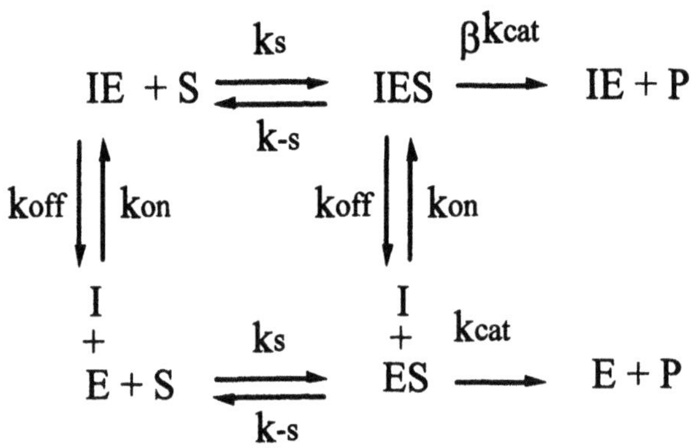

where I, S, and E respectively denote free inhibitor, substrate, and enzyme; ES is the productive enzyme-substrate complex with a rate constant k_{cat}; IE is the enzyme-inhibitor (peripheral site) complex; β is the fraction of residual activity reflecting the partial productivity of IES; the ratio of rate constants, k_{off}/k_{on} = K_i, is the reversible dissociation constant of the IE complex. This scheme assumes that the right and left k_{on} and k_{off}'s are the same, so that I combines equally well with E and ES, as does S with E and IE (the same applies to the top and bottom k_s and k_{-s}'s). A more general scheme, where substrate in excess interacts with the peripheral site and competes with fasciculin binding, may be drawn (cf. Radic et al., 1994, 1995).

Activity determination

Spectrophotometer data output: $vi = \Delta A_{412nm}\ min^{-1}$
In terms of concentration: $\Delta C = \Delta A_{412nm} / \varepsilon\, l$ ($\varepsilon = 13{,}600\ M^{-1}\ cm^{-1}$ is the extinction coefficient of the coloured anion 5-thio-2-nitobenzoic acid, and l = 1 cm).

Example for 20 pM AChE: $V_m = 0.042\ \Delta A_{412nm}\ min^{-1}$
$V_m = 0.042 / 13{,}600 = 3.08\ 10^{-6}\ M\ min^{-1}$
$V_m = k_{cat}\ [E]$; $[E] = 20\ pM$
$k_{cat} = V_m / [E] = 3.08\ 10^{-6} / 20\ 10^{-12} = 154\ 000\ min^{-1}$

Rates of fasciculin interaction with AChE

The interaction between fasciculin and AChE is analysed by combining, at zero time, AChE (vo) and fasciculin and by removing aliquots at specified time intervals to measure residual AChE activity (vi). When using high fasciculin concentrations, fasciculin is added simultaneously to the substrate at a concentration high enough to be not appreciably reduced during the reaction interval, and the decreasing rate of substrate hydrolysis is measured continuously. The rate of approach to the new equilibrium is described by:

$$\frac{(vo - vi)}{vo} = \frac{k_{on}[I]}{k_{on}[I] + k_{off}}\left(1 - e^{-(k_{on}[I] + k_{off})t}\right)$$

GraphPad PRISM equation:

kobs = kon * I + koff
Y = [(kon * I) (1 – exp (-kobs * X))] / kobs
; vo is the activity at initial time (control)
; vi is the activity at intermediate time after addition of inhibitor
; X = t is the time (min)
; Y = (vo – vi) / vo is the fraction of enzyme inhibition at time t (0 – 1)
; kon is the second-order rate constant for formation of the complex (M^{-1} min^{-1}, x 10^{+9})
; koff is the first-order rate constant for dissociation of the complex (min^{-1})
; I is the inhibitor concentration (nM)

Notes:

- k_{on} [I] and k_{off} should be of the same magnitude, i.e., [±] ~ K_i.

Enzyme inhibition at equilibrium (Ackermann-Potter plot)

A tight binding inhibitor-enzyme complex with K_i in the picomolar range, reflected by a small k_{off}, implies that the inhibitor interaction should be measured by prolonged (usually over night) incubation with various inhibitor concentrations prior to measuring fractional inhibition. Since k_{off} is small, the residual enzyme activity gives a measure of the fraction of enzyme in interaction with the inhibitor. The reversible inhibition constant is determined at equilibrium by fitting ratio viβ/vo to the following equation:

$$\frac{vi\beta}{vo} = \frac{vi}{[E]}\left(1 + \frac{\beta[I]}{(K_i + v)}\right) \text{ where}$$

$$vi = \frac{\left([E] - K_i - [I] + \sqrt{\left((K_i + [I] + [E])^2 - 4[I][E]\right)}\right)}{2}$$

where [E] and [I] should be of the same magnitude and between 0.1 and 10 K_i (Ackermann and Potter, 1949; Radic et al., 1995).

GraphPad PRISM equation:

vi = (E – Ki – X + sqrt ((Ki + X + E)^2 – 4 * X * E)) / 2
Y = (vi / E) * (1 + beta * X / (Ki + vi))
; vo is the activity at initial time (control)
; viβ is the residual activity after addition of inhibitor (at equilibrium)
; X = [I] is the inhibitor concentration (pM)
; Y = viβ/vo is the enzyme activity at time t as a fraction of the control (1 – 0)
; E is the enzyme concentration (pM)
; Ki is the reversible inhibition constant for I (Ki = koff / kon) (pM)
; β is the fraction of residual activity reflecting partial productivity of IES

Notes:

- Without considering any residual activity of IES (i.e., β = 0):
  Y = (E – Ki – X + sqrt((Ki + X + E)^2 – 4 * X * E)) / 2 * E
- For a tight binding, noncompetitive inhibitor like fasciculin, K_i is best determined from the above-described nonlinear regression analysis. However, the initial parameter estimates required for this procedure can be found by the graphical method of Dixon (1972) as detailed by Greco and Hakala (1979).

5 ^{125}I-fasciculin Binding Assays

5.1 Radioiodination of fasciculins

Lactoperoxidase-catalysed iodination results in full retention of both the binding and anticholinesterase activities of fasciculins. This enzyme substitutes an iodine from a hydrogen on phenol rings of tyrosines that are solvent-accessible. Thus, Fas1 and Fas3 can be iodinated on both Tyr47 and Tyr61 and Fas2 on Tyr61 only, since Tyr4 and Tyr23 are buried inside the molecule (Marchot et al., 1993).

Prepare:

- **Iodination buffer**
  50 mM $NaPO_4$, pH 7.4
- **BSA buffer**
  50 mM $NaPO_4$, pH 7.4
  BSA 0.1 mg ml^{-1}
  store at –20°C
- **Lactoperoxidase** at 0.25 mg ml^{-1} in water (control concentration spectrophotometrically);
  store aliqoted at –20°C
- **one 1.5 ml tube with 100** µl spin-dried Dowex 1 x 8 resin equilibrated in iodination buffer

Protocol 5 Radioiodination

1. At room temperature, add successively to a 1.5-ml tube, 5 µg Fas2 (0.73 nmol as 2–10 µl depending on the stock concentration), 10 µl lactoperoxidase, 1 mCi ^{125}I-Na (10 µl, to be adjusted depending on stock concentration and titration date), and iodination buffer to a total volume of 80 µl. Mix gently.
2. Prepare (now, not earlier) a 60 µM H_2O_2 solution by serial dilution of the commercial solution with iodination buffer (if commercial solution is 3 M, dilute 10 µl in 1 ml; 10 µl in 1 ml; 200 µl in 1 ml), then with 1-min intervals and gentle mixing add 4 x 5 µl of the H_2O_2 solution to the iodination mixture, close the tube, vortex, spin down.

Prepare counting sample #1 (total ^{125}I -Na added):

- Dilute 20,000-fold 5 µl of the iodination mixture with BSA-buffer (5 µl in 1 ml; 10 µl in 1 ml); transfer 50 µl to a counting tube and keep for further counting.

Removal of free iodide:

3. Add 100 µl ice-cold iodination buffer to the iodination mixture, mix, spin down.
4. Transfer the total 195 µl sample to the spin-dried Dowex-containing tube, mix gently, incubate for 3–5 min.
5. Add 200 µl ice-cold BSA-buffer, mix well, spin down, transfer supernatant to a pretagged 1.5 ml storage tube (measure volume, should be 395 µl).
6. Rinse the resin with 150 µl ice-cold BSA-buffer, mix, spin down, transfer supernatant (measure volume, should be 150 µl), repeat.

   Notes:
   The storage tube now contains 695 µl of the ^{125}I-fasciculin stock solution.
7. Mix well, spin down, and keep at 4°C in a tagged lead container.

Prepare counting sample #2 (incorporated ^{125}I):

- Take 5 µl from the stock solution and dilute as for sample #1. Count samples #1 and #2 for γ-radioactivity.

Quality control by paper chromatography:

8. Place 5–10 ml of an NaI-methanol mixture (a pinch of NaI in 100 ml of methanol) in a 200-ml graduated glass cylinder.
9. Cut two 20 x 1 cm bands of Whatman 3M paper and label for easy identification; on first band, load (as a spot, 2–2.5 cm above bottom end) 10–50 µl of a 1:5 dilution of sample #1 first dilution tube (20–50 x 10^3 cpm); on second band, load 10–50 µl of sample #2 first dilution tube (again, 20–50 x 10^3 cpm).
10. Let the spots dry, then immerse the bottom ends of the bands in the NaI-methanol mixture (keep loading spots out of the liquid), while maintaining tops of the bands with paper tape; allow liquid to absorb up to 15 cm (30–60 min).
11. Remove the bands from the cylinder, mark liquid levels with a pencil, allow to dry, slice the bands every centimeter, place in tubes (keep the right order!) and count for γ-radioactivity.

    Notes:
    Results for the two bands should differ: for the first band, 90% of the deposited cpm should be at the loading spot (this is ^{125}I-fasciculin) and 5–10% at the top of the band (this is free iodide); for the second band, all cpm should be at the loading spot (^{125}I-fasciculin only).

Data processing:

- Count #1 (before the Dowex step) is the total ^{125}I added to the iodination mixture, of which a few percent do not incorporate in fasciculin and are removed by charge retention on the resin.
- Count #2 (after the Dowex step) is the total ^{125}I incorporated into fasciculin.
- Paper chromatography reveals how much of the added ^{125}I was not incorporated into fasciculin (the 10% cpm at the top of first band) and whether this nonincorporated ^{125}I was efficiently removed upon the Dowex step (no more cpm should remain at the top of second band).

Count #1 should be about 40,000 cpm, which is 1.6 x 10^9 cpm (2.5 x 10^9 dpm for a usual counter yield of 65% or 1.1 mCi, since 2.2 x 10^9 dpm = 1 mCi) in the 100-µl iodination mixture which initially contained 0.73 nmol of fasciculin; the specific radioactivity (SRA) is thus 1.5 mCi $nmol^{-1}$. The nonincorporated cpm (free ^{125}I), as determined from first paper chromatography band (for example, 10%), has to be subtracted from this value; thus SRA = 1.4 mCi $nmol^{-1}$. Count #2 should be about 3500 cpm; thus 1 x 10^9 cpm, 1.5 x 10^9 dpm or 0.7 mCi (same as above) in the 695-µl ^{125}I-fasciculin stock, or 1 mCi ml^{-1}; thus the ^{125}I-fasciculin stock solution is 0.7 nmol ml^{-1} (since SRA = 1.4 mCi nmol^{-1}).

Notes:
- Use the right buffer at the right time: using the BSA-buffer for iodination or dilution of H_2O_2 would yield ^{125}I-BSA; using the iodination buffer for dilution of counting samples would result in erroneous counts.
- All reagents for iodination must be kept at room temperature and at the correct pH: lactoperoxidase catalysis is temperature- and pH-dependent. In contrast, use ice-cold buffers for the Dowex steps.
- ^{125}I-fasciculin can be stored at 4°C for 3 weeks with no loss of its binding properties towards AChE or anti-fasciculin serum; however, one should take into account the radioactivity decay. This method is set up for a SRA of 1500 Ci mmol^{-1}. Should a lower value be needed, one can either use less ^{125}I-Na (radioiodination product will be a mixture of ^{125}I-fasciculin and unlabelled fasciculin), or an isotopic dilution with ^{127}I-Na (radioiodination product will be a mixture of ^{125}I-fasciculin and iodinated, but not radiolabelled, ^{127}I-fasciculin).

5.2 Binding on membrane-bound AChE

Binding of ^{125}I-fasciculin to membrane-bound AChE can be examined using the centrifugation technique. The method described below was set up for synaptosomal rat brain AChE (P_2) but can be used with any membrane-bound AChE preparation of a low specific AChE activity (Marchot et al., 1993).

Binding assays are performed at 25°C in 1.5-ml tubes.

Protocol 6 Binding assay

1. To 10 mM Tris-HCl, pH 7.5, 100 mM NaCl, 0.1 mg/ml BSA buffer, add successively ^{125}I-fasciculin (various concentrations) and P_2 (0.1 mg/ml) to a final volume of 250 µl.

   Notes:
   For determination of nonspecific binding (parallel assay), add unlabelled fasciculin to a 1000-fold excess over the highest ^{125}I-fasciculin concentration.

2. After incubation, tubes are centrifuged at 11,000 x g for 5 min; the supernatant is kept for quantification of the free (unbound) ^{125}I-fasciculin.

3. The pellet is washed twice with 1 ml of ice-cold buffer and then counted for γ-radioactivity.

   Notes:
   Data points should correspond to duplicates that differ by less than 10% and from which nonspecific binding has been subtracted.

Notes:

- The centrifugation technique requires that the membrane pellet be detected at the bottom of the tube; small size, totally transparent tubes should be used.
- Should a high activity preparation be available, it might be worth diluting the pellet with other membranes; in this case, however, check carefully for nonspecific binding.
- To avoid dissociation of the ^{125}I-fasciculin-AChE complex during the procedure, be sure to washing pellets with ice-chilled buffer and as quickly as possible (do not mix or incubate); as much as possible, use a refrigerated centrifuge (or place it in the coldroom).

5.3 Binding on soluble AChE

Binding of ^{125}I-fasciculin to soluble, or solubilised AChE cannot be examined by the centrifugation technique; instead, a rapid filtration assay must be used. The choice of a filtration membrane that permits both a minimal binding of the free (unbound) ^{125}I-fasciculin (nonspecific binding) and an optimal recovery of the ^{125}I-fasciculin-AChE complex is critical. A positively charged membrane will fulfil the need of not binding the cationic fasciculin, but other specifications of the membrane must be carefully investigated depending on the chemical characteristics of the AChE species in use. Thus, a cationic polysulfone-based membrane initially designed for oligonucleotide blots was found to satisfactorily bind the soluble, recombinant form of AChE from mouse (mAChE), of pI 6.0–6.5 (Marchot et al., 1995); however, this membrane may yield poor results with an AChE species of a higher pI value or of a different hydrophobicity.

The technical principle of the filtration assay is the same as for the centrifugation assay. Briefly, after incubation of ^{125}I-fasciculin and AChE (or, for determination of nonspecific binding: ^{125}I-fasciculin, unlabelled fasciculin, and AChE), the liquid in each tube is quickly filtered (under vacuum) through the premoistened (ice-cold buffer) membrane; the filtrate is kept for determination of free ^{125}I-fasciculin; still under vacuum, and using the incubation tube, the membrane is washed several times (optimal number of washes to be determined previously) with ice-cold buffer, dried, and counted for γ-radioactivity.

Notes:

- Presoaking the filtration membrane in the assay buffer complemented with BSA at 1.0 mg ml^{-1} may help reduce nonspecific binding.

5.4 Liquid-phase radioimmunoassay (RIA)

Protocol 7 Liquid-phase radioimmunoassay

1. In a 1.5-ml tube, incubate for 90 min at 37°C:
   - ^{125}I-fasciculin (10,000 cpm in 50 μl) and the anti-fasciculin serum at its titre (added as 50 μl of a 10-fold-the-titre solution) in 200 μl of 50 mM $NaPO_4$, pH 7.5, 0.1 mg/ml BSA.
   - Dilute to 500 μl with buffer and competitor, incubate overnight at 4°C.
2. For double immunoprecipitation at 4°C of the antigen-antibody complex add successively:
   - Ice-cold normal rabbit serum (50 μl of a 1/50 dilution) and goat anti-rabbit serum (50 μl of a 1/20 dilution); incubate for 90 min.
   - 400 μl of a 20% (w/v) PEG 8000 solution (extemporaneously prepared), incubate for 90 min.
   - Tube is then centrifuged (10,000 x g, 25 min, 4°C), the supernatant sucked out, and the pellet counted for γ-radioactivity.

   Notes:
   The serum titre is the dilution which binds 50% of the ^{125}I-fasciculin added to the sample; this value is to be determined with a fixed concentration of ^{125}I-fasciculin and a series of concentrations of the serum.

5.5 Data processing

Saturation binding to a single site (Scatchard analysis)

Scatchard transformations of binding data bias the experimental error; it is therefore worth analysing the data by nonlinear regression to fit the equation:

$$Y = \frac{B_{max}X}{K_d + X}$$

GraphPad PRISM equation:

Y = Bmax * X / (Kd + X)
; Y is in fmol mg^{-1}
; X, the free ^{125}I-fasciculin, and K_d are in nM

However, once the data are processed by nonlinear regression, it is useful to use a Scatchard plot of the data (B / F versus B) to show a change in B_{max} or K_d more clearly.

Dissociation process

Initially, ^{125}I-fasciculin and AChE are allowed to associate, although not necessarily to equilibrium, then further binding is blocked by dilution of the incubation mixture or by adding a very large excess of unlabelled fasciculin. Binding is then measured at various times and analysed as a one-phase exponential decay according to the equation

$$B = B_{max}e^{-k_{off}t} + B\infty$$

GraphPad PRISM equation:

Y = Span * exp(-koff * X) + Plateau
; Y = B is in fmol mg^{-1}
; X = t is time in min
; Plateau = $B\infty$ is the new equilibrium reached
; Starts at Span + Plateau and decays to Plateau with a first-order rate constant k_{off} in min^{-1}; The half-life is Log2 / k_{off} = 0.69 / k_{off}

Association process

Association data can be analysed as a single exponential according to equation

$$B = B_{max}\left(1 - e^{-kt}\right)$$

where B and B_{max} are the amounts of ^{125}I-fasciculin specifically bound at all times and at equilibrium (limit), respectively, and k is the pseudo-first-order rate constant of the association process.

GraphPad PRISM equation:

Y = Ymax * (1 – exp(-k * X))
; Y = B is in fmol mg^{-1}
; X = t is time in min
; Starts at zero and ascends to B_{max} with a rate constant k in min^{-1}
; Half time is Log2 / k = 0.69 / k.

k is related to the second-order rate constant of association process (k_{on}), the ^{125}I-fasciculin concentration, and the first-order rate constant of dissociation process (k_{off}); thus, the data from destruct experiments, performed with varying [I], can be plotted as:

$$k = k_{on}[I] + k_{off}$$

Competitive binding

Competition experiments are performed by quantifying, with the use of ^{125}I-fasciculin, the number of binding sites which are *not* occupied by the competitor (another peripheral site ligand, cf. Marchot et al., 1993). The competitor is first allowed to react for 30 min with AChE, then the initial binding rate of ^{125}I-fasciculin (v, cpm min^{-1}) is determined and compared with that of a control experiment performed in the absence of competitor (vo). Assuming that the time required for the determination of the initial binding rate of ^{125}I-fasciculin is short enough (<10 min) not to disturb the competitor equilibrium binding to a mutually exclusive site on AChE, the competitor concentration yielding the half-effect (EC_{50}) is the dissociation constant (K_d).

Data can be analysed as an equilateral hyperbola.

GraphPad PRISM equation:

Y = Bottom + (Top – Bottom) / (1 + 10^(X-LogEC$_{50}$))
; X = log(Concentration)
; Y = v / vo

RIA

Normalised data (B / Bo) can be analysed as an equilateral hyperbola.

GraphPad PRISM equation:

Y = Bottom + (Top – Bottom) / (1 + 10^(X – LogEC$_{50}$))
; X = log(Concentration)
; Y = B / Bo

Acknowledgements

The authors are grateful to Dr. Palmer Taylor (University of California at San Diego, La Jolla) and Dr. Evert Karlsson (Biomedical Center, Uppsala) for critical reading of the manuscript and helpful suggestions.

References

Ackermann WW, Potter VR (1949) Enzyme inhibition in relation to chemotherapy. Proc. Soc. Exp. Biol. Med. 72: 1–9

Adem A, Åsblom A, Johansson G et al (1988) Toxins from the venom of the green mamba Dendroaspis angusticeps that inhibit the binding of quinuclidinyl-benzilate to muscarinic acetylcholine receptors. Biochem. Biophys. Acta 968: 340–345

Albrand JP, Blackledge MJ, Pascaud F et al (1995) NMR and restrained molecular dynamics study of the three-dimensional solution structure of toxin FS2: a specific blocker of the L-type calcium channel, isolated from black mamba venom. Biochemistry 34: 5923–5937

Augustinsson KB, Eriksson H, Faijersson Y (1978) A new approach to determining cholinesterase activities in samples of whole blood. Clin. Chim. Acta 89: 239–252

Bougis P, Rochat H, Piéroni G, Verger R (1981) Penetration of phospholipid monolayers by cardiotoxins. Biochemistry 20: 4915–4920

Bougis PE, Marchot P, Rochat H (1986) Characterization of Elapidae snake venom components using optimized reverse-phase high performance liquid chromatographic conditions and screening assays for α-neurotoxin and phospholipase A_2 activities. Biochemistry 25: 7235–7243

Bourne Y, Taylor P, Marchot P (1995) Acetylcholinesterase inhibition by fasciculin: crystal structure of the complex. Cell 83: 503–512

Cerveñansky C, Dajas F, Harvey AL, Karlsson E (1991) Fasciculins, anticholinesterase toxins from mamba venoms: biochemistry and pharmacology. Snake Toxins (Harvey AL, ed) pp 303–321: New York: Pergamon Press

Cerveñansky C, Engström A, Karlsson E (1994) Study of structure-activity relationship of fasciculin by acetylation of amino groups. Biochem. Biophys. Acta 1199: 1–5

Cerveñansky C, Engström A, Karlsson E (1995) Role of arginine residues for the activity of fasciculin. Eur. J. Biochem. 229: 270–275

Cerveñansky C, Duran R, Karlsson E (1996) Fasciculin: modification of carboxyl groups and discussion of structure-activity relationship. Toxicon 34: 718–721

Changeux JP, Kasai M, Lee CY (1970) The use of a snake venom toxin to characterize the cholinergic receptor protein. Proc. Natl. Acad. Sci. USA 67: 1241–1247

Chiappinelli VA (1983) Kappa-bungarotoxin: a probe for the neuronal nicotinic receptor in the avian ciliary ganglion. Brain Res. 277: 9–22

Cousin X, Bon S, Duval N et al (1996) Cloning and expression of acetylcholinesterase from Bungarus fasciatus venom. A new type of COOH-terminal domain; involvement of a positively charged residue in the peripheral site. J. Biol. Chem. 271: 15099–15108

Cousin X, Hotelier T, Giles K et al (1997) The α/β-fold family of protein database and the cholinesterase gene server ESTHER. Nucleic Acids Res. 25: 143–146

Cousin X, Hotelier T, Giles K et al (1998) aCHEdb: the database system for ESTHER, the α/β fold family of proteins and the Cholinesterase gene server. Nucleic Acids Res. 26: 226–228

de Weille JR, Schweitz H, Maes P et al (1991) Calciseptine, a peptide isolated from black mamba venom, is a specific blocker of the L-type calcium channel. Proc. Natl. Acad. Sci. USA 88: 2437–2440

Dixon M (1972) The graphical determination of K_M and K_i. Biochem. J. 129: 197–202

Dufton MJ, Hider RC (1991) The structure and pharmacology of elapid cytotoxins. Snake Toxins (Harvey AL, ed) pp 259–302: New York: Pergamon Press

Duran R, Cerveñansky C, Dajas F, Tipton KF (1994) Fasciculin inhibition of acetylcholinesterase is prevented by chemical modification of the enzyme at a peripheral site. Biochem. Biophys. Acta 1201: 381–388

Duran R, Cerveñansky C, Karlsson E (1996) Effect of fasciculin on hydrolysis of neutral and choline esters by butyrylcholinesterase, cobra venom and chicken acetylcholinesterases. Toxicon 34: 959–963

Eastman J, Wilson EJ, Cerveñansky C, Rosenberry TL (1995) Fasciculin 2 binds to the peripheral site on acetylcholinesterase and inhibits substrate hydrolysis by slowing a step involving proton transfer during enzyme acylation. J. Biol. Chem. 270: 19694–19701

Ellman GL, Courtney KD, Andres V Jr, Featherstone RM (1961) A new and rapid colorimetric determination of acetylcholinesterase activity. Biochem. Pharmacol. 7: 88–95

Endo T, Tamiya N (1991) Structure-function relationships of postsynaptic neurotoxins from snake venoms. Snake Toxins (Harvey AL, ed) pp. 165–222: New York: Pergamon Press

Frobert Y, Créminon C, Cousin X et al (1997) Acetylcholinesterases from Elapidae snake venoms: biochemical, immunological, and enzymatic characterization. Biochem. Biophys. Acta 1339: 253–267

Futerman AH, Fiorini R-M, Roth E et al (1985) Physicochemical behavior and structural characteristics of membrane-bound acetylcholinesterase from Torpedo electric organ. Biochem J. 226: 369–377

Grassi J, Frobert Y, Lamourette P, Lagoutte B (1988) Screening of monoclonal antibodies using antigens labeled with acetylcholinesterase: application to the peripheral protein of photosystem 1. Anal. Biochem. 168: 436–450

Gray EG, Whittaker VP (1962) J. Anat. Lond. 96: 79–88

Greco WR, Hakala M (1979) Evaluation of methods for estimating the dissociation constant of tight binding enzyme inhibitors. J. Biol. Chem. 254: 12104–12109

Harel M, Kleywegt GJ, Ravelli RBG et al (1995) Crystal structure of an acetylcholinesterase-fasciculin complex: interaction of a three-fingered toxin from snake venom with its target. Structure 3: 1355–1366

Harvey AL, Anderson AJ, Mbugua PM, Karlsson E (1984) Toxins from mamba venoms that facilitate neuromuscular transmission. J. Toxicol.-Toxin Reviews 3: 91–137

Heukeshaven, J, Dernick, R (1988) Increased sensitivity for coomassie staining of sodium dodecyl sulfate-polyacrylamide gels using PhastSystem development unit. Electrophoresis 9: 60–61

Israel M, Frachon P (1970) Achives d'Anatomie Microscopique 59: 383

Joubert FJ, Taljaard N (1978) The complete primary structure of toxin C from Dendroaspis polylepis polylepis (black mamba) venom. S. Afr. J. Chem. 31: 107–110

Karlsson E, Mbugua PM, Rodriguez-Ithurralde D (1984) Fasciculins, anticholinesterase toxins from the venom of the green mamba Dendroaspis angusticeps. J. Physiol. Paris 79: 232–240

Karlsson E, Cerveñansky C, Jolkkonen M (1990) Chemistry of mamba toxins. Toxicon 28: p. 154

Le Du MH, Marchot P, Bougis PE, Fontecilla-Camps JC (1989) Crystals of fasciculin 2 from green mamba snake venom: preparation and preliminary X-ray analysis. J. Biol. Chem. 264: 21401–21402

Le Du MH, Marchot P, Bougis PE, Fontecilla-Camps JC (1992) 1.9-Å resolution structure of fasciculin 1: an anti-acetylcholinesterase toxin from green mamba snake venom. J. Biol. Chem. 267: 22122–22130

Le Du MH, Housset D, Marchot P et al (1996) Structure of fasciculin 2 from green mamba snake venom: evidence for unusual loop flexibility. Acta Crystallogr. D52: 87–92

Lee CY, Chang CC (1966) Modes of action of purified toxins from elapid venoms on neuromuscular transmission. Mem. Inst. Butantan 33: 555–572

Lin WW, Lee CY, Carlsson FHH, Joubert FJ (1987) Anticholinesterase activity of angusticeps-type toxins and protease inhibitors homologues from mamba venoms. Asia Pacific J. Pharmacol. 2: 79–85

Marchot P (1986) Contribution à l'étude de la purification et du mode d'action des toxines des venins de serpents Elapidae. Thèse de Doctorat de l'Université d'Aix-Marseille II, Spécialité Neurosciences

Marchot P, Khélif A, Ji Y-H et al (1993) Binding of ^{125}I-fasciculin to rat brain acetylcholinesterase. The complex still binds diisopropyl fluorophosphate. J. Biol. Chem. 268: 12458–12467

Marchot P, Camp S, Radic Z et al (1995) Structural determinants of fasciculin specificity for acetylcholinesterase. Enzymes of the Cholinesterase Family (Quinn DM, Balasubramanian AS, Doctor BP, Taylor P, eds) pp 197–202: New York: Plenum

Marchot P, Ravelli, RBG, Raves ML et al (1996) Soluble monomeric acetylcholinesterase from mouse: Expression, purification, and crystallization in complex with fasciculin. Protein Sci. 5: 672–679

Marchot P, Prowse CN, Kanter J et al (1997) Expression and inhibitory activity of mutants of fasciculin, a peptidic acetylcholinesterase inhibitor from mamba venom. J. Biol. Chem. 272: 3502–3510

Marchot P, Bourne Y, Prowse CN et al (1998) Crystal structure of the fasciculin-mouse acetylcholinesterase complex and mutagenesis of fasciculin. Structure and Function of Cholinesterases and Related Proteins (Doctor BP, Quinn DM, Rotundo RL, Taylor P, eds) New York: Plenum (in press)

Massoulié J, Bon S (1976) Affinity chromatography of acetylcholinesterase. The importance of hydrophobic interactions. Eur. J. Biochem. 68: 531–539

Massoulié J, Pezzementi L, Bon S et al (1993) Molecular and cellular biology of cholinesterases. Progr. Neurobiol. 41: 31–91

McDowell RS, Dennis MS, Louie A et al (1992) Mambin, a potent glycoprotein IIb-IIIa antagonist and platelet aggregation inhibitor structurally related to the short neurotoxins. Biochemistry 31: 4766–4772

Morel N, Israel M, Monaranche R, Mastour-Frachon P (1977) Isolation of pure cholinergic nerve endings from Torpedo electric organ. J. Cell. Biol. 75: 43–55

Oswald RE, Sutcliffe MJ, Bamberger M et al (1991) Solution structure of neuronal bungarotoxin determined by two-dimensional NMR spectroscopy: sequence-specific assignments, secondary structure, and dimer formation. Biochemistry 30: 4901–4909

Puu G, Koch M, (1990) Comparison of kinetic parameters for acetylcholine, soman, ketamine, and fasciculin towards acetylcholinesterase in liposomes and in solution. Biochem. Pharmacol. 40: 2209–2214

Radic Z, Duran R, Vellom DC et al (1994) Site of fasciculin interaction with acetylcholinesterase.J. Biol. Chem. 269: 11233–11239

Radic Z, Quinn DM, Vellom DC et al (1995) Allosteric control of acetylcholinesterase catalysis by fasciculin. J. Biol. Chem. 270: 20391–20399 (correction in J. Biol. Chem. N52: 31414)

Radic Z, Kirchhoff P, Quinn DM et al (1997) Electrostatic influence on the kinetics of ligand binding to acetylcholinesterase: distinctions between active center ligands and fasciculin. J. Biol. Chem. 272: 23265–23277

Reed JK, Raftery MA (1976) Properties of the tetrodotoxin binding component in plasma membranes isolated from Electrophorus electricus. Biochemistry 15: 944–953

Rémy MH, Frobert Y, Grassi J (1995) Characterization of monoclonal antibodies that strongly inhibit Electrophorus electricus acetylcholinesterase. Eur. J. Biochem. 231: 651–658

Rodriguez-Ithurralde D, Silveira R, Barbeito L, Dajas F (1983) Fasciculin, a powerful anticholinesterase polypeptide from Dendroaspis angusticeps venom. Neurochem. Int. 5: 267–274

Rosenberry TL, Rabl C-R, Neumann E (1996) Binding of the neurotoxin fasciculin 2 to the acetylcholinesterase peripheral site drastically reduces the association and dissociation rate constants for N-methylacridinium binding to the active site. Biochemistry 35: 685–690

Segalas I, Roumestand C, Zinn-Justin S et al (1995) Solution structure of a green mamba toxin that activates muscarinic acetylcholine receptors, as studied by nuclear magnetic resonance and molecular modeling. Biochemistry 34: 1248–1260

Stadler H, Tashiro T (1979) Isolation of synaptosomal plasma membranes from cholinergic nerve terminals and a comparison of their proteins with those of synaptic vesicles. Eur. J. Biochem. 101: 171–178

Strydom DJ (1976) Snake venom toxins. Purification and properties of low-molecular-weight polypeptides of Dendroaspis polylepis polylepis (black mamba) venom. Eur. J. Biochem. 69: 169–176

Sussman JL, Harel M, Frolow F et al (1988) Purification and crystallization of a dimeric form of acetylcholinesterase from Torpedo californica subsequent to solubilisation with phosphatidylinositol-specific phospholipase C. J. Mol. Biol. 203: 821–823

Sussman JL, Harel M, Frolow F et al (1991) Atomic structure of acetylcholinesterase from Torpedo californica: a prototypic acetylcholine-binding protein. Science 253: 872–879

Sutcliffe MJ, Jaseja M, Hyde EI et al (1994) Three-dimensional structure of the RGD-containing neurotoxin homologue dendroaspin. Nature Struct Biol 1: 802–807

Taylor P, Radic Z (1994) The cholinesterases: from genes to proteins. Annu. Rev. Pharmacol. Toxicol. 34: 281–320

Van den Born HKL, Radic Z, Marchot P et al (1995) Theoretical analysis of the structure of the peptide fasciculin and its docking to acetylcholinesterase. Protein Sci. 4: 703–715

Viljoen CC, Botes DP (1973) The purification and amino acid sequence of toxin F_{VII} from Dendroaspis angusticeps venom. J. Biol. Chem. 248: 4915–4919

17 Dendrotoxins and BPTI-like Proteins

Alan L. Harvey and Edward G. Rowan

Contents

1 Introduction

The dendrotoxins are a family of homologous proteins that are isolated from mamba snake venoms: namely, α-dendrotoxin (α-Dtx), β_1-Dtx, β_2-Dtx, γ-Dtx, and δ-Dtx from the venom of the Eastern green mamba (*Dendroaspis angusticeps*), two closely related peptides, toxin I and toxin K, from the venom of the black mamba (*Dendroaspis polylepis polylepis*), and Dv14 from Western green mamba (*Dendroaspis viridis*) (Harvey and Karlsson, 1980; 1982; Joubert and Taljaard, 1980; Strydom, 1973; Benishin *et al.*, 1988; Harvey and Anderson, 1991). They consist of 57–60 amino acid residues in single polypeptide chains, containing three intramolecular disulfide bonds. The crystal structure of α-dendrotoxin has been solved (Skarzynski, 1992). Dendrotoxins are highly homologous to Kunitz-type serine proteinase inhibitors, such as bovine pancreatic trypsin inhibitor (BPTI or aprotinin) (see Fig. 1). However, BPTI does not have K^+ channel-blocking properties (Marshall and Harvey, 1992), and

Methods and Tools in Biosciences and Medicine
Animal Toxins, ed. by H. Rochat and M.-F. Martin-Eauclaire

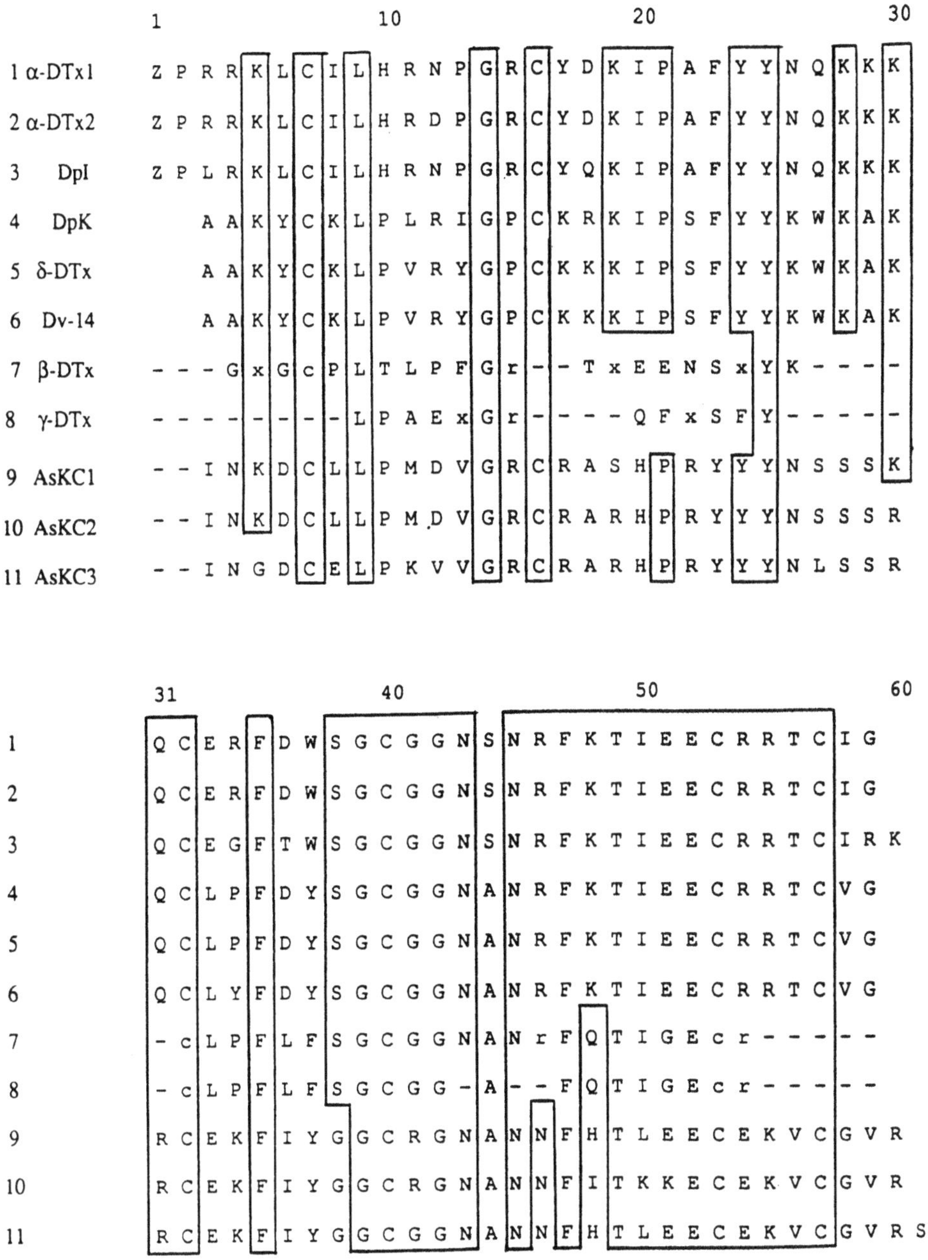

Figure 1 Amino acid sequences of dendrotoxins and kalicludines. α-DTx1 and 2 refer to α-dendrotoxins from Eastern green mamba (*Dendroaspis angusticeps*) venom – see Danse *et al.* (1994).

DpI and DpK are from black mamba (*Dendroaspis polylepis*) venom; Dv-14 is from Western green mamba (*Dendroaspis viridis*) venom. δ-DTx, β-DTx, and γ-DTx are from *D. angusticeps* venom; blanks or "x" indicate uncertainties in the sequence determination (Benishin *et al.*, 1988). AsKC1, 2 and 3 are the kalicludines from the sea anemone *Anemone sulcata* (Schweitz *et al.*, 1995).

although dendrotoxins inhibit some proteinases (Marshall and Harvey, 1990), this may be unrelated to their ability to block K^+ channels. More recently, three dendrotoxin homologues, the kalicludines, were isolated from the sea anemone *Anemonia sulcata* (Schweitz *et al.*, 1995). They are several orders of magnitude less active than toxin I, and they may be more closely related to proteinase inhibitors.

The dendrotoxins from mamba venom appear to be equivalent in their actions at the neuromuscular junction (Harvey and Karlsson, 1982), but there are at least two subgroups based on their effects on different components of $^{86}Rb^+$ flux from synaptosomes (Benishin *et al.*, 1988). Additional differences in binding properties have also been found (Muniz *et al.*, 1990; Awan and Dolly, 1991).

Dendrotoxins have been shown to block some types of K^+ channels but not others (for reviews see Moczydlowski *et al.*, 1988; Harvey and Anderson, 1991). In general, they block some of the voltage-dependent outward K^+ currents (mainly on neurones), but are inactive on inward rectifying K^+ channels or on Ca^{2+}-activated K^+ channels. They are active in the nanomolar range, and high-affinity binding sites have been detected in the brain. They are not very toxic on systemic administration, but their toxicity increases 10,000-fold on direct injection into brain. The dendrotoxins induce repetitive firing of neurons and also enhance transmitter release.

Facilitation of transmitter release was the first effect of dendrotoxins to be characterized (Harvey and Karlsson, 1980, 1982; Harvey *et al.*, 1984; Anderson 1985; Anderson and Harvey 1988), following observations of this effect of green mamba (*Dendroaspis angusticeps*) venom (Barrett and Harvey, 1979). Subsequently, dendrotoxins were found to specifically block some neuronal K^+ channels. In hippocampal pyramidal cells, dendrotoxin inhibits a portion of the transient K^+ current (Dolly *et al.*, 1984; Halliwell *et al.*, 1986). An equivalent effect has been demonstrated by measuring $^{86}Rb^+$ efflux from synaptosomes (Benshin *et al.*, 1988) and by extracellular recording from nerve terminals in rat olfactory cortex (McGivern *et al.*, 1993). In nodes of Ranvier of frog sciatic nerves, dendrotoxin prolonged the duration of action potentials by blocking a fraction of the K^+ current (Weller *et al.*, 1985), which resulted from a specific block of the f_1 component of the K^+ current (Benoit and Dubois, 1986; Brau *et al.*, 1990). This has been confirmed by single-channel recording experiments on *Xenopus* axons (Jonas *et al.*, 1989), where the single-channel conductance blocked by dendrotoxin was 23pS. The paranodal regions of rat motor nerve fibres also have similar K^+ currents that are very sensitive to dendrotoxin (Corrette *et al.*, 1991).

K^+ channels have been cloned and expressed in *Xenopus* oocytes or mammalian cell lines, and some of the neuronally derived channels are sensitive to dendrotoxin (see, e.g., Stuhmer *et al.*, 1989; Grissmer *et al.*, 1994). Dendrotoxin blocks some K^+ currents (Kv 1.1; Kv 1.2 and Kv 1.6: $K_d < 20$ nM) derived from complementary DNA (cDNA) (Werkman *et al.*, 1992; Grissmer *et al.*, 1994), whereas other channels (Kv1.5, Kv3.1) are insensitive to micromolar concentrations of dendrotoxin.

Radiolabelled dendrotoxin has also been used to isolate binding proteins from brain membranes (e.g., Black *et al.*, 1988; Rehm and Lazdunski, 1988;

Parcej and Dolly, 1989; Sorensen and Blaustein, 1989). The N-terminal sequence of an isolated dendrotoxin-binding protein is highly homologous to the sequence of rat cortex K^+ channel 5 (RCK5, now known as rKv 1.2) deduced from the cDNA cloned from rat brain (Scott *et al.*, 1990). Proteins isolated on a dendrotoxin affinity column were reconstituted in lipid bilayers (Rehm *et al.*, 1989a). The resulting K^+ channel activity was blockable by dendrotoxin. Additionally, antibodies against parts of the sequence of the mouse brain K^+ channel 1 (MBK1 or mKv 1.1) cross-react with dendrotoxin-binding proteins (Rehm *et al.*, 1989b). By comparing sequences of dendrotoxin-sensitive and -insensitive clones expressed in oocytes and by making specific mutations, the binding site for dendrotoxin can be localized on the K^+ channel polypeptide (Hurst *et al.*, 1991; Tytgat *et al.*, 1995). Mutations in the H5 loop between transmembrane domains S5 and S6 reduced the binding affinity of dendrotoxin. The particular amino acids involved were Ala 352, Glu 353, and Tyr 379. All four subunits in the K^+ channel appear to interact simultaneously with dendrotoxin, and only one toxin molecule is required to block the channel.

Dendrotoxins, therefore, are valuable tools for the study of voltage-dependent K^+ channels in neurones.

In this chapter, we discuss the commercially available dendrotoxins and give protocols for

(i) competition binding studies with radioiodinated dendrotoxin, (ii) assessment of neuromuscular facilitation using the chick biventer cervicis nerve-muscle preparation, and (iii) assessment of effects on motor nerve terminals using the mouse triangularis sterni nerve-muscle preparation.

2 Materials and Chemicals

2.1 Buffers and reagents

- **Krebs-Henseleit solution (Solution 1)**
  - 118.4 mM NaC1
  - 4.7 mM KC1
  - 1.2 mM $MgSO_4$
  - 1.2 mM KH_2PO_4
  - 2.5 mM $CaC1_2$
  - 25 mM $NaCHO_3$
  - 11.1 mM glucose
- **High Mg/low Ca recording solution (Solution 2)**
  - 118.4 mM NaC1
  - 4.7 mM KC1
  - 2.0–6.0 mM $MgSO_4$ (as required)
  - 1.2 mM KH_2PO_4
  - 0.5 mM $CaC1_2$
  - 25 mM $NaCHO_3$
  - 11.1 mM glucose

- **HEPES physiological salt solution (Solution 3)**
  154 mM NaC1
  5.0 mM KC1
  1.2 nM $MgC1_2$
  2.5 mM $CaC1_2$
  11.1 mM glucose
  5.0 (pH 7.4) mM Na-HEPES
- **Synaptosomal buffer (Solution 4)**
  130 mM NaC1
  3 mM KC1
  2 mM $CaC1_2$
  2 nM $MgC1_2$
  20 (pH 7.4) mM Tris HCl
- **Homogenisation buffer (Solution 5)**
  320 mM sucrose
  2 (pH 7.4) mM Tris HCl
- **Lysis buffer (Solution 6)**
  5 mM Tris HCl (pH 8.1)
  Acetylcholine chloride: 10^{-1} M in water
  Carbachol chloride: 10^{-3} M in water
  KC1: 1 M in water
  Tubocurarine chloride: 10^{-3} M in water

2.2 Chemicals

Acetylcholine chloride	Sigma	A6625
Carbachol chloride	Sigma	C4382
μ-Conotoxin GIIIB	Peptide Institute	4217-v
Dendroaspis angusticeps venom	Sigma	V7750
	Latoxan	L13 07
Dendroaspis polylepis venom	Sigma	V8000
	Latoxan	L13 09
Dendroaspis viridis venom	Latoxan	L13 10
α-Dendrotoxin	Alomone Laboratories	D-350
	Latoxan	L81 06
	Calbiochem	253704
β-Dendrotoxin	Alomone Laboratories	D-360
	Calbiochem	253706
γ-Dendrotoxin	Alomone Laboratories	D-370
	Calbiochem	253711
δ-Dendrotoxin	Alomone Laboratories	D-380
	Calbiochem	253708;
Dendrotoxin I	Alomone Laboratories	D-390
	Calbiochem	253703
^{125}I α-Dendrotoxin	Amersham International	IM 267
Sylgard 184 silcone elastomer base and curing agent	Dow Corning	
Tubocurarine chloride	Sigma	T2379

3 Commercially Available Dendrotoxins from Mamba Venoms

Originally, dendrotoxins had to be isolated from mamba venoms by individual laboratories. Currently, several dendrotoxins are available from commercial sources. These include the four dendrotoxins from *D. angusticeps* (green mamba) venom and dendrotoxin I from *Dendroaspis polylepis* (black mamba) venom. Synthetic dendrotoxin I is available from the Peptide Institute of Osaka.

With all toxins, care has to be taken about the purity of the toxin and about the concentration of toxin in solution. Apart from the product from the Peptide Institute, all commercially available dendrotoxins have been isolated from venoms. This means that there is the potential for some contamination of the dendrotoxin with small amounts of other components. Purity of about 99% is possible after purification using HPLC, and the effects of any contaminating material may not be significant. However, it is possible for traces of curaremimetic α-neurotoxins to mask some of the facilitatory effects of dendrotoxins in experiments at the neuromuscular junction. This is more likely to be a problem with dendrotoxins isolated from black mamba venom than with dendrotoxins from green mamba venom, because the former venom is much richer in α-neurotoxins than is the latter.

As with all freeze-dried protein toxins, the dendrotoxins may absorb water on storage, and they may also contain varying amounts of salts from the freeze-drying process. Therefore, the concentrations of toxins in solution should not be calculated from the weight of dried toxin that has been dissolved. It is more reliable to measure the absorbance of the toxin-containing solution and calculate the concentration of toxin from its molar absorbance. At 280 nm, the absorbances of 1 mg/ml solutions are 1.48 for both α-dendrotoxin and dendrotoxin I.

Solutions of dendrotoxins in water are very stable, particularly when the concentration of toxin is more than 0.1 mg/ml. Such solutions can be stored reliably either frozen or at 4°C.

Mamba venoms are available from a number of commercial sources. Dendrotoxins can be isolated by use of standard chromatographic procedures, as described by Harvey and Karlsson (1980) and by Sorensen and Blaustein (1992).

4 Competition Binding Studies with Radioiodinated Dendrotoxin

Protocol 1 Preparation of synaptosomal membranes

1 Rats are killed by cervical dislocation, and decapitated.

2 Rapidly remove brains (apart from the cerebellum) and rinse in ice-cold synaptosomal buffer.

3 Homogenise the brains in 10 ml ice-cold homogenisation buffer (Solution 5) (8 strokes at 1300 rpm on a Heidolph homogeniser) at 4°C.

4 Centrifuge the suspension at 900 × g for 10 min.

5 Recover the supernatant.

6 Centrifuge the supernatant for 15000 x g for 12min.

7 Recover the pellet.

8 Resuspend the pellet in 10 ml lysis buffer (Solution 6) and incubate for 30 min at 4°C.

9 Centrifuge the suspension at 48000 x g for 20 min at 4°C.

10 Recover the pellet.

11 Resuspend the pellet in 25 ml synaptosomal buffer (Solution 4).

12 Remove an aliquot for protein determination, and store the remainder as 5-ml aliquots at –20°C until needed. They will be approximately 1.85 mg protein/ml.

Competitive binding studies

Radiolabelled ^{125}I-α-dendrotoxin is available at 2000 Ci/mmol from Amersham International; its K_d is 0.7 nM. Specific binding to potassium channels on synaptosomal membranes can be determined (e.g., Black *et al.*, 1988; Rehm and Lazdunski, 1988; Parcej and Dolly 1989; Sorensen and Blaustein, 1989), and interactions with unlabelled compounds can be studied using the simple centrifugation-based assay described below. Each assay point is prepared in duplicate or triplicate.

Protocol 2 Competition binding studies

1. Add 20 pM ^{125}I-α-dendrotoxin (2000 Ci/mmol) to 100 μl synaptosomal membrane suspension in 1-ml Eppendorf tubes, and make up to 0.5 ml with synaptosomal buffer (Solution4) containing 0.1% bovine serum albumin Notes: *this will represent 100% binding.*
2. Prepare similar tubes including cold α-dendrotoxin at 2 nM *Notes: these will estimate nonspecific binding*
3. Prepare similar tubes containing various concentrations of the cold competing ligand.
4. Incubate for 30 min at room temperature on a rotatory agitator.
5. Centrifuge on a microcentrifuge at 13,000 rpm for 3 min.
6. Remove the supernatant and wash the pellet with 0.5 ml synaptosomal buffer (Solution 4) containing BSA.
7. Place the tubes in a calibrated gamma counter to count the membrane-bound radioactivity.

Data from the experiments can be analysed using a variety of curve-fitting software, such as the Equilibrium Binding Data Analysis program (Elsevier-Biosoft).

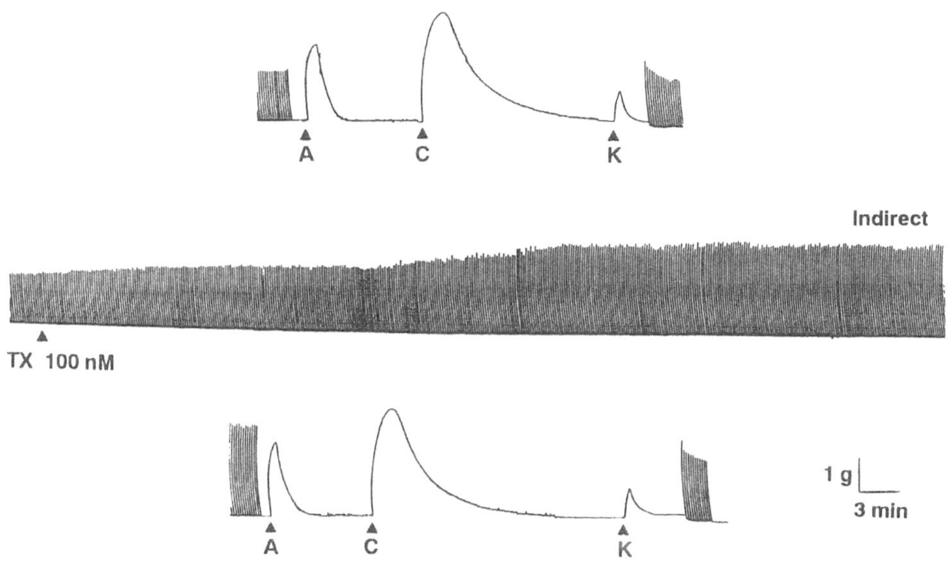

Figure 2 Effect of dendrotoxin on a chick biventer cervicis nerve-muscle preparation.

Upper panel: Control twitch responses to indirect stimulation at 0.1 Hz, and (in the absence of nerve stimulation) control responses to acetylcholine (A, 1 mM), carbachol (C, 40 μM), and KCl (K, 20 mM). Middle panel: Twitch responses are augmented following the addition of the toxin (at TX). Lower panel: After maximum twitch augmentation had been reached, responses to acetylcholine, carbachol, and KCl were unchanged from control.

Protocol 4 Experimental setup

1. Mount pairs of muscles and associated nerves in 2-ml, 5-ml, or 10-ml tissue baths containing Krebs-Henseleit solution (pH 7.3) **(Solution1)** continuously bubbled with oxygen containing 5% CO_2.

   Notes:
   The solution is maintained at a fixed temperature (either 37°C or 32–33°C for more stable responses over a longer period).
2. Place a ring electrode around the tendons and connect the long threads to a force-displacement transducer.
3. Evoke twitches by stimulating the motor nerve at the desired frequency (routinely 0.1 Hz) with pulses of 0.2 ms duration and a voltage greater than that which produces a maximal twitch.
4. Set the resting tension initially at 1 g, but adjust it for each preparation so that nerve stimulation produces the maximum force from the muscle.
5. In the absence of nerve stimulation, obtain contractions to submaximal concentrations of exogenously applied acetylcholine (1–3 x 10^{-3} M), carbachol (1–4 x 10^{-5} M), and KCl (1–4 x 10^{-2} M) prior to the addition of the toxin and at the end of the experiment (Fig. 2). Allow acetylcholine and KCl to remain in contact with the preparations for 30 s and carbachol for 60 s. Wash the preparations by overflow for 15 sec at between 5 and 10 ml/s.
6. After obtaining control responses to agonists, allow preparations to stabilize for about 30 min with continuous nerve stimulation (0.1 Hz) before the addition of toxin. Twitches and contractures are recorded isometrically on suitable ink-writing polygraphs.

If required, preparations can be stimulated directly. Neuromuscular transmission is first abolished by, e.g., tubocurarine (10–20 μM), and the electrodes are placed around the belly of the muscle. Direct muscle stimulation can be evoked by increasing the pulse duration to 1–2 ms and by increasing the strength of the stimulus. As the multiply innervated fibers can give a slowly developing contracture during direct stimulation, it is difficult to achieve a maximal twitch.

6 Assessment of Effects on Motor Nerve Terminals using the Mouse Triangularis Sterni Nerve-Muscle Preparation

For electrophysiological studies on neuromuscular transmission, it is best to have a thin muscle so that end-plate regions can be located by visual identification. Additionally, fibers that are large enough to allow stable impalements for at least 1 h are necessary for experiments with dendrotoxins because effects are often slow in onset. Several preparations have been used: for example, frog sartorius and frog cutaneous pectoris (Dreyer and Peper, 1974), mouse or rat diaphragms, mouse triangularis sterni (McArdle *et al.*, 1981), and mouse levator auris longus (Angaut-Petit *et al.*, 1987). We will describe the triangularis sterni preparation of the mouse.

Dissection

Experiments are performed on the left triangularis sterni (TS) nerve-muscle preparation (McArdle *et al.*, 1981) isolated as described below from mice weighing about 20–25 g. As different strains of mice from different suppliers seem to vary in the pattern of innervation of the TS, it may be necessary to select a strain that provides convenient access to the three nerves innervating the TS. The complete dissection of the muscle with its three nerves is performed under continuous perfusion at a rate of 15–20 ml per min with Krebs-Henseleit solution (aerated with oxygen containing 5% carbon dioxide). A dissecting microscope with transillumination is used to aid dissection.

The TS muscle is located on the inner surface of the rib cage. It is a trapezoidal sheet originating at the sternum and inserting at the intercalations of the ribs. The muscle is bounded by the second and seventh intercostal spaces, which form its inner and outer limits.

Protocol 5 Dissection of triangularis sterni

1. A mouse is killed by cervical dislocation and exsanguination, and the entire rib cage is removed.
2. Pin the rib cage, with the inner ribs uppermost, to a Sylgard-lined petri dish.
3. Carefully clear the inner area of rib of all excess tissue, ensuring that the exposed TS is not damaged.
4. Turn the rib cage over and pin it to the Sylgard base, with the outer ribs uppermost. To ease the dissection, the ribs are fanned out and secured with pins. At this point, two preparations can be made by carefully dividing the preparation longitudinally at the sternum, although we do not normally do this.
5. Carefully remove the muscles at each intercostal space (starting at the abdominal end) in layers until the TS (or usually the nerve branch innervating the muscle) is visible.
6. Remove the remaining overlying muscle a few millimeters proximal to the nerve branch; repeat at each intercostal space.
7. Cut the ribs free of the sternum and cut a few millimeters proximal to the nerve branch, and then lift the ribs free of the preparation. The surface of the TS can be cleaned of any remaining tissue that may cause obstructions.

8. Isolate the separate nerve trunks running parallel to the rib; at least 1 cm of nerve can be dissected out.
9. For experiments, pin the TS nerve-muscle preparation, thoracic side downward, to the base of a 2- to 3-ml Sylgard-lined tissue bath, which is perfused at a rate of 5–10 ml/min with Krebs-Henseleit solution or HEPES-buffered solution. Twitching is abolished by addition of tubocurarine, μ-conotoxin, or by using a high Mg^{2+}/low Ca^{2+} physiological salt solution, as appropriate for the particular experiment.

Electrophysiological recording

Intact nerve-muscle preparations are observed using a binocular compound microscope with a movable stage. To assist the localization of superficial motor nerve endings for electrode placement, preparations are observed at a magnification of ×400–600 using a high-magnification (x40), long working distance objective (Zeiss or Olympus).

Microelectrodes used for intracellular and extracellular recording are prepared from 1.2-mm (external diameter) borosilicate glass capillaries containing internal glass filaments (Clark Electromedical Instruments, type GCl20f). The microelectrodes are pulled on an electrode puller and filled with 2 M NaC1 (resistance 5–15 MΩ for extracellular recording, and 3 M KCl (resistance 10–20 MΩ for intracellular recording.

The potential difference between a silver/silver chloride reference electrode (pellet type) in the bath and the recording electrode is measured by a high-impedance unity gain electrometer, displayed and amplified on a dualbeam storage oscilloscope and simultaneously stored on FM tape or videotape, as described below under *Data Analysis*.

For extracellular recording of nerve terminal action potentials, the intercostal nerves are stimulated via a suction electrode every 2 s with pulses of 50 μs duration and superamaximal voltage. Recording electrodes are placed inside the perineural sheath (near end-plate areas) of one of the branches of an intercostal nerve (see Mallart, 1985; Penner and Dreyer, 1986; Anderson *et al.*, 1988). Usually 20–25 waveforms are recorded at each time period. As the shape of the waveform recorded is directly dependent on the electrode position, waveforms are monitored for 20 min prior to the start of the experiment and continuously from the same site throughout application of toxins and drugs. If the waveform changes by more than 10% during the initial 20 min monitoring period, the electrode position is adjusted or a new site is tried. The microelectrode should be as close as possible to the preterminal branching of the motor nerve. In general, the second negative waveform should be as large as possible (see Fig. 3). The first negative deflection is caused by the Na^+ current, and the second negative deflection reflects the K^+ current at the terminals. This is due to the asymmetrical distribution of Na^+ and K^+ channels in the mammalian motor nerve (see Mallart, 1985; Penner and Dreyer, 1986; Anderson *et al.*, 1988). Sodium ion channels are not present at high density in the nonmyelinated terminal regions; hence, the active depolarization at the last nodes of Ranvier and the heminodal region caused by the inward Na^+ current of the action potential gives rise to a local circuit current that appears to be outward at the terminal regions. Conversely, K^+ channels are found at the terminals, but not at the nodes of Ranvier. The active, outward K^+ current at the terminals creates a local circuit current that is inward at the heminode and final nodes of Ranvier.

We routinely use perineural recording in preference to focal recording from the terminals because the signal is larger, has a better signal-to-noise ratio, and is more stable. This last point is important when working with toxins that have a slow and progressive action on acetylcholine release.

Although a stable Na^+ spike is necessary in order to be confident about changes in the second negative deflection being caused by the toxin rather than change in electrode position, there are other problems of interpretation. For example, low concentrations of tetrodotoxin or μ-conotoxin can cause an apparently selective block of terminal "K^+currents" (Braga *et al.*, 1992). Higher concentrations reveal the expected block of the main Na^+ spike. The explanation is presumably because of a loss of some of the driving force for nerve terminal depolarization.

Dendrotoxins cause little overall change to perineural K^+ currents, but they induce repetitive firing of the motor nerve (Fig. 3).

For intracellular recording, the microelectrode is inserted into a muscle fiber at an end-plate region. Cells are rejected if the initial membrane potential is less than –70 mV, if the membrane potential is not stable, or if the rise time of the end-plate potential or miniature end-plate potential is greater than 1 ms (see Fig. 4). Recordings are made continuously from one end-plate before and during application of toxins. Dendrotoxins induce an increase in quantal content of end-plate potentials and repetitive firing of end-plate potentials (Fig. 4).

Physiological solution (10–20 ml) containing the toxin at the desired concentration is perfused onto the preparation and recycled after aeration.

Most experiments are performed at room temperature, although it should be remembered that dendrotoxins have greater effects in mammalian preparations at body temperature (Anderson and Harvey, 1988). For experiments performed at 30 or 37°C, the perfusing solution is passed through a constant-temperature hot water jacket before entering the tissue bath. To reduce the temperature drop across the preparation, the tissue bath is fitted into a water jacket mounted on the microscope stage. Hot water at the required temperature is pumped through the water jacket in the direction opposite to the flow of the physiological salt solution. This procedure allows countercurrent heat exchange to take place and thus prevent a temperature gradient in the tissue bath.

We have found that nerve terminal function is not well maintained in mouse TS preparations at 37°C in Krebs-Henseleit solution unless flow rate is rapid. Consequently, we often use a HEPES-buffered solution.

Data analysis

All signals collected on the FM or video tape recorder are amplified to approximately 4 V and then digitized by an analog-to-digital converter (CED 502). Each digitized event consists of 512 points collected every 40 μs. These events are then analyzed by an IBM PC computer. The analysis program (SCAN or CDR, written by John Dempster, Department of Physiology and Pharmacology, University of Strathclyde, and available on request) is used to visualize the digitized event and either accept it for further analysis or reject it. The accepted events are used by the computer to generate histograms and averaged records.

For intracellular records, the analysis program determines the amplitude, rise time, and decay time constant of the digitized signals. It can also determine quantal content by three methods (the direct method, by failures, or by analysis of variance).

The amplitude of an event is measured as the difference between 1 ms of preceding baseline and the point of greatest amplitude during the event. The rise time is calculated as the time between points at 10 and 90% of the peak amplitude derived from a straight line fitted to the points between 10 and 90% of the peak. The decay time constant is estimated as the inverse of the slope of a single exponential fitted to the points between 95 and 5% of the peak amplitude on the decay phase of the event.

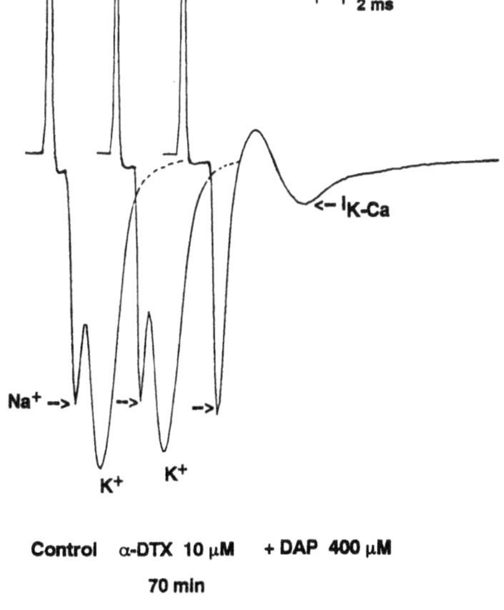

Figure 3 Perineural recordings from a mouse triangularis sterni preparation.

Even at very high concentrations (10 μM), α-dendrotoxin did not produce a marked reduction in the second negative deflection that is associated with K^+ currents at motor nerve terminals (although it caused repetitive firing of action potentials – not shown). Subsequent addition of 3,4-diaminopyridine eliminated the second negative deflection and revealed a slower waveform associated with Ca^{2+} -activated K^+ currents. All recordings were made from the same site; they are shown slightly displaced for clarity. *, stimulus artefact. From an unpublished experiment of Dr. H. Vatanpour.

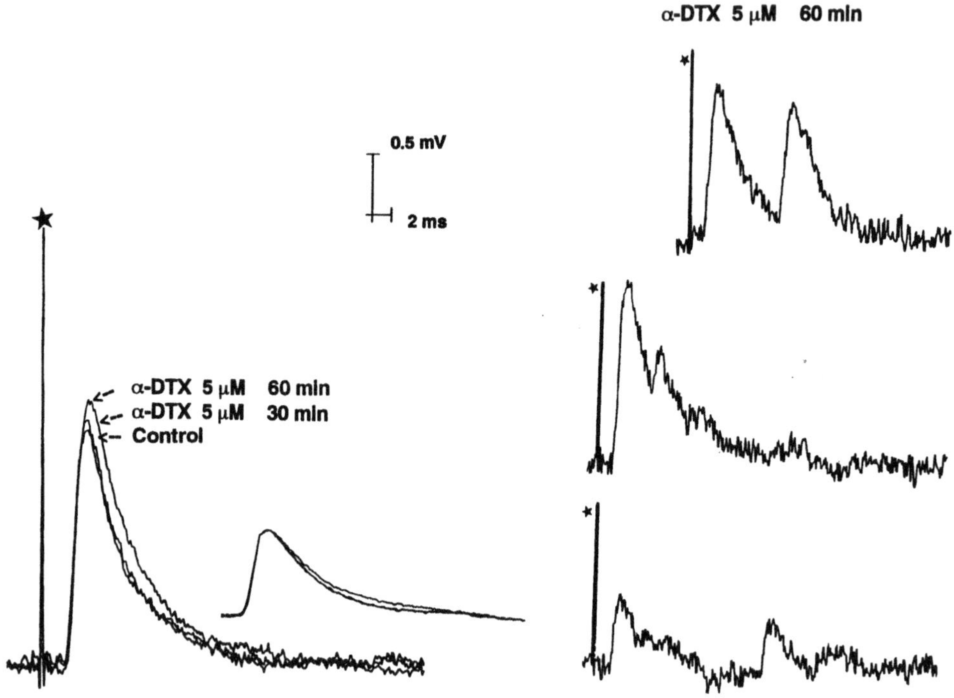

Figure 4 Recordings of end-plate potentials from a mouse triangularis sterni preparation. Left: α-Dendrotoxin produces a small increase in the amplitudes of end-plate potential without changing the amplitude of miniature end-plate potentials (inset). Right: α-Dendrotoxin induces repetitive firing of end-plate potentials. *, stimulus artefact. From an unpublished experiment of Dr. H. Vatanpour.

References

Anderson, A.J. (1985). The effects of protease inhibitor homologues from mamba snake venoms on autonomic neurotransmission. Toxicon 23: 947–954

Anderson, A.J. and Harvey, A.L. (1988). Effects of the potassium channel blocking dendrotoxins on acetylcholine release and motor nerve terminal activity. Br. J. Pharmacol. 93, 215–221

Anderson, A.J., Harvey, A.L., Rowan, E.G. and Strong, P.N. (1988). Effects of charybdotoxin, a blocker of Ca^{2+} -activated K^+ channels, on motor nerve terminals. Brit. J. Pharmacol. 95, 1329–1335

Angaut-Petit, D., Molgo, J., Connold, A.L. and Faille, L. (1987). The levator auris longus muscle of the mouse: a convenient preparation for studies of short- and long-term presynaptic effects of drugs or toxins. Neurosci. Lett. 82, 83–88

Awan, K.A. and Dolly, J.O. (1991). K^+ channel sub-types in rat brain: characteristic locations revealed using β-bungarotoxin, α- and β-dendrotoxins. Neuroscience 40, 29–39

Barrett, J.C. and Harvey, A.L. (1979). Effects of the venom of the green mamba Dendroaspis angusticeps on skeletal muscle and neuromuscular transmission. Br. J. Pharmacol. 167, 199–205

Benishin, C.G., Sorensen, R.G., Brown, W.E., Krueger, B.K. and Blaustein, M.P. (1988). Four polypeptide components of green mamba venom selectively block certain potassium channels in rat brain synaptosomes. Mol. Pharmacol. 34, 152–159

Benoit, E. and Dubois, J.M. (1986). Toxin I from the snake Dendroaspis polylepis polylepis: a high specific blocker of one type of potassium channel in myelinated nerve fibre. Brain Res. 37, 374–377

Black, A.R., Donegan, C.M., Denny, B.J. and Dolly, J.O. (1988). Solubilization and physical characterization of acceptors for dendrotoxin and β-bungarotoxin from synaptic membranes of rat brain. Biochemistry 27, 6814–6820.

Braga, M.F.M., Anderson, A.J., Harvey, A.L. and Rowan, E.G. (1992). Apparent block of K^+ currents in mouse motor nerve terminals by tetrodotoxin, μ-conotoxin and reduced external sodium. Brit. J. Pharmacol. 106, 91–94.

Brau, M.E., Dreyer, F., Jonas, P., Repp, H. and Vogel, W. (1990). A K^+ current in Xenopus nerve fibres selectively blocked by bee and snake toxins: binding and voltage-clamp experiments. J. Physiol. 420, 365–385.

Corrette, B.J., Repp, H., Dreyer, F. and Schwarz, J.R. (1991). Two types of fast K^+ channels in rat myelinated nerve fibres and their sensitivity to dendrotoxin. Pflugers Arch. 418, 408–416.

Danse, J.M., Rowan, E.G., Gasparini, S., Ducancel, F., Vatanpour, H., Young, L.C., Poorheidari, G. Lajeunesse, E., Drevet, P., Menez, R., Pinkasfeld, S., Boulain, J.-C., Harvey A.L. and Menez, A. (1994). On the site by which α-dendrotoxin binds to voltage-dependent potassium channels: site-directed mutagenesis reveals that the lysine triplet 28–30 is not essential for binding. FEBS Lett. 356, 153–158.

Dolly, J.O., Halliwell, J.V., Black, J.D., Williams, R.S., Pelchen-Matthews, A., Breeze, A.L., Mehraban, F., Othman, I.B. and Black, A.R. (1984). Botulinum neurotoxin and dendrotoxin as probes for studies on transmitter release. J. Physiol. (Paris), 79, 280–303.

Dreyer, F. and Peper, K. (1974). A monolayer preparation of innervated skeletal muscle fibres of M. cutaneous pectoris of the frog. Plugers Arch. 348, 257–262.

Ginsborg, B.L. and Warriner, J. (1960). The isolated chick biventer cervicis nerve-muscle preparation. Brit. J. Pharmacol. 15, 410–411.

Grissmer, S., Nguyen, A.N., Aiyar, J., Hanson, D.C., Mather, R.J. and Gutman, G.A. (1994). Pharmacological characterization of 5 cloned voltage-gated K+ channels, types KV1.1, KV1.2, KV1.3, KV1.5, and KV3.1, stably expressed in mammalian cell lines. Mol. Pharmacol. 45, 1227–1234.

Halliwell, J.V., Othman, I.B., Pelchen-Matthews, A. and Dolly, J.O. (1986). Central action of dendrotoxin: selective reduction of transient K conductance in hippocampus and binding to localized acceptors. Proc. Natl. Acad. Sci. USA 83, 493–497.

Harvey, A.L. and Anderson, A.J. (1991). Dendrotoxins: snake toxins that block potassium channels and facilitate neurotransmitter release. Snake Toxins (Harvey AL, ed) pp 131–164: New York Pergamon Press

Harvey, A.L. and Karlsson, E. (1980). Dendrotoxin from the venom of the green mamba, Dendroaspis angusticeps. A neurotoxin that enhances acetylcholine release at neuromuscular junctions. Naunyn-Schmiedebergs Arch. Pharmacol. 312, 1–6.

Harvey, A.L. and Karlsson, E. (1982). Protease inhibitor homologues from mamba venoms: facilitation of acetylcholine release and interactions with prejunctional blocking toxins. Br. J. Pharmacol. 77, 133–161.

Harvey, A.L., Anderson, A.J. and Karlsson, E. (1984). Facilitation of transmitter release by neurotoxins from snake venoms. J. Physiol. (Paris) 79, 222–227.

Hurst, R.S., Busch, A.E., Kavanaugh, M.P., Osborne, P.B., North, R.A. and Adelman, J.P. (1991). Identification of amino acid residues involved in dendrotoxin block of rat voltage-dependent potassium channels. Mol. Pharmacol. 140, 572–576.

Jonas, P., Brau, M.E., Hermsteiner, M. and Vogel, W. (1989). Single-channel recording in myelinated nerve fibres reveals one type of Na channel but different K channels. Proc. Nat. Acad. Sci. USA 86, 7238–7242.

Joubert, F.J. and Taljaard, N. (1980). The amino acid sequence of two proteinase inhibitor homologues from Dendroaspis angusticeps venom. Hoppe Seylers Z. Physiol. Chem. 361, 661–674.

Mallart, A. (1985). Electric current flowinside perineurial sheaths of mouse motor nerves. J. Physiol. 368, 565–575.

Marshall, D.L. and Harvey, A.L. (1990). Reexamination of the protease inhibitor activities of the dendrotoxins from mamba venoms. Toxicon 28, 157–158.

Marshall, D.L. and Harvey, A.L. (1992). Protease inhibitor homologs of dendrotoxin do no bind to dendrotoxin acceptors on synaptosomal membranes or facilitate neuromuscular transmission. Biol. Chem. Hoppe-Seyler 373, 707–714.

McArdle, J.J., Angaut-Petit, D., Mallart, A., Bournaud, R., Faille, L. and Brigant, J.L. (1981). Advantages of the triangularis sterni muscle for investigation of synaptic phenomena. J. Neurosci. Methods 4, 109–116.

McGivern, J., Scholfield, C.H. and Dolly, J.O. (1993). Action of α-dendrotoxin on K^+ currents in nerve terminal regions of axon in rat olfactory cortex. Br. J. Pharmacol. 109, 535–538.

Moczydlowski, E., Lucchesi, K. and Ravindran, A. (1988). An emerging pharmacology of peptide toxins targeted against potassium channels. J. Membrane Biol. 105, 95–111.

Muniz, Z.M., Diniz, C.R. and Dolly, J.O. (1990). Characterisation of binding sites for δ-dendrotoxin in guinea-pig synaptosomes: relationship to acceptors for the K^+ channel probe α-dendrotoxin. J. Neurochem. 54, 343–346.

Parcej, D.N. and Dolly, J.O. (1989). Dendrotoxin acceptor from bovine synaptic plasma membranes. Biochem. J. 257, 899–903.

Penner, R. and Dreyer F. (1986). Two different presynaptic calcium currents in mouse motor nerve terminals. Pflugers Arch. 406, 190–197.

Rehm, H. and Lazdunski, M. (1988). Purification and subunit structure of a putative K^+-channel protein identified by its binding properties for dendrotoxin I. Proc. Natl. Acad. Sci. USA 85, 4919–4923.

Rehm, H., Pelzer, S., Cochet, C., Chambaz, E., Tempel, B.L., Trautwein, W., Pelzer, D. and Lazdunski, M. (1989a). Dendrotoxin-binding brain membrane protein displays a K^+ channel activity that is stimulated by both cAMP-dependent and endogenous phosphorylations. Biochemistry 28, 6455–6460.

Rehm, H., Newitt, R.A. and Tempel, B.L. (1989b). Immunological evidence for a relationship between the dendrotoxin-binding protein and the mammalian homologue of the Drosophila shaker K^+ channel. FEBS Lett. **249**, 224–228.

Schweitz, H., Bruhn, T., Guillemare, E., Moinier, D., Lancelin, J.-M., Beress, L. and Lazdunski, M. (1995). Kalicludines and kaliseptine. Two different classes of sea anemone toxins for voltage-sensitive K^+ channels. J. Biol. Chem. 270, 25121–25126.

Scott, V.E.S., Parcej, D.N., Keen, J.N., Findlay, J.B.C. and Dolly, J.O. (1990). a-Dendrotoxin acceptor from bovine brain is a K^+ channel protein. Evidence from the N-terminal sequence of its larger subunit. J. Biol. Chem. 265, 20094–20097.

Skarzynski, T. (1992). Crystal-structure of alpha-dendrotoxin from the green mamba venom and its comparison with the structure of bovine pancreatic trypsin-inhibitor. J. Mol. Biol. 224, 671–683.

Sorensen, R.G. and M.P. Blaustein (1989). Rat brain dendrotoxin receptors associated with voltage-gated potassium channels: dendrotoxin binding and receptor solubilization. Mol. Pharmacol. 36, 689–698.

Sorensen, R.G. and M.P. Blaustein (1992). Dendrotoxin acceptor sites: identification and labelling of brain potassium channels. Methods in Neurosci. 8, 235–246.

Strydom, D.J. (1973). Protease inhibitors as snake venom toxins. Nature New Biol. 243, 88–89.

Stuhmer, W., Ruppersberg, J.P., Schroter, K.H., Sakmann, B., Stocker, M., Giese, K.P., Perschke, A., Baumann, A. and Pongs, O. (1989). Molecular basis of functional diversity of voltage-gated potassium channels in mammalian brain EMBO J. 8, 3295–3244.

Tytgat, J., Debont, T., Carmeliet and Daenens, P. (1995). The α-dendrotoxin footprint on a mammalian potassium channel. J. Biol. Chem. 270, 24776–24781,

Weller, U., Bernhardt, Y., Siemen, D., Dreyer, F., Vogel, W. and Habermann, E. (1985). Electrophysiological and neurobiochemical evidence for the blockade of a potassium channel by dendrotoxin. Naunyn-Schmiedeberg's Arch. Pharmacol. 330, 77–83

Werkman, T.R., Kawamura, T., Yokoyama, S., Higashida, H., and Rogawski, M.A. (1992). Charybdotoxin, dendrotoxin and mast cell degranulating peptide block the voltage-activated K^+ current of fibroblast cells stably transfected with NGK1 (KV1.2) K^+-channel. Neuroscience 50, 935–946.

18 Sarafotoxins: Cloning of mRNAs Encoding Sarafotoxin Precursors

Frédéric Ducancel, Jean-Claude Boulain and André Ménez

Contents

1 Introduction

When applied to animal toxins, molecular biology technologies offer several interests, including the possibility to elucidate the structure of toxin precursors (Ducancel *et al.*, 1991). In general, purified messenger RNAs (mRNAs) are converted into stable complementary copies of DNA (cDNAs) which are then cloned into recombinant vectors and used to transform bacteria. Two screening strategies can be considered to identify clones containing the precursor nucleic information. One of them consists in performing *in situ* hybridizations (Sambrook *et al.*, 1989), using labeled nucleic acid probes that are complementary to a portion of the precursor sequence. The other strategy consists in using cloning/expression vectors such as λgt11 or pBluescript, and in screening

Methods and Tools in Biosciences and Medicine
Animal Toxins, ed. by H. Rochat and M.-F. Martin-Eauclaire

with precursor-specific antibodies (Mierendorf *et al.*, 1987). The nucleic informations contained in the screened clones are then elucidated by DNA sequencing (Sanger *et al.*, 1977). Irrespective of the strategy, an appropriate cDNA library should:

- be large enough to contain representatives of all possible interesting sequences
- contain a limited number of clones with small cDNA inserts
- include nearly full-length cDNA copies of the mRNAs from which they are derived

Numerous kits are now available to perform the different steps required for preparation of cDNA libraries. In this chapter we will present the protocols and kits that have been used to construct a cDNA library containing clones encoding the long precursor of sarafotoxins from the venom glands of the burrowing asps of the genus *Atractaspis engaddensis* (Ducancel *et al.*, 1993, 1995). The three steps that will be considered are: (i) extraction and purification of mRNAs from venom glands, (ii) cDNA synthesis and plasmid cloning, and (iii) screening of positive clones using a ^{32}P-labeled oligonucleotide probes.

2 Materials and Chemicals

Buffers and reagents

- **Denaturing solution for total RNAs extraction (Solution 1)**
  4 M guanidinium thiocyanate
  25 mM sodium citrate, pH 7
  0.5% sarcosyl
  0.1 M β-mercaptoethanol
  prepared as following: 250 g guanidinium thiocyanate are dissolved directly in the manufacter's bottle with 293 ml H_2O, 17.6 ml of a 0.75 M solution of sodium citrate, pH 7, and 26.4 ml of 10% of pre-warmed sarcosyl at 65% (Chomczynski and Sacchi, 1987). This solution is stable for several months at room temperature. Just before use add 0.36 ml of β-mercaptoethanol.
- **10x Tris-Borate-EDTA (TBE) solution (Solution 2)**
  1 M Tris-HCl
  1 M boric acid
  20 mM EDTA
  pH adjusted to 8.3

- **10x T4 polynucleotide kinase buffer (Solution 3)**
  0.5 M Tris-HCl, pH 7.6
  0.1 M $MgCl_2$
  50 mM DTT
  1 mM spermidine
  1 mM EDTA
- **TE solution**
  10 mM Tris-HCl pH 8
  1 mM EDTA
- **20x SSC**
  3 M NaCl
  300 mM sodium citrate, pH 7.0
- **50x Denhardt's reagent (Solution 4)**
  1% bovine serum albumin (BSA)
  1% Ficoll
  1% polyvinylpyrrolidone
- **Polyacrylamide gel electrophoresis (PAGE) solutions for DNA sequence analysis**
  46% urea
  acrylamide/bisacrylamide 30/2%
  1x TBE
  H_2O
- **Blocking solution for DNA transfer (Solution 5)**
  200 µg/ml of heat-denaturated single-stranded herring sperm DNA
- **Solid Luria-Bertani/ampicillin medium per liter**
  10 g of BactoTryptone
  5 g of NaCl
  5 g of yeast extract
  12 g of Bacto-Agar, pH 6.8–7.0
  After sterilization, add 100–200 µg/ml of ampicillin
- **Solutions for bacteria transfer and DNA fixation on nitrocellulose membranes**
  sodium dodecyl sulfate (SDS) 10%
  0.5 N NaOH
  1.5 M NaCl
  0.5 M Tris-HCl pH 8
  1.5 M NaCl
  2x SSC

Chemicals

Diethyl pyrocarbonate	Sigma	D-5758
Guanidinium thiocyanate	Fluka	50980
N-Laurylsarcosine	Sigma	L-5125
Tri-Sodium Citrate, dihydrate analytical reagent	Prolabo	27 833.294

mRNA Purification Kit	Pharmacia	27-9258-01
SuperScript Plasmid System for cDNA synthesis and Plasmid Cloning	Gibco BRL	18090-019
T4 polynucleotide kinase	Gibco BRL	18248-013
γ^{32}P-ATP (3000 Ci/mmol)	Amersham	PB. 10168
BactoTryptone	Difco	0123-17.3
Yeast Extract	Difco	0127-17.9
Bacto Agar	Difco	0140-01
^{T7}Sequencing Kit	Pharmacia	27-1682-01
α^{35}S-dCTP (1000 Ci/mmol)	Amersham	AG. 0135

Equipment

- Ultra-Turrax homogenizer
- horizontal and vertical DNA gel electrophoresis systems
- water-bath
- sequencing gel drying

3 Extraction and Purification of mRNAs

The construction of a cDNA library began with the isolation of the largest possible amount of pure and undegraded mRNAs from total RNA fraction. However, venom gland tissues contain active ribonucleases (RNAses) which may affect the yield of extraction. Thus, to perform efficient extraction of total RNAs from RNase-rich tissues, we minimized the activity of RNAses by wearing and changing frequently gloves, using RNase-free materials and solutions, and by using strong protein denaturants.

For preparation and storage of RNAs, we used sterile and RNAase-free disposable plasticwares. Laboratory glasswares were baked at 180°C at least 8 h. Polypropylene plasticwares and rubber items (tube stoppers) were rinsed 5–10 min with chloroform. We used DEPC (diethyl pyrocarbonate)-treated water and new chemicals to inhibit the enzymatic activity of RNAses. We added 0.1% of DEPC to sterile water, waited for 2–3 h at 37°C, and heated the mixture at 100°C for 15 min or autoclaved it to remove traces of DEPC. Because DEPC is suspected to be a carcinogenic compound, it should be used with caution.

3.1 Extraction of total RNAs

We adopted the following stategy to extract total RNA fractions from venom glands of *Atractaspis engaddensis*.

Protocol 1 Extraction of total RNAs

1. Decapitate snake heads and immediately immerse them in liquid nitrogen.
2. Dissect venom glands by placing frozen snake heads on dry ice during the operation.
3. Then, drop frozen venom glands in 10 ml of denaturing solution in a 25-ml Corex tube.
4. The tissue are homogenized 2 × 60 s at full speed at room temperature, using an Ultra-Turrax homogenizer.
5. Add sequentially to the homogenate 1 ml of 2 M sodium acetate, pH 4, 10 ml of phenol (water-saturated), and 2 ml of chloroform-iso-amyl alcohol mixture (49:1), and mix the solution by inversion after addition of each reagent. The final suspension should be vigorously shaken for 15 s, cooled on ice for 15 min, and then centrifuged at 10,000 t/min at 4°C, for 20 min.
6. Transfer the aqueous phase containing the RNAs to a fresh RNAse-free Corex tube, mix with 10 ml of isopropanol and place the tube at –20°C for at least 1 h.
7. Sedimentation of the precipitated total RNAs is performed at 10,000 t/min at 4°C for 20 min.
8. Dissolve the resulting wet RNAs pellet in 0.3 ml of denaturing solution, transfer the solution into a 1.5-ml Eppendorf tube, and perform a second RNA precipitation with one volume of isopropanol at –20°C for 1 h.
9. An RNA pellet is obtained after centrifugation at 10,000 t/min at 4°C for 10 min.
10. Wash it with 0.5 ml 75% ethanol.
11. Drain the pellet by inverting the tube on a tissue paper for 10 min.

3.2 Purification of poly (A^+) RNAs

Most newly synthesized eukaryotic mRNAs are characterized by the presence of 3' polyadenylated tracts, which should further constitute the sites of initiation of first strand cDNA synthesis using a complementary polythymidine primer (see below). Thus, purification of poly (A^+) RNAs contained within the RNA pellet constituted a key step. Poly (A^+) RNAs contained in two *Atractaspis engaddensis* venom glands were collected by oligo (dT) affinity chromatography using the mRNA Purification Kit purchased by Pharmacia. Typically, 5 µg of purified mRNAs were obtained from 400 µg of extracted total RNAs.

4 cDNA Synthesis and Plasmid Cloning

To prepare the DNA double-stranded copies of purified mRNAs, we carefully followed the procedures described in the technical manual of the kit "Superscript Plasmid System for cDNA Synthesis and Plasmid Cloning" obtained from Gibco BRL. In brief, from 1.25 μg of purified mRNAs we obtained 0.135 μg of double-stranded cDNAs. Orientated ligation into the plasmid vector pSPORT 1 was performed using 10 ng of size-fractionated cDNA (>500 pb) whose extremities were successively modified by addition of *Sal*I adapters, then by *Not*I digestion.

5 Screening of Positive Clones Using a Specific ^{32}P-Labeled Oligonucleotide Probe

Using the whole ligation mixture and NM 522 competent cells prepared according to the method of Hanahan (1983), we generated a library of 7000 transformants. In most cases, such a size is sufficient to contain the majority of toxin precursors synthesized by the cells forming the venom glands. However, before the screening, we evaluated the quality of the library by characterizing the size of 24 random cDNA clones, using the endonuclease restriction enzyme *Mlu* I. Thus, we noticed the presence of a majority of large inserts (>1.6 kbp) and the absence of empty vector, indicating the success of previous steps.

5.1 Transfer of transformed colonies

Protocol 2 Transfer of transformed colonies

1. Number and asymetrically mark sterile nitrocellulose filters (Schleicher and Schuell BA-S 85, 0.45 μm) with a ballpoint pen.
2. Using sterile forceps, lay a filter marked side up on the surface of a master petri-dish plate containing about 300 isolated clones, and wait for 1 min.

   Notes:
   20 different plates, i.e., about 6000–7000 recombinant clones were thus replicated.
3. Remove the filters from the plates and invert them on a day-old LB/ampicillin plate marked side down.
4. Incubate them at 37°C until small colonies appear with the size of a pinhead.

5.2 Fixation of DNA to filters

Protocol 3 Fixation of DNA to filters

1. Leave the replicate filters (colonies up) 5 min on a large sheet of Whatman 3MM paper previoulsly wet with 10% SDS solution.
2. Tranfer them on a second 3MM paper wet with a DNA denaturing solution (0.5 M NaOH, 1.5 M NaCl) for 15 min.
3. Neutralize the filters 10 min by transferring them onto a third 3MM paper wet with 0.5 M Tris-HCl, pH 8, 1.5 M NaCl.
4. Place the filters on a 3MM paper wet with a 2 x SSC transfer buffer for 5 min.
5. Dry the filters 1 h at room temperature.
6. Finally, place the filters between two Whatman papers and then for 0,5 h at 80°C in a vacuum oven.

5.3 Hybridization of the DNA using ^{32}P-labeled probe

Since comparison of sarafotoxin isoforms revealed a high degree of sequence identity, we selected recombinant bacterial clones using a ^{32}P-labeled oligonucleotide probe specific for the conserved C-terminal part of sarafotoxins: 5' ATTTTTGTCATCAGGATGTG/T ATG/TTGG 3'. Its melting temperature (T_m) from the target sequence was estimated to be 64°C, assuming that nondegenerated positions occupied by A or T and G or C nucleotide correspond to 2 or 4°C, respectively. The temperature used for hybridization was 40°C.

Protocol 4 ^{32}P labelling of the oligonucleotide probe

1. Labeling of the purified probe was performed as following: 20 pmol (1 µl) of oligonucleotide were mixed with 20 pmol of γ^{32}P-ATP (10 µl), 2 µl of 10 x T4 polynucleotide kinase buffer, 1 µl of T4 polynucleotide kinase (10 units), and 6 µl H_2O. Incubate sucessively 1 h at 37°C, then 10 min at 68°C.
2. Precipitate the labeled DNA by adding 230 µl of 5 M ammonium acetate and 750 µl of absolute ethanol (–20°C).
3. Mix well and let the mixture for at least 1 h at –70°C. Centrifuge 20 min at 13,000 t/min at 4°C, rinse with 0.5 ml of 70% ethanol (–20°C), decant, and dry at room temperature.
4. Resuspend the radioactive pellet in 2 or 3 ml of TE solution, and filter it using a syringe with a 0.2-µm filter.

Notes:
Typically, 300,000–500,000 cpms/10 µl were obtained, a value compatible with hybridization conditions. Thus, for 10 filters we currently carried on hybridization by adding 10^6 cpms per/ml of radioactive/filtered DNA probe, in a final volume of 20 ml.

Protocol 5 Screening of specific cDNA clones

1. First, immerse the filters in a crystallizing dish containing 20 ml of 5x SSC, 5x Denhardt's, and 0.1% SDS, supplemented with 150 µg/ml of heat-denaturated (5 min at 100°C) herring sperm DNA, as saturating compound.
2. Incubate them at least 2 h at 40°C in a water-bath.
3. Add 2.10^7 cpms of radiolabelled probe and hermetically close the crystallizing dish.
4. Perform the hybridization overnight at 40°C, under gentle agitation.
5. Transfer the radioactive filters into a second crystallizing dish containing a nonradioactive 2x SSC washing solution.
6. Remove the radioactivity by washing the filters 3–4 times at room temperature under gentle agitation for periods of 5 min.
7. Reduce the radioactivity associated with the filters by two or three repetitive washing steps of 10 min each, at room temperature, first in 1x SSC, 0.1% SDS, then in 0.1x SSC, 0.1% SDS solutions.

   Notes:
   If the radioactivity does not significantly decrease, repeat the previous washing steps at higher temperatures, i.e., increased by successive steps of 5°C from the initial temperature of hybridization.
8. When 50% of the initial bound radioactivity is eluted from the filters, expose them at –80°C.

6 Characterization of Selected Clones

Radioactive footprint autoradiographs obtained after SRTX-specific probe screening revealed 14 significantly labelled spots. Once identified on the master petri-dish plates, the corresponding recombinant bacterial clones were isolated on freshly prepared LB/ampicillin plates.

6.1 Size characterization of selected clones

We evaluated the size of the cDNA inserts contained in the cloning plasmid pSPORT 1 by using the restriction enzyme *Mlu* I. Indeed, two *Mlu* I restriction sites surround cloned cDNAs: one of them is located within added adapters at the 5' extremity, whereas the other one belongs to pSORPT 1, upstream of the *Not* I site used to covalently link the 3' extremity. Surprisingly, two of them contained large inserts of 2.0 kb, whereas the others displayed sizes ranging from 0.8 to 1.2 kb. Such a result was unexpected regarding both the size of SRTXs (21 amino acids) but also the size of other snake toxin precursors encoding larger mature toxic proteins (Ducancel *et al.*, 1991).

6.2 Strategy to sequence selected clones

To confirm the selection of cDNAs encoding SRTXs, we sequenced their 5' and 3' extremities using double-stranded DNA templates. All displayed a poly (A^+) tract, preceded by an identical 3' untranslated region of 154 pb, upstream of

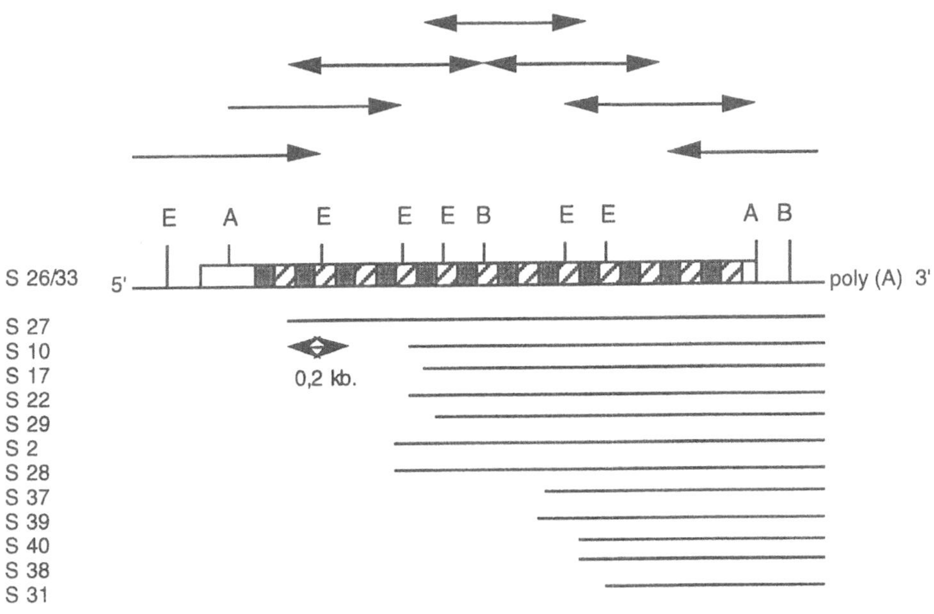

Figure 1 Structure and partial restriction map of cDNAs encoding precursors of sarafotoxins, and strategies for nucleotide sequence determination.

The number of each specific clone is indicated on the left of the figure. The large and open rectangle shown in clones S26 and S33 represents their ORF (nucleotides 162–1791). Untranslated 5' and 3' regions are indicated by continuous lines. Restriction enzymes used in mapping include A (*AlwN* I), B (*Bgl* I), E (*Ear* I). Arrows indicate both direction and length of nucleotide sequencing, using universal or internal oligonucleotidic probes.

which we identified the C-terminal part of an open reading frame (ORF) containing a sequence encoding a sarafotoxin-a isoform. In contrast, the 5' part of these 14 clones revealed variable nucleotide sequences within SRTX coding sequences. Together, these results (i) confirmed us that we cloned cDNAs encoding SRTXs precursors, and (ii) showed that SRTXs are produced from large precursors containing about 2,000 bp. The short length of some clones probably reflects incomplete *in vitro* cDNA synthesis of the 5' region, since all selected cDNAs displayed identical 3' untranslated regions.

Surprinsingly, when the screening probe (complementary to the C-terminal part of SRTXs) was used as an internal sequencing primer to determine the nucleotide sequence of the two larger cDNAs, mixed sequences appeared, suggesting multiple priming events. Consequently, we looked for rare internal restriction sites to subclone fragments of these large precursors. However, analysis of their restriction maps (i) showed the absence of most of the classical restriction sites, and (ii) revealed a repetitive organization of the mRNA since even the rare *Ear* I site yielded multiple fragments (Fig. 1). Fortunately, restriction of the longest cDNAs with *Bgl* I generated two DNA fragments of 1.0 and 0.9 kb that we subcloned in both orientations into M13mp 18 and 19 cloning vectors, respectively. Thus, we successfully carried out the complete nucleotide sequence elucidation of the two larger cDNA clones, following the strategy described on the upper part of Figure 1.

Finally the SRTX precursors were identified as large pre-pro-polypeptide chains of 543 amino acids, containing 12 successive and similar streches of 40 residues. Each stretch encoded one SRTX isoform preceded by a highly invariant sequence of 18 or 19 amino acid residues. Thus, we noticed that six different SRTX isoforms are encoded by a complete precursor. In conclusion, we anticipated that the overproduction of SRTXs in venom glands of the snakes *Atractaspis engaddensis* results from the processing of this unusual repetitive “rosary-type” precursor.

7 Troubleshooting

Low abundance or absence of mRNAs:

- Snake heads may have not been treated correctly. Avoid keeping the heads unfrozen prior to venom gland extraction.
- Perform extraction of venom glands as quickly as possible on dry ice. During extraction, do not hesitate to regularly drop the heads into liquid nitrogen. Maintain venom glands in liquid nitrogen until used.
- The solutions, the vessel, and/or the material may not be RNAse-free. Start again using new reagents; prepare the solutions with DEPC-treated water.

Poor yield of cDNA synthesis:

- Synthesis of the first-strand cDNA is mainly dependent upon the quality and/or the purity of the starting mRNAs but also on RNase contamination. Repeat oligo (dT) chromatography twice to maximally enrich poly (A^+) RNAs. Use at least 1–2 µg of purified mRNAs. Use new, sterile, and choloroform-treated Eppendorf tubes and Pipetman tips. Use the control mRNA to be certain that the kit components are properly active.
- To obtain an appropriate amount of the second strand, carefully respect the dilution procedure of the first-strand solution before performing the second-strand reaction.

Low yield of cDNA cloning and reduced number of bacterial recombinant clones:

- Addition of adapters, endonuclease digestion, and cDNA column chromatography constitute successive difficulties which may influence the quality of a library. As a consequence, it is recommended to carefully respect the described experimental procedures. Take care to properly remove residual adapters, *Not* I fragments, and small cDNA inserts.
- Low transformation efficiency, i.e., $<10^4$ clones per µg of cDNA may reflect incorrect ligation conditions. Check the ligation mixtures in the presence or absence of cDNAs, using comparative agarose gel electrophoresis, and/or increase the amount of cDNAs in the ligation reaction. One may also increase the number of transformants by introducing the ligated cDNAs mixtures into *Escherichia coli* by electroporation. If so, the ligation mixture absolutely must be concentrated by ethanol precipitation, and the pellet rinsed twice and resuspended in distilled water to reduce the salt concentration as much as possible.

References

Chomczynski P, Sacchi N (1987) Single-step method of RNA isolation by acid guanidinium thiocyanate-phenol-chloroform extraction. Anal. Biochem. 162: 156–159

Ducancel F, Bouchier C, Tamiya T et al (1991) Cloning and expression of cDNAs encoding snake toxins. Snake Toxins (Harvey, A.L, ed) pp 385–414: New York: Pergamon Press

Ducancel F, Matre V, Dupont C et al (1993) Cloning and sequence analysis of cDNAs encoding precursors of sarafotoxins. J. Biol. Chem, 268: 3052–3055

Ducancel F, Matre V, Dupont C et al (1995) Structure and function of mRNAs encoding sarafotoxins precursors. Frontiers in Endocrinology: Endothelins in Endocrinology, New advances, vol 15 (Baldi E, Maggi M, Cameron IT, Dunn MJ, eds) pp 1–8: Ares Serono Symposia

Hanahan D (1983) Studies on transformation of *E. coli* with plasmids. J. Mol. Biol. 166: 557–580

Mierendorf RC, Percy C, Young RA (1987) Gene isolation by screening λgt11 libraries with antibodies. Methods Enzymol. 152: 458

Sambrook J, Fritsch EF, Maniatis T (1989) Molecular Cloning: A Laboratory Manual, 2nd edn. Cold Spring Harbor, New York: Cold Spring Harbor Laboratory Press

Sanger F, Nicklen S, Coulson AR (1977) DNA sequencing with chain-terminating inhibitors. Proc. Natl. Acad. Sci, USA 74: 5463–5467

19 Snake Venom Phospholipase A_2 Enzymes in Cell Biology

R. Manjunatha Kini

Contents

1 Introduction

Phospholipase A_2 (PLA_2) (EC 3.1.1.4) enzymes are esterolytic hydrolases which catalyze the hydrolysis of 3-*sn*-phosphoglycerides. They specifically hydrolyze and release fatty acids from the second position (Fig. 1). These enzymes are ubiquitous in nature, and they occur as both intracellular (cellular type or $cPLA_2$ enzymes) and extracellular (secretory type or $sPLA_2$) forms. In general, $cPLA_2$ enzymes are found in small quantities, and hence their purification is relatively difficult. $sPLA_2$ enzymes, on the other hand, are found in larger quantities in nature, and snake venoms and mammalian pancreas are particularly rich sources of these enzymes. Since purification of $sPLA_2$ enzymes is easier, several hundreds of these enzymes have been purified and characterized. Mammalian PLA_2 enzymes play important roles in the maintenance of the cellular phospholipid pools and membrane repair through deacylation/reacylation pathways (Verheij *et al.*, 1981; Waite, 1987; Dennis *et al.*, 1991; Kudo *et al.*, 1993; Dennis, 1994) and biosynthesis of prostaglandins and leukotrienes. They have crucial roles in various physiological and pathological processes (Glaser *et al.*, 1993; Dennis, 1994; Mukherjee *et al.*, 1994). The importance of PLA_2 enzymes in fertilization (Fry *et al.*, 1992), cell proliferation (Arita *et al.*,

Methods and Tools in Biosciences and Medicine
Animal Toxins, ed. by H. Rochat and M.-F. Martin-Eauclaire

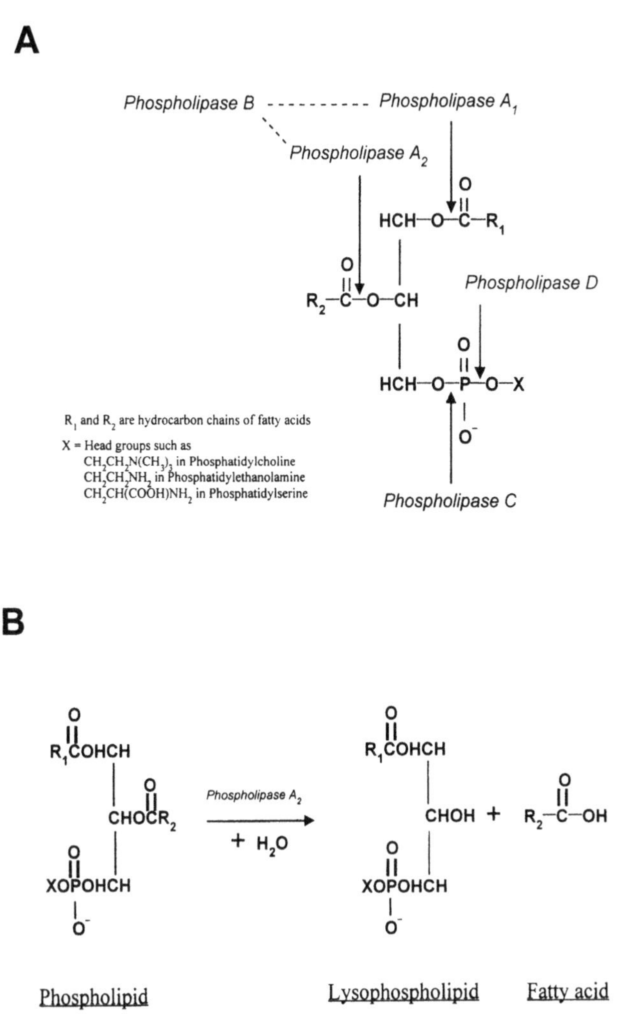

Figure 1 Reaction catalyzed by phospholipases.

A. Sites of action of various phospholipases

B. Reaction catalyzed by PLA_2

1991), smooth muscle contraction (Vadas and Pruzanski, 1986; Nakajima *et al.*, 1992; Sommers *et al.*, 1992), and hypersensitization and chronic inflammatory diseases (Vadas and Pruzanski, 1986; Vadas *et al.*, 1993) have been recognized. They are also important in cellular functions such as signal transduction and membrane homeostasis (Kudo *et al.*, 1993).

Despite structural and catalytic similarities with the mammalian enzymes, venom PLA_2 enzymes induce a wide variety of pharmacological effects and show a range of toxicities. Neurotoxic, cardiotoxic, myotoxic, anticoagulant, antiplatelet, convulsive, hemolytic and hemorrhagic effects are some of the most important pharmacological effects induced by snake venom PLA_2 enzymes. Although a variety of pharmacological effects are induced by PLA_2 enzymes, not all the effects are exhibited by all PLA_2 enzymes. Each enzyme exhibits a specific effect. For example, β-bungarotoxin, a PLA_2-containing toxin, induces a presynaptic effect (Strong *et al.*, 1976) but fails to show any postsynaptic effect (Yang, 1978). This toxin affects only nicotinic acetylcholine transmission, but not glutamate and adrenergic transmission even at high concentrations (Abe *et al.*, 1971). It also fails to induce myotoxicity (Fraenkel-Conrat, 1982–83) and anticoagulant effects (Boffa *et al.*, 1980; Verheij *et al.*, 1980a). Thus β-bungarotoxin affects only the presynaptic site and hence is "targeted" to the presynaptic site. The example of *Vipera russelli* isoenzymes provides a stronger indication of "targeting". When the isoenzymes are injected intraperitoneally, they exhibit various effects: one specifically induces hemorrhage in liver and kidney (Vishwanath *et al.*, 1985), the second damages the lungs causing hemorrhage (Kasturi and Gowda, 1989), whereas the third kills the mice through neurotoxic symptoms (T. V. Gowda, personal communication). This indicates the susceptibility of a specific system, tissue or organ to a particular PLA_2 enzyme. Similar specificity of attack on a particular tissue was also found in *Trimeresurus flavoviridis* PLA_2 isoenzymes (Vishwanath *et al.*, 1987). Such susceptibility can be explained by the presence of specific "target sites" on the surface of target cells or tissues (Kini and Evans, 1989). These target sites are recognized by specific "pharmacological sites" on the PLA_2 molecule. These pharmacological sites are independent of, but sometimes overlapping with, the active site of the enzyme (Rosenberg, 1986). It was proposed that the target sites and pharmacological sites are complementary to each other in terms of charges, hydrophobicity, and van der Waal's contact surfaces, hence the high affinity (Kini and Evans, 1989). High affinity between the target and pharmacological sites determines the specific pharmacological effects of PLA_2 enzymes.

The proposed target sites should be located on the surface of the target cells. Therefore these target sites could be either membrane lipids or proteins (glycoproteins). Since PLA_2 enzymes bind to and hydrolyze phospholipids, phospholipids could act as target sites. But the ubiquitous presence of phospholipids, the susceptibility of only mature mice to β-bungarotoxin, and resistance of immature muscle cells to myotoxic PLA_2 enzymes argue against this possibility. On the other hand, proteins (or glycoproteins) could act as specific target sites for PLA_2 enzymes (for details of this model, see Kini and Evans, 1989). Thus the specific pharmacological effects of PLA_2 enzymes are most likely due to high-affinity interaction between PLA_2 protein and membrane protein in the specific target tissue. However, phospholipids near the interaction sites may also contribute to these specific interactions.

In addition, two PLA_2 enzymes affecting the same cell may bind either to two separate sites of the same acceptor/receptor protein or two separate unrelated acceptor/receptor proteins and induce similar, if not identical, symptoms. For example, although presynaptic neurotoxins (PLA_2 enzymes or their complexes) such as notexin, crotoxin, and taipoxin show similar effects on the neuromuscular transmission, they show synergistic effects and mutual potentiation. It is also important to note that there is lack of competition for binding between these neurotoxins (Tzeng *et al.*, 1986). These factors make PLA_2

Table 1 Comparison of Phospholipase A_2 Assays

Method	Detection Equipment	Detection Limit[1]	Monitoring Capability[2]	Natural Substrate	Remarks[3]
Titrametric	pH stat	20 nmol	Kinetics	Yes	Inexpensive equipment; simple and straightforward; most common; *less sensitive*
Acidimetric	pH stat	100 pmol	Kinetics	Yes	Inexpensive equipment; *pH varies continuously; sensitivity determined by pH*
Radiometric	Scintillation counter	1 fmol–1 pmol	End-point	No	Sensitive assay; most common; *laborious; requires TLC or HPLC methods to separate products*
Nuclear magnetic resonance (NMR)	NMR spectrometer	1 µmol	End-point	Yes	*Expensive equipment; less sensitive; not suitable for routine assays*
Monolayer	Monolayer trough	—	Kinetics	Yes	Very sensitive method; activity at constant surface pressure; *specialized equipment; difficult study near or above* K_m *values*
Spectrophotometric	Spectrophotometer	200 pmol-25 nmol	Kinetics/ End-point	Yes/No	Several different assays; simple to laborious
Fluorometric	Fluorometer	1 pmol	Kinetics/ End-point	Yes/No	Sensitive methods; *mostly substrates are fluorescently labeled*; ADIFAB method is recommended[4] (see text)
Electron spin resonance (ESR)	ESR spectrometer	1 nmol	Kinetics	No	Sensitive method; *expensive and specialized equipment*

Modified from Reynolds *et al.* (1991). For in-depth analyses of various methods see Reynolds *et al.* (1991) and Hendrickson (1994).

1. Describes general sensitivity of the assays. No adjustments were made for assay volumes.
2. Ability of the method to monitor the activity by either kinetics (continuously) or end-point (fixed time point) assays.
3. Remarks in normal letters are merits of the method, whereas those in italicized letters indicate demerits.
4. This method is most suitable for cellular PLA_2 enzymes. However, for the study of snake venom enzymes, titrametric methods are most convenient and simple to use.

enzymes attractive tools in cell biology: (i) they specifically interact with a particular cell type, and hence one can study their effects on various important cells or tissues; (ii) several PLA_2 enzymes can bind to different cell surface acceptor/receptor proteins and alter their (or neighboring protein's) functions.

In this chapter, I have described general methods used in the purification and characterization of snake venom PLA_2 enzymes. A pure and thoroughly characterized enzyme will provide a good research tool. A few examples of these enzymes which are used as tools are also described. For more details on structure, function, and mechanism of snake venom PLA_2 enzymes, readers are recommended to consult the recent book (Kini, 1997).

2 Purification of PLA_2 Enzymes

Snake venoms are rich sources for PLA_2 enzymes. The amount of PLA_2 enzymes in venoms could be as high as 20–40% of the total venom by weight. However, there are some problems in the purification of these enzymes:

- **Number of isoenzymes.** Snake venom contains multiple forms of PLA_2 enzymes. For example, *Naja naja, Vipera russelli, Trimeresurus flavoviridis,* and *Pseudechis australis* venoms contain more than 10 isoenzymes (Braganca and Sambray, 1967; Shiloah *et al.*, 1973; Vishwanath *et al.*, 1987, 1988; Takasaki *et al.*, 1990a,b; Ogawa *et al.*, 1992). In several cases, although isoenzymes are closely related, they differ in their pharmacological effects. This poses a problem in the purification of a specific PLA_2 enzyme. Coelution could be a problem in the purification of these isoenzymes.
- **Self-aggregation.** Protein-protein interactions between PLA_2 isoenzymes and self-aggregation can contribute significantly to the problem of purification (R. M. Kini and H. J. Evans, unpublished observations).
- **Interaction with other venom proteins.** In addition, PLA_2 enzymes also interact with other snake venom proteins, such as cardiotoxins (Klibansky *et al.*, 1968). These interactions contribute significantly to the enzymatic and, sometimes, the pharmacologic activity of PLA_2 enzymes. In some cases, PLA_2 enzymes form multimeric complexes, and such complex formation is important as the components show synergistic effect on the pharmacological potency and toxicity of PLA_2 enzymes.

2.1 Purification strategy

Several different strategies are used in the purification of PLA_2 enzymes. Since each PLA_2 isoenzyme differs in its physical properties and elution characteristics and since each snake venom is distinct from others in its composition, there is no single procedure which can be considered as an ideal method. However, in general, a first step of gel filtration chromatography is recommended. This step is particularly useful when purifying PLA_2 enzymes from elapid snake venoms; it will help separate the group of PLA_2 enzymes from smaller polypeptides, such as cardiotoxins and neurotoxins. This is important because of the synergistic interactions between PLA_2 enzymes and cardiotoxins. If ion-exchange chromatographic methods are used as a first step, the increasing concentrations of the salt used in eluting the proteins enhances the hydrophobic interactions between these two groups of proteins. Gel filtration could be performed on Sephadex (G-50 or G-75), Bio Gel (P-10 or P-30), Superdex 75, or similar columns. In snake venoms, there are only a few other proteins which have molecular mass similar to PLA_2 enzymes. Thus gel filtration is a very good first step to separate PLA_2 enzymes from other groups of proteins.

Ion-exchange chromatography is a logical second step in the purification of PLA_2 isoenzymes. In this step, PLA_2 enzymes are separated based on their charge differences. Most of the isoenzymes can be separated on any of the cationic exchange columns (either commercially available or packed in-house). Gradients of buffers ranging in pH values between 6 and 8 along with salt concentrations up to 1 M are most suitable for separation of various PLA_2 isoenzymes. At least in some cases, a third and final step of reverse-phase HPLC may be required to separate the PLA_2 enzymes from contaminating proteins based on the differences in their hydrophobicity.

If PLA_2 enzymes are needed for nonspecific applications, generally it is easy to isolate and purify an enzyme from snake venom. However, if a specific PLA_2 enzyme is needed, it can sometimes be tedious and cumbersome. In the latter case, it is critical to purify the enzymes to homogeneity to avoid nonspecific effects of the contaminating proteins.

2.2 Test for homogeneity

Similar molecular size, closer isoelectric points, and even identical amino terminal sequences between the isoenzymes (Evans *et al.*, 1980; Kini *et al.*, 1986; Takasaki *et al.*, 1990b) lead to difficulty in ascertaining the purity of the preparation. Therefore more rigorous methods should be considered in determining homogeneity of PLA_2 enzymes. Sometimes even HPLC methods may not be suitable to define the homogeneity of the preparation. Sophisticated methods including capillary electrophoresis and mass spectrometry will be useful in determining the homogeneity. The large number of theoretical plates/cm (several hundred thousand in capillary electrophoresis compared with a few thousand in HPLC) which provides a much higher resolution and higher sensitivity can help in identifying any contaminating proteins. Mass spectrometric methods using electrospray ionization (ESI; also called ionspray) or matrix-associated laser desorption ionization (MALDI) can play a significant role in determining the homogeneity of the preparation.

3 Assay of Enzymatic Activity

Several assay procedures to determine the catalytic activity are available. These methods are based on various principles including titrametric, radiometric, spectrophotometric, and fluorometric measurements. The methods differ from each other in their sensitivity, usefulness in continuous monitoring, and utility in studying hydrolysis of natural substrates (Reynolds *et al.*, 1991). The radiometric method is the most sensitive (1 fmol of hydrolyzed substrate/min), but is not useful for continuous monitoring or for natural substrates (Reynolds *et al.*, 1991). Titrametric methods using pH stats are less sensitive (20 nmol/min), but are suitable for continuous monitoring and for natural substrates. Since most of the snake venom PLA_2 enzymes are highly active, with the exception of Lys-49 enzymes, pH stat methods are suitable for studying venom enzymes. It is recommended that the reader consult a reference (Dennis, 1991) for a detailed account of the advantages and disadvantages of various methods.

Fluorometric methods appear to be ideal in terms of their sensitivity (1 pmol/min or less) and their suitability for continuous monitoring (Reynolds *et al.*, 1991). In most of these methods fluorescently modified substrates are used (for a recent review, see Hendrikson, 1994) and thus most are not suitable for studying the hydrolysis of natural phospholipids. Recently, a new fluorescent probe, acrylodated intestinal fatty acid binding protein (ADIFAB) has been developed (Richieri *et al.*, 1992; Richieri and Kleinfeld, 1995). This probe can determine the aqueous-phase monomeric concentration of fatty acid as the fluorescence of ADIFAB shifts from 432 to 505 nm upon binding a fatty acid. The ratio of 505 and 432 nm fluorescence gives a direct measure of the bound fraction of ADIFAB. Absolute free fatty acid concentrations are calculated based on this ratio and the measured fatty acid binding constants. Richieri and Kleinfeld (1995) have developed an assay method for continuous measurement of PLA_2 activity. This method can be used to study the hydrolysis of natural phospholipids.

4 Cell Biology and PLA_2 Enzymes

Similar to other toxins described in this book, PLA_2 enzymes can also be used as research tools in cell biology. Although specificity of PLA_2 enzymes towards a particular cell type or organs was well documented, this group of proteins was not used as probes as other toxins. Several pharmacological effects of PLA_2 enzymes were thought to be due to nonspecific and indiscriminate destruction of plasma membranes. This is because their inherent phospholipolytic activity and the specificity was related to their affinity towards specific phospholipids or phospholipid domains. However, studies in the last 15 years have indicated that PLA_2 enzymes bind to various cell types because of their high-affinity towards target proteins found on the surface of the cells. The affinity between PLA_2 and its target protein is in the low nanomolar range, whereas the binding between PLA_2 and phospholipids is in the high micromolar range. Thus there is an affinity difference of four to six orders of magnitude between the protein-protein interaction and the protein-phospholipid interaction. Because of this high degree of specificity of pharmacologically active PLA_2 enzymes, they can be used as research tools in cell biology.

4.1 Identification of target proteins

Identification of the target proteins for PLA_2 enzymes on cells serves two distinct purposes:

- it clarifies the physiological role of the target protein and describes the susceptibility of a specific cell. This is particularly useful in developing new research tools for cell biology or pharmaceutical agents for specific disorders; and
- it contributes to our understanding of the mechanism of toxicity/pharmacological effects of PLA_2 enzymes.

As mentioned above, this field is still in the developing stages. In recent years, several laboratories have shown specific binding of PLA_2 enzymes to protein targets in the target cell or tissue, and have identified and isolated the target proteins (for a recent review, see Lambeau *et al.*, 1997). The structural and functional characteristics of some of these membrane-bound receptors/acceptors are described below.

Presynaptic neurotoxic PLA_2 enzymes

All known presynaptic neurotoxins isolated from snake venoms are either PLA_2 enzymes or contain PLA_2 enzymes as an integral part of the toxin complex. These toxins specifically bind to presynaptic sites and inhibit neurotransmission. Although their overall effect on the presynaptic site is inhibiting the release of neurotransmitter, there is a possibility of their binding to either different sites of the same acceptor (or target) protein or different target proteins. The acceptor protein for β-bungarotoxin, a presynaptic toxin complex isolated from *Bungarus multicinctus* venom, has been identified as a potassium channel protein (Breeze and Dolly, 1989; Awan and Dolly, 1991; Schmidt *et al.*, 1988). It is also known that OS_2, a toxic PLA_2 enzyme from *Oxyuranus scutellatus* venom, binds to specific proteins of molecular weights 34, 48, and 82 kDa (Lambeau *et al.*, 1989, 1991). Several presynaptic neurotoxic PLA_2 enzymes have been shown to compete with OS_2 (Lambeau *et al.*, 1989). However, this N-type receptor for OS_2 or any other acceptor proteins for presynaptically active PLA_2 enzymes have not been isolated and characterized. It is also not clear at a molecular level how this group of PLA_2 enzymes show their presynaptic effects. The possible role in phosphorylation of proteins and inhibition of recycling of presynaptic vesicles are being investigated (Ueno and Rosenberg, 1990, 1992, 1995). Several groups have been working on the isolation and characterization of acceptor (target) proteins (Tzeng, 1993; Krizaj *et al.*, 1994) and on the molecular mechanisms by which this group of PLA_2 enzymes induce their effects. These enzymes are thus useful tools to understand the biology of neurons, particularly at the presynaptic site of neurons.

Myotoxic PLA_2 enzymes

This group of PLA_2 enzymes induce depolarization, contraction, and finally lysis of muscle cells. A receptor (or acceptor) protein, aptly named M-type receptor, has been isolated from rabbit skeletal muscle (Lambeau *et al.*, 1990). This protein is responsible for the high-affinity binding of OS_1 and OS_2, PLA_2 enzymes from *O. scutellatus* venom, to muscle cells. This receptor protein is homologous with maltose-binding protein and contains eight carbohydrate recognition domains (CRDs) (Lambeau *et al.*, 1994; Ancian *et al.*, 1995). The fifth CRD appears to be crucial for interaction with PLA_2 enzymes (Nicolas *et al.*, 1995). The residues in the calcium binding loop of PLA_2 enzymes have been shown to participate in the recognition of the M-type receptor (Lambeau *et al.*, 1995). Interestingly, a similar, if not identical, receptor protein has also been identified and characterized for pancreatic PLA_2 enzymes (Hanasaki and Arita, 1992; Ishizaki *et al.*, 1994; Higashino *et al.*, 1994). Upon binding to this receptor, pancreatic enzyme stimulates mitogenesis, relaxation of arteries and migration of vascular smooth muscle cells (Arita *et al.*, 1991; Kishino *et al.*, 1992; Nakajima *et al.*, 1992; Kanemasa *et al.*, 1992). The role of M-type receptor protein in neither the function of muscle cells nor the myolysis (or myotoxicity) is clear. Further studies are also needed to uncover cellular events involved the action of these PLA_2 enzymes. These enzymes help us in our understanding of Ca^{2+} movement and biology of muscle cells.

PLA_2 enzymes and platelet aggregation

Several snake venom PLA_2 enzymes affect platelet aggregation: one group initiates aggregation, a second group inhibits aggregation, whereas a third group of enzymes initiates as well as inhibits platelet aggregation (Kini and Evans, 1997). These effects of PLA_2 enzymes in some cases are dependent on the enzymatic activity, but in other cases are independent of phospholipolytic activity. In addition, antiplatelet PLA_2 enzymes inhibit platelet aggregation by several different mechanisms (Kini *et al.*, unpublished observations) and most likely they interact with different platelet membrane acceptor/receptor proteins. No acceptor protein on platelets has been identified, and the details of molecular events are also scanty. These enzymes are suitable to understand cellular events in thrombosis and hemostasis.

4.2 Methods to identify target proteins

Although different approaches can be used to identify a target protein, the following step-by-step description and associated precautions may be useful. The first step for identification of a target protein is to study the specific binding of the radiolabeled PLA_2 enzyme to the specific cells or tissue (see Lambeau *et al.*, 1997 and references therein). The following are some of the precautions one should take avoid erroneous results and interpretations:

- The radiolabeled PLA_2 should have a pharmacological potency similar to the unlabeled, native PLA_2. This modification becomes important if the radiolabelling occurs at the pharmacological site.
- PLA_2 enzymes are targeted to surface proteins (according to the model). Therefore it is critical to avoid the use of nonspecific proteinases during the preparation of cells. For example, instead of trypsinization, a treatment with collagenase may be more suitable for harvesting the cells from cell culture.
- Generally, the affinity constants between the PLA_2 enzyme and its protein target range from sub-nM to nM. Observation of low-affinity binding to cells, in the range of low mM, is more likely due to binding of the PLA_2 to phospholipids rather than the specific target protein. Generally, numbers of the low-affinity binding sites are larger than the high-affinity sites.
- Cross-linking studies of radiolabeled PLA_2 to the target protein are used in the second stage to identify the size as well as the specificity of binding. This can be achieved by glutaraldehyde or bifunctional cross-linking agents.

In some cases, photoaffinity-labelling techniques can also be used. In this method, results can change drastically depending on the position of the label, because the cross-linking occurs with the topologically nearest reactive group. Therefore it is possible to get a cross-link with two different proteins by having the photolabile group in different locations of the molecule; hence care must be taken during the interpretation of these studies. In both cases, suitable controls and binding inhibition by excess unlabeled PLA_2 are essential. It should be noted that in some cases, inherent enzymatic activity of PLA_2 protein may compromise the binding data. Therefore it is recommended to perform the binding studies with catalytically inactive conditions (either chemically modified protein or in the presence of an inhibitor). Isolation and further characterization strategies depend on the nature of the target protein.

5 Role of Enzymatic Activity and Its Effect on Target Cells

Upon binding to their target proteins, PLA_2 enzymes induce their effects by mechanisms either dependent on or independent of their catalytic activity. This binding process and subsequent catalytic damage near the acceptor/receptor most probably results in the loss of specific function of the acceptor/receptor protein or its neighboring protein(s) due to change in the environment of these membrane-bound proteins. This in turn could and does affect the normal physiological status of the target cell or tissue. In some cases, the released products, such as lysophospholipids and fatty acids, can cause the pharmacological effect (Kini and Evans, 1989). In other cases, binding to the target protein can cause the pharmacological effect by acting as an agonist or an antagonist, or by blocking the interaction of the target protein with its normal ligands. However, even in the latter case, the enzymes would hydrolyze the phospholipids in the neighborhood of their target proteins, but this hydrolysis is inconsequential to the observed pharmacological effect. This is exemplified by the mechanism of inhibition of the prothrombinase complex by a strongly anticoagulant PLA_2 enzyme, CM-IV from *Naja nigricollis* venom (Stefansson *et al.*, 1990). Finally, there are cases where the effect is due to both enzymatic and nonenzymatic mechanisms (Kini and Evans, 1995). Thus, upon binding to the target protein, the enzymes induce their effects either dependent on or independent of their enzymatic activity (Kini and Evans, 1989). Therefore, it is critical to determine the role of enzymatic activity of PLA_2 enzymes in inducing the pharmacological effects.

One of the simplest ways to determine the role of enzymatic activity is to increase the (pre)incubation time of the PLA_2 enzyme with the "complete" system and determine whether the pharmacological effect increases with the (pre)incubation time (Kini and Evans, 1988). When phospholipid hydrolysis contributes to the effect, the potency of the effect increases with the incubation time, since more and more phospholipids are hydrolyzed. On the other hand, in nonenzymatic mechanisms, the effect will not increase with the (pre)incubation times. The presence or absence of (fatty acid-free) serum albumin (such as BSA) may help in resolving whether it is the hydrolysis of phospholipids or the released products that cause the pharmacological effects (for details, see Fletcher and Rosenberg, 1997). The main advantage is that no additional chemicals or reagents are added to the system. However, if the binding interaction between the PLA_2 enzyme and its target protein is slow, but tight, the results may lead to erroneous interpretations.

PLA_2 enzymes require Ca^{2+} for their activity. Thus inclusion of chelators, such as EDTA, may help determine the role of enzymatic activity (for example, see Kini and Evans, 1988). The replacement of the metal ion by other metal ions, such as Ba^{2+} and Sr^{2+}, is also used (Kelly *et al.*, 1975). This is particularly important in cases where the removal of Ca^{2+} ions affects the normal functions of the system. Sr^{2+} substitution to some extent supports the normal functions of the system, but not the phospholipolytic activity of the PLA_2 enzyme in question (Kelly *et al.*, 1975). Nonenzymatic mechanisms are not affected by EDTA or other metal ions, whereas the enzymatic contributions to the effect are compromised significantly. Care must be taken to insure that the additional components do not themselves contribute to the effect. The above methods, however, are suitable only in the *in vitro* systems. Additional studies on the hydrolysis of specific phospholipids may assist in clarifying the role of enzymatic activity.

The histidine residue at position 48 is conserved in PLA_2 enzymes and plays a significant role in phospholipid hydrolysis by these enzymes (Scott, 1997). Alkylation of this His residue leads to complete loss of enzymatic activity. Thus His-modified PLA_2 enzymes are suitable for both *in vitro* and *in vivo* systems for studying the role of enzymatic activity. The active site His can be easily alkylated using *p*-bromophenacyl bromide (*p*-BPB) or methyl *p*-nitrobenzenesulfonate (Yang and King, 1980; Verheij *et al.*, 1980b). Although the alkylation does not significantly affect the conformation of the PLA_2 enzyme and its ability to bind to phospholipids, it could alter the ability of the enzyme to interact with specific proteins or ligands. More details on the pitfalls in the methods for determining the role of enzymatic activity and on the correlation between the enzymatic activity and pharmacological effects are described elsewhere (Fletcher *et al.*, 1997; Rosenberg, 1997).

Acknowledgements

This work is supported by a financial grant from the Economic Development Board of Singapore and Academic research grants (RP 940325, RP 950377 and RP 960304) from the National University of Singapore.

References

Abe T, Alema S, Miledi R (1977) Isolation and characterization of presynaptically acting neurotoxins from the venom of Bungarus snakes. Eur. J. Biochem. 80: 1–12

Ancian P, Lambeau G, Mattéi MG, Lazdunski M (1995) The human 180-kDa receptor for secretory phospholipases A_2. Molecular cloning, identification of a secreted soluble form, expression, and chromosomal localization. J. Biol. Chem. 270: 8963–70

Arita H, Hanasaki K, Nakano T et al (1991) Novel proliferative effect of phospholipase A_2 in Swiss 3T3 cells via specific binding sites. J. Biol. Chem. 266: 19139–41

Awan KA, Dolly JO (1991) K^+ channel subtypes in rat brain: characteristic locations revealed using β-bungarotoxin, α- and δ-dendrotoxins. Neuroscience 40: 29–39

Boffa MC, Rothen C, Verheij HM et al (1980) Correlation of enzymatic activity and anticoagulant properties of phospholipase A_2. Natural Toxins (Eaker D, Wadstrom T, eds) pp 131–8: Oxford: Pergamon Press

Braganca BM, Sambray YM (1967) Multiple forms of cobra venom phospholipase A. Nature 216: 1210–1

Breeze AL, Dolly JO (1989) Interactions between discrete neuronal membrane binding sites for the putative K^+-channel ligands β-bungarotoxin, dendrotoxin and mast-cell-degranulating peptide. Eur. J. Biochem. 178: 771–8

Dennis EA (ed) (1991) Phospholipases. Methods Enzymol. 197 San Diego: Academic Press

Dennis EA (1994) Diversity of group types, regulation, and function of phospholipase A_2. J. Biol. Chem. 269: 13057–60

Dennis EA, Rhee SG, Billah MM, Hannun YA (1991) Role of phospholiase in generating lipid second messengers in signal transduction. FASEB J. 5: 2068–77

Evans HJ, Franson R, Qureshi GD, Moo-Penn WF (1980) Isolation of anticoagulant proteins from cobra venom (Naja nigricollis). Identity with phospholipases A_2. J. Biol. Chem. 255: 3793–7

Fletcher JE, Rosenberg P (1997) The cellular effects and mechanisms of action of presynaptically acting phospholipase A_2 toxins. Venom Phospholipase A_2 Enzymes: Structure, Function, and Mechanism (Kini RM, ed) pp 417–58: Chichester: Wiley

Fletcher JE, de Araujo HSS, Ownby CL (1997) Molecular events in the myotoxic action of phospholipases. Venom Phospholipase A_2 Enzymes: Structure, Function, and Mechanism (Kini RM, ed) pp 459–501: Chichester: Wiley

Fraenkel-Conrat H (1982–83) Snake venom neurotoxins related to phospholipase A_2. J. Toxicol. Toxin Rev. 1: 205–21

Fry MR, Ghosh SS, East JM, Franson RC (1992) Role of human sperm phospholipase A_2 in fertilization: effects of a novel inhibitor of phospholipase A_2 activity on membrane perturbations and oocyte penetration. Biol. Reprod. 47: 751–9

Glaser KB, Mobilio D, Chang JY, Senko N (1993) Phospholipase A_2 enzymes: regulation and inhibition. Trends Pharmacol. Sci. 14: 92–8

Hanasaki H, Arita H (1992) Purification and characterization of a high-affinity binding protein for pancreatic-type phospholipase A_2. Biochim. Biophys. Acta 1127: 233–41

Hendrickson HS (1994) Fluorescence-based assays of lipases, phospholipases and other lipolytic enzymes. Anal. Biochem. 219: 1–8

Higashino K, Ishizaki J, Kishino J et al (1994) Structural comparison of phospholipase A_2-binding regions in phospholipase A_2 receptors from various mammals. Eur. J. Biochem. 225: 375–82

Ishizaki J, Hanasaki K, Higashino K et al (1994) Molecular cloning of pancreatic group I phospholipase A_2 receptor. J. Biol. Chem. 269: 5897–904

Kanemasa T, Hanasaki H, Arita H (1992) Migration of vascular smooth muscle cells by phospholipase A_2 via its specific binding sites. Biochim. Biophys. Acta 1125: 210–4

Kasturi S, Gowda TV (1989) Purification and characterization of a major phospholipase A_2 from Russell's viper (Vipera russelli) venom. Toxicon 27: 229–37

Kelly RB, Oberg SG, Strong PN, Wagner GM (1975) β-Bungarotoxin, a phospholipase that stimulates transmitter release. Cold Spring Harb. Symp. Quant. Biol. 40: 117–25

Kini RM (ed) (1997) Venom Phospholipase A_2 Enzymes: Structure, Function, and Mechanism, Chichester: Wiley

Kini RM, Evans HJ (1988) Correlation between the enzymatic activity, anticoagulant and antiplatelet effects of phospholipases A_2 isoenzymes from Naja nigricollis venom. Thrombos. Haemos. 60: 170–3

Kini RM, Evans HJ (1989) A model to explain the pharmacological effects of snake venom phospholipases A_2. Toxicon 27: 613–35

Kini RM, Evans HJ (1995) The role of enzymatic activity in inhibition of the extrinsic tenase complex by phospholipases A_2 isoenzymes from Naja nigricollis venom. Toxicon 33: 1585–90

Kini RM, Evans HJ (1997) Effects of phospholipase A_2 enzymes on platelet aggregation. Venom Phospholipase A_2 Enzymes: Structure, Function, and Mechanism (Kini RM, ed) pp 373–91: Chichester: Wiley

Kini RM, Kawabata SI, Iwanaga S (1986) Comparison of amino terminal region of three isoenzymes of phospholipases A_2 (TFV PL-Ia, TFV PL-Ib, TFV PL-X) from Trimeresurus flavoviridis (habu snake) venom and the complete amino acid sequence of the basic phospholipase, TFV PL-X. Toxicon 24: 1117–29

Kishino J, Tohkin M, Arita H (1992) Proliferative effect of phospholipase A_2 in rat chondrocyte via its specific binding sites. Biochem. Biophys. Res. Commun. 186: 1025–31

Klibansky C, London Y, Frenkel A, de Vries A (1968) Enhancing action of synthetic and natural basic polypeptides on erythrocyte-ghost phospholipid hydrolysis by phospholipase A. Biochim. Biophys. Acta 150: 15–23

Krizaj I, Dolly JO, Gubensek F (1994) Identification of the neuronal acceptor in bovine cortex for ammodytoxin C, a presynaptically phospholipase A_2. Biochemistry 33: 13938–45

Kudo I, Murakami M, Hara S, Inoue K (1993) Mammalian non-pancreatic phospholipase A_2. Biochim. Biophys. Acta 1170: 217–31

Lambeau G, Barhanin J, Scweitz H et al (1989) Identification and properties of very high-affinity brain membrane-binding sites for a neurotoxic phospholipase from taipan venom. J. Biol. Chem. 264: 11503–10

Lambeau G, Scmid-Alliana A, Lazdunski M, Barhanin J (1990) Identification and purification of a very high-affinity binding protein for toxic phospholipases A_2 in skeletal muscle. J. Biol. Chem. 265: 9526–32

Lambeau G, Barhanin J, Lazdunski M (1991) Identification of different receptor types of toxic phospholipases A2 in rabbit skeletal muscle. FEBS Lett. 293: 29–33

Lambeau G, Ancian P, Barhanin J, Lazdunski M (1994) Cloning and expression of a membrane receptor for secretory phospholipases A_2. J. Biol. Chem. 269: 1575–8

Lambeau G, Ancian P, Nicolas JP et al (1995) Structural elements of secretory phospholipases A_2 involved in the binding to M-type receptors. J. Biol. Chem. 270: 5534–40

Lambeau G, Cupillard L, Lazdunski M (1997) Membrane receptors for venom phospholipases A_2. Venom Phospholipase A_2 Enzymes: Structure, Function, and Mechanism (Kini RM, ed) pp 393–416: Chichester: Wiley

Mukherjee AB, Miele L, Pattabhiraman N (1994) Phospholipase A_2 enzymes: regulation and physiological role. Biochem. Pharmacol. 48: 1–10

Nakajima M, Hanasaki H, Ueda M, Arita H (1992) Effect of pancreatic type phospholipase A_2 on isolated porcine cerebral arteries via its specific binding sites. FEBS Lett. 309: 261–4

Nicolas JP, Lambeau G, Lazdunski M (1995) Identification of the binding domain for secretory phospholipases A_2 on their M-type 180-kDA membrane receptor. J. Biol. Chem. 270: 28869–73

Ogawa T, Oda N, Nakashima KI, Sasaki H et al (1992) Unusually high conservation of untranslated sequences in cDNAs for Trimeresurus flavoviridis phospholipase A_2 isozymes. Proc. Natl. Acad. Sci, USA 89: 8557–61

Reynolds LJ, Washburn WN, Deems RA, Dennis EA (1991) Assay strategies and methods for phospholipases. Methods Enzymol. 197: 3–23

Richieri GV, Kleinfeld AM (1995) Continuous measurement of phospholipase A_2 activity using fluorescent probe ADIFAB. Anal. Biochem. 229: 256–63

Richieri GV, Ogata RT, Kleinfeld AM (1992) A fluorescently labeled intestinal fatty acid binding protein. Interactions with fatty acids and its use in monitoring free fatty acids. J. Biol. Chem. 267: 23495–501

Rosenberg P (1986) The relationship between enzymatic activity and pharmacological properties of phospholipases in natural poisons. Natural Toxins (Harris JB, ed) pp 129–74: Oxford: Oxford University Press

Rosenberg P (1997) Pitfalls to avoid in the study of correlations between enzymatic activity and pharmacological properties of phospholipase A_2 enzymes. Venom Phospholipase A_2 Enzymes: Structure, Function, and Mechanism (Kini RM, ed) pp 159–87: Chichester: Wiley

Schmidt RR, Betz H, Rehm H (1988) Inhibition of β-bungarotoxin binding to brain membranes by mast cell degranulating peptide, toxin I, and ethylene glycol bis(β-aminoethyl ether)-N,N,N',N'-tetraacetic acid. Biochemistry 27: 963–7

Scott DL (1997) Phospholipase A_2: structure and catalytic properties. Venom Phospholipase A_2 Enzymes: Structure, Function, and Mechanism (Kini RM, ed) pp 101–32: Chichester: Wiley

Shiloah J, Klibansky C, De Vries A (1973) Phospholipase isoenzymes from Naja naja venom – I. Purification and partial characterization. Toxicon 11: 481–90

Sommers CD, Bobbitt JL, Bemis KG, Snyder DW (1992) Porcine phospholipase A_2-induced contractions of guinea pig lung pleural strips. Eur. J. Pharmacol. 216: 87–96

Stefansson S, Kini RM, Evans HJ (1990) The basic phospholophase A_2 from naja nigricollis venom inhibits the prothrombinase complex by a novel nenenzymatic mechanism. biochemistry 29: 7742–7746.

Strong PN, Goerke J, Oberg SG, Kelly RB (1976) β-Bungarotoxin, a presynaptic toxin with enzymatic activity. Proc. Natl. Acad. Sci USA 73: 178–82

Takasaki C, Suzuki J, Tamiya N (1990a) Purification and properties of several phospholipases A_2 from the venom of Australian king brown snake (Pseudechis australis). Toxicon 28: 319–27

Takasaki C, Yutani F, Kajiyashiki T (1990b) Amino acid sequences of eight phospholipases A_2 from the venom of Australian king brown snake (Pseudechis australis). Toxicon 28: 329–39

Tzeng MC (1993) Interaction of presynaptically toxic phospholipases A_2 with membrane receptors and other binding sites. J. Toxicol. Toxin Rev. 12: 1–62

Tzeng MC, Hseu MJ, Yang JH et al. (1986) Specific binding of three neurotoxins with phospholiphase $_2$ activity to synaptosomal membrane preparations from the guinea pig brain. J. Prot. Chem. 5: 221–228

Ueno E, Rosenberg P (1990) Inhibition of phosphorylation of rat synaptosomal proteins by snake venom phospholipase A_2 neurotoxins (β-bungarotoxin, notexin) and enzymes (Naja naja atra, Naja nigricollis). Toxicon 28: 1423–1437

Ueno E, Rosenberg P (1992) Inhibition of phosphorylation of synapsin I and other synaptosomal proteins by β-bungarotoxin, a phospholipase A_2 neurotoxin. J. Neurochem. 59: 2030–9

Ueno E, Rosenberg P (1995) β-bungarotoxin blocks phorbol ester-stimulated phosphorylation of MARKS, GAP-43 and synapsin I in rat brain synaptosomes. Toxicon 33: 747–62

Vadas P, Pruzanski W (1986) Role of secretory phospholipase A_2 in the pathobiology of disease. Lab. Invest. 55: 391–404

Vadas P, Browning J, Edelson J, Pruzanski W (1993) Extracellular phospholipase A_2 expression and inflammation: the relationship with associated disease states. J. Lipid Mediators 8: 1–30

Verheij HM, Boffa MC, Rothen C et al (1980a) Correlation of enzymatic activity and anticoagulant properties of phospholipase A_2. Eur. J. Biochem. 112: 25-32

Verheij HM, Volwerk JJ, Jansen EJM et al (1980b) Methylation of histidine-48 in pancreatic phospholipase A2. Role of histidine and calcium ion in the catalytic mechanism. Biochemistry 19: 743–750

Verheij HM, Slotboom AJ, De Haas GH (1981) Structure and function of phospholipase A_2. Rev. Physiol. Biochem. Pharmacol. 91: 91–203

Vishwanath BS, Kini RM, Gowda TV (1985) Purification of an edema inducing phospholipase A_2 enzyme from Vipera russelli venom and its interaction with aristolochic acid. Toxicon 23: 617

Vishwanath BS, Kini RM, Gowda TV (1987) Characterization of three edema-inducing phospholipase A_2 enzymes from habu (Trimeresurus flavoviridis) venom and their interaction with the alkaloid aristolochic acid. Toxicon 25: 501–515

Vishwanath BS, Kini RM, Gowda TV (1988) Purification and partial biochemical characterization of an edema inducing phospholipase A_2 from Vipera russelli (Russell's viper) snake venom. Toxicon 26: 713–20

Waite M (ed) (1987) The Phospholipases. New York: Plenum Press

Yang CC (1978) Chemistry and biochemistry of snake venom neurotoxins. Toxicon 17 (Suppl 1) 261–92

Yang CC, King K (1980) Chemical modification of histidine residue in basic phospholipases A_2 from the venom of Naja nigricollis. Biochim. Biophys. Acta 614: 373–88

Yen CH, Tzeng MC (1991) Identification of a new binding protein for crotoxin and other neurotoxic phospholipase A_2s on brain synaptic membranes. Biochemistry 30: 11473–7

20 Polypeptides and Proteins Active in the Coagulation Process

Jüri Siigur and Ene Siigur

Contents

Methods and Tools in Biosciences and Medicine
Animal Toxins, ed. by H. Rochat and M.-F. Martin-Eauclaire

1 Introduction

Snake venoms are known to contain a number of factors affecting blood coagulation. These venoms have been found to have potent effects on coagulation through both pro- and anticoagulant mechanisms (reviews: Kornalik, 1985, 1990; Seegers and Ouyang, 1979; Tu, 1977; Stocker, 1990a,b; Markland, 1991; Siigur and Siigur, 1992; Ouyang *et al.*, 1992).

The present review will concentrate on the effects of Viperidae venom components affecting mainly the blood coagulation system and platelet function.

To avoid involvement in some of the uncertainties in the classification of venomous snakes, we use here the term "Viperidae snakes" to indicate the following genera: *Bitis, Cerastes, Echis, Eristicophis, Vipera.*

Studies of the blood coagulation process using snake venom components are important not only for understanding blood coagulation mechanisms but also for development of new agents for the treatment of thrombic diseases.

2 Coagulant Action

The blood coagulation process is rather complicated. Here, a simplified coagulation scheme is reported to better understand the following text.

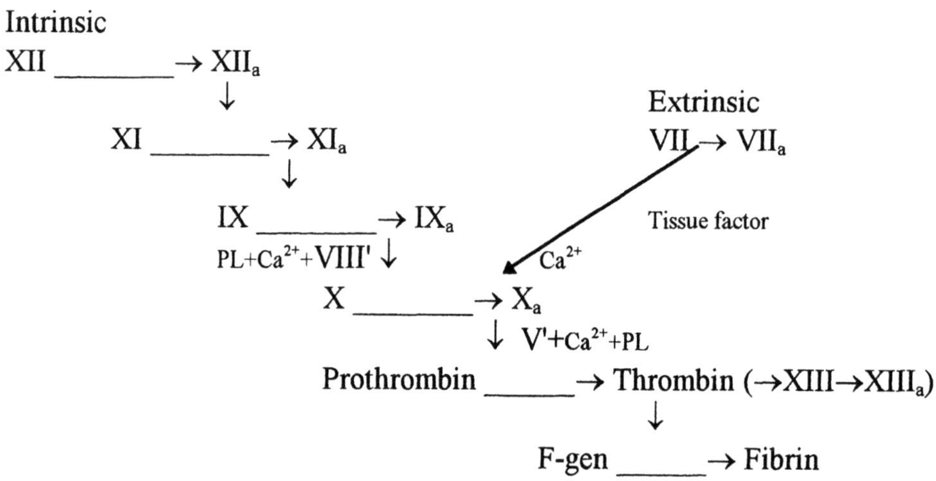

Figure 1 Blood coagulation process (principal scheme).

2.1 Fibrinogen-converting (thrombin-like) enzymes

Conversion of fibrinogen to fibrin occurs via liberation of fibrinopeptides A and B from the fibrinogen N-terminal disulfide knot. The arising fibrinogen monomers aggregate to form a firm fibrin clot:

$$n(A\alpha B\beta\gamma)_2 \rightarrow 2nA + 2nB + n(\alpha\beta\gamma)_2 \rightarrow [(\alpha\beta\gamma)_2]_n$$

firm clot (Stocker, 1980)

Such a behavior is characteristic to thrombin, and to a *Bitis gabonica* venom thrombin-like enzyme called gabonase. This enzyme, isolated by Pirkle *et al.* (1986), is a single-chain glycoprotein (20.6% carbohydrates) with an M_r of 30,600 and pI of 5.3. It is a serine proteinase, and TAME is the best synthetic substrate for gabonase (Pirkle *et al.*, 1986).

Thrombin-like enzymes with different specificity have been isolated from the venoms of Cerastes: cerastobin from *C. vipera* (Farid and Tu, 1989) and proteinase RP34 from *C.cerastes* venom (Laraba-Djebari *et al.*, 1992). Proteinase RP34 splits off some fibrinopeptide of A-type, forming a weak clot:

$$n(A\alpha B\beta\gamma)_2 \rightarrow 2nA + n(\alpha B\beta\gamma)_2 \rightarrow [(\alpha B\beta\gamma)_2]_n$$

weak clot (Stocker, 1980)

Cerastobin is a single-chain protein with an M_r of 38,000, 348 amino acid residues, pI 7.7, and no carbohydrates (Farid and Tu, 1989). It has very broad substrate specificity, hydrolyzing fibrinogen, fibrin-monomer, factor X (no activation), antithrombin III, esters and amides of arginine (but not BAPNA), releases bradykinin from bradykininogen, and causes uterine contractions. In addition, cerastobin has potent platelet aggregatory activity (Farid et al., 1990).

Proteinase RP34 has an M_r of 97,000 (nonreduced) or 48,500 (reduced form) on SDS-PAGE and a pI of 3.75. From the same venom (*C. cerastes*) another thrombin like enzyme – afaâcytin – has been isolated (Laraba-Djebari *et al.*, 1995). It had an M_r of 40,000 in nonreduced form and yielded three distinct bands under reducing conditions [with M_r of 43,000 (α), 35,500 (β), and 10,200 (β')].

Afaâcytin has broad substrate specificity, splitting casein, fibrinogen, arginine esters and amides, etc. (Laraba-Djebari *et al.*, 1995).

The N-terminal sequences of gabonase, cerastobin, proteinase RP34, and afaâcytin α- and β-chains were highly homologous.

2.2 Prothrombin activation by viperidae venom enzymes

Prothrombin is unique among the clotting proteases because, after activation, other proteases remain bound to the phospholipid surface, whereas active thrombin is released and is able to reach all its substrates in the environment (Kornalik, 1985).

In blood coagulation cascade, thrombin is formed from prothrombin by the so-called prothrombinase complex, consisting of activated factors X and V, phospholipids, and Ca^{2+}. Prothrombin is activated also by some components of snake venoms (Rosing and Tans, 1992). An active component found in the venom of *Echis carinatus* – ecarin – was first partially purified by Schieck *et al.* (1972) and later isolated and characterized by a number of other investigators (Morita and Iwanaga, 1978; Rhee *et al.*, 1982; Fortova *et al.*, 1983). According to Rosing and Tans (1992) ecarin belongs to the first group of prothrombin activators which do not require added metal ions or phospholipids or factor Va. These activators are only capable of hydrolyzing the Arg 322 – Ile 323 bond in prothrombin.

The product of prothrombin activation by ecarin is meizothrombin:

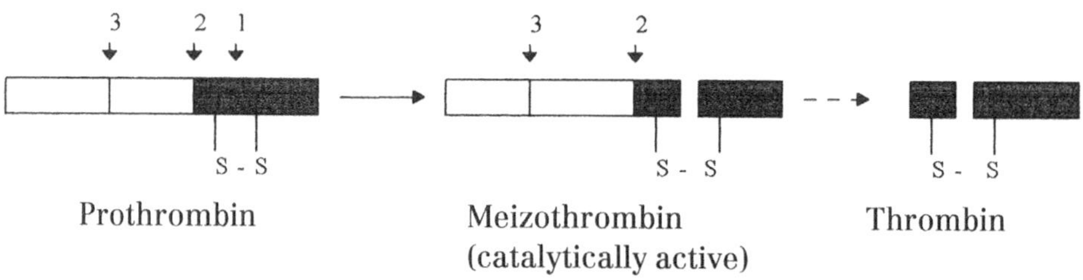

In some cases prolonged incubation will result in thrombin formation (mainly with human prothrombin) probably by autocatalytic cleavage of bond 2 by meizothrombin (Rhee *et al.*, 1982).

Ecarin is a single-chain glycoprotein with a molecular weight of 55,000–63,000 (Morita and Iwanaga, 1978; Rhee *et al.*, 1982; Fortova *et al.*, 1983) and pI of 4.5 ± 0.1 (Morita and Iwanaga, 1978). It contains 16.6% of carbohydrates (Rhee *et al.*, 1982). Ecarin has very narrow substrate specificity: it does not activate factor X or IX, plasminogen, prekallikrein, trypsinogen, has no effect on casein, arginine or lysine esters or amides (Morita and Iwanaga, 1978). The prothrombin activation can be monitored with chromogenic peptides that release pNA proportionally to the amount of generated thrombin derivative (Stocker *et al.*, 1986a). Ecarin is relatively unstable in acidic ($pH < 5$) or basic ($pH > 10$) media. Activity is lost by heating 10 min at 60°C and by metalloproteinase and thiol inhibitors (EDTA, 2-mercaptoethanol, DTT) (Morita and Iwanaga, 1978).

In 1995, the complete amino acid sequence of ecarin was published, deduced from the nucleotide sequence of a complementary DNA (cDNA) clone isolated by screening a venomous gland cDNA library of Kenyan *Echis carinatus* (Nishida *et al.*, 1995). The cDNA sequence with 2379 bp encodes an open reading frame of 616 amino acids. Ecarin is translated as a precursor protein, which may be processed posttranslationally. The processed mature protein consists of 426 amino acid residues, showing the strongest sequence similarity with that of Russell's viper venom factor X activator heavy chain (64% identity). Like RVV-X, ecarin contains metalloproteinase, disintegrin-like, and cysteine-rich domains (Nishida *et al.*, 1995).

An enzyme similar to ecarin – ecamulin – was isolated from the venom of Central Asian *E. multisquamatus* (Solovyov *et al.*, 1996). The enzyme gives a doublet with M_r of 67,000 and 27,000 on SDS-PAGE under nonreducing conditions and produces three distinct bands in the presence of DTT or 2-mercaptoethanol, with M_r of 67,000, 14,000 and 13,000. The pI of ecamulin is in the pH range of 4.3–4.5. It does not hydrolyze esters and amides of arginine, except the chromogenic substrates for kallikreins – S 2302 and S2266. It is a Zn enzyme. The enzymatic activity is inhibited by EDTA and DTT but not by DFP or PMSF. The activation process of prothrombin is identical to that of ecarin and does not require Ca^{2+}, phospholipids, or factor V (Solovyov *et al.*, 1996).

Recently a novel prothrombin activator with a unique catalytic mechanism was discovered in the venom of *E. carinatus* – carinactivase (CA-1) (Yamada *et al.*, 1996). On SDS-PAGE two bands (60,000/62,000 doublet plus 25,000) were obtained under nonreducing conditions, and three bands (62,000/64,000, 17,000 and 14,000) under reducing conditions. The 60/62-kDa chain had high N-terminal sequence homology to ecarin, and this subunit had metalloproteinase activity. The two chains of the 25-kDa subunit resembled one another, and were rather similar to those of snake venom factor IX/factor X-binding protein (Atoda *et al.*, 1991).

The specificity of CA-1 is as strict as that of ecarin. The main difference between CA-1 and ecarin was CA-1's requirement for Ca^{2+} ions for the activation of prothrombin. As a result of many different experiments, the authors came to the conclusion that the unique structure of CA-1 explains the unique mechanism of its activation of prothrombin; the 25,000-Da subunit first recognizes the N-terminal Gla domain of prothrombin in a Ca^{2+}-dependent fashion, and then the 62,000-Da subunit cleaves the bond between the A-chain and the B-chain (Yamada *et al.*, 1996).

In addition to *E. carinatus* prothrombin activator, a procoagulant with M_r 125,000 (reduced form 66,000) and pI of 4.3 has been purified from *Vipera aspis aspis* venom which might participate in the activation of prothrombin (Komori *et al.*, 1993).

The enzyme has relatively weak dimethyl casein hydrolyzing activity, but no proteolytic activity; it does not cleave esters and amides of arginine. Incubation of the aspis procoagulant with prothrombin in the presence of factor V, phospholipids, and Ca^{2+} decreased the clotting time of fibrinogen remarkably, at the same time increasing the esterase activity of the mixture which could be explained by thrombin generation from prothrombin. The absence of factor V, phospholipids, and Ca^{2+} drastically decreased esterolytic activity of the reaction mixture. This may be the indirect proof of prothrombin activation by *V. aspis* procoagulant. As the process is inhibited by DFP, the enzyme is serine proteinase, and thus the *V. aspis* venom procoagulant might react with prothrombin as factor X_a from plasma.

2.3 Factor V activator

Factor V is a protein cofactor that accelerates the conversion of prothrombin to thrombin by activated factor X. In plasma its activation was shown to occur by discrete proteolysis by thrombin to give a multiple subunit protein (Kisiel, 1991). Factor V can also be activated by a component of Russell's viper venom.

RVV-V is a glycoprotein with M_r of 28,000–29,000 (Smith and Hanahan, 1976; Kisiel, 1979) containing 6% carbohydrate. It has weak amidase activity but readily hydrolyzes TAME. The hydrolysis is inhibited by DFP, indicating the serine proteinase nature of RVV-V. Although RVV-V appeared homogeneous on SDS-PAGE, Tokunaga *et al.* (1988) resolved it into three fractions (RVV-Vα, RVV-Vβ and RVV-Vγ), by reversed-phase HPLC on Cosmosil SC_4 300 columns. The three isomers were present in the final preparation in a ratio of 2:1:6, respectively. The complete amino acid sequences of RVV-Vα and RVV-Vγ have been determined (Tokunaga *et al.*, 1988). Both enzymes consist of 236 residues with 6 disulfide bridges and differ in only six amino acid residues in the molecule. The N-terminal sequences are homologous to other viper venom serine proteinases, such as gabonase (Pirkle *et al.*, 1986) and cerastobin (Farid and Tu, 1989).

Although thrombin-catalyzed activation of factor V occurs in three cleavages in the factor V molecule (Kisiel, 1991), RVV-V cleaves only one peptide bond (Arg 1545 – Ser 1546) in factor V (Kisiel, 1991). The same specificity might be characteristic to partially purified factor V activator from *V. aspis* venom (Boffa and Boffa, 1974).

2.4 Factor X activators

Factor X is a glycoprotein that plays a central role in blood coagulation. It is activated by two physiological pathways called the intrinsic and extrinsic systems. Factor X_a is a serine proteinase which in turn activates prothrombin in the presence of factor V, calcium ions, and phospholipids. Factor X can be also activated by components of some snake venoms, the best-known among them is RVV-X – *V. russellii* venom factor X activator.

The conversion of factor X to factor X_a by factor IX_a and factor VIII' as well as by a proteinase from Russell's viper venom is due to the cleavage of the peptide bond between Arg-51 and Ile-52, giving rise to a new N-terminal isoleucine and a decrease in the molecular weight of the zymogen from 55,100 to 45,300 ($X_{a,\alpha}$). In the second step, the peptide bond between Arg-290 and Gly-291 is cleaved by factor $X_{a\alpha}$ to yield factor $X_{a\beta}$. Cleavage of the second bond does not alter the specific activity of the factor X_a (Fujikawa *et al.*, 1975).

RVV-X is isolated in several different laboratories using gel filtration and ion-exchange chromatography methods (Furukawa *et al.*, 1976; Kisiel *et al.*, 1976; Teng *et al.*,1984; Jayanthi and Gowda, 1990), immunoaffinity with monoclonal antibodies (Pukrittayakamee *et al.*, 1983; Durkee *et al.*, 1993), or affinity chromatography on columns of Sepharose-factor X equilibrated with lanthanide ions (Furie and Furie, 1975). The physicochemical properties of isolated enzymes are quite conflicting.

Takeya *et al.* (1992) determined the primary structure of the RVV-X heavy chain and one of the light chains (LC1). These chains were separated after reduction and S-pyridylethylation on the reversed-phase HPLC. The heavy chain consists of 427 amino acid residues with four asparagine-linked sugar chains at positions 28, 69, 163, and 183. Light-chain LC1 consists of 123 amino acid residues containing an asparagine-linked sugar chain at position 24. The heavy chain contains metalloproteinase, disintegrin-like, and cysteine-rich domains. The light chain shows sequence homology similar to that found in the so-called C-type lectins (Takeya *et al.*, 1992).

RVV-X potently inhibits collagen- and ADP-induced platelet aggregations in a dose-dependent manner, probably due to its disintegrin-like domain [although this domain contains an Arg-Asp-Glu-(RDE) sequence instead of Arg-Gly-Asp-(RGD)]. Platelet aggregation inhibitory activity could, however, be due to some of the proteolytic activity or the lectin-like activity, since whole RVV-X was used for that experiment (Takeya *et al.*, 1992).

Gowda *et al.* (1994) consistently obtained by SDS-PAGE, under nonreducing conditions, apparent M_r values of 93,000 for RVV-X and 78,000 for the completely deglycosylated molecule. They also observed that the two distinct light chains are present in equal proportions on SDS-PAGE of the reduced molecule [M_r: 62,000, heavy chain (α-chain); 21,000, β-chain; 18,000, γ-chain]. After deglycosylation, the M_r of reduced forms were 48,000, α-chain; 17,000, β-chain; 14,000, γ-chain.

The α-chain N-terminal microsequence corresponded to the sequence of Takeya *et al.* (1992), except for Ala in the third cycle. The amino acid sequence of the γ-chain appeared to be identical to that reported by Takeya *et al.* (1992) for LC1. The NH_2-terminal amino acid sequence of the β-chain was not present within the complete sequences of either the α- or γ-chain determined by Takeya *et al.* (1992). As a conclusion, they found that (i) RVV-X is a 92,880-kDa glycoprotein and consists of three disulfide-linked glycosylated polypeptide chains, and (ii) all three polypeptide chains of RVV-X contain asparagine-linked oligosaccharides. The α-chain contains four oligosaccharides; β- and γ-chains contain one oligosaccharide each (Gowda *et al.*, 1994).

Gowda *et al.* (1996) investigated the role of carbohydrates in the structure and functional activity of RVV-X and found that peripheral carbohydrate moieties were not involved in interactions with factor X. Removal of whole oligosaccharide chains with N-glycanase caused an almost total loss of the ability of RVV-X to activate factor X to factor X_a. Significant changes in the secondary structure of the protein also appeared.

Enzymes activating factor X are also isolated from other snake venoms. Factor X activator from *V. aspis aspis* venom was purified by Komori *et al.* (1990). It had an M_r of 75,000 (nonreduced, SDS-PAGE) and pI 4.6. Reduced protein gave two bands in the ratio of ~2:3, with M_r of 16,000 and 14,000 on SDS-PAGE. The amino terminal sequence of reduced and pyridylethylated derivatives of the activator was different from that of RVV-X. It is interesting that all the derivatives possessed the same sequence, so the *aspis* activator is suspected to consist of several peptide chains with M_r of 14,000–16,000.The cleavage site of factor X by *aspis* activator has not been defined (Komori *et al.*, 1990).

An activator of factor X was found in the venom of the horned viper *Cerastes cerastes* by Franssen *et al.* (1983). This activator resembles RVV-X both in structural and catalytic properties. The nonreduced protein gives a doublet with M_r of 76,000 and 67,000 on SDS-PAGE; the reduced form contains a heavy chain of M_r 58,000 and two light chains of M_r 17,700 and 15,000. The activator appears to cleave the same bond in factor X that is cleaved by factor IX_a. Factor X activation is strongly stimulated by Ca^{2+} (Franssen *et al.*, 1983).

A special low-molecular weight factor X activator has been purified from *Cerastes vipera* (Sahara sand viper) venom (Farid *et al.*, 1993). It had an M_r of 12,500 on SDS-PAGE both in reduced and nonreduced conditions, and a pI of 4.4. Activation of factor X was studied on SDS-PAGE and using a chromogenic substrate for factor X_a (S-2222, Bz-Ile-Glu-Gly-Arg pNAHCl). The activation effects were inhibited by PMSF, suggesting that the *C. vipera* activator is a serine proteinase. Their observations indicated that the venom activator is able to convert factor X to either factor X_a or an intermediate having the same amidolytic and coagulant activities (Farid *et al.*, 1993).

In 1995 a medium molecular weight factor X activator was isolated from *Vipera berus berus* venom (Samel and Siigur, 1995). This enzyme had a M_r of 38,000 both in reduced and nonreduced form on SDS-PAGE. The pI of the enzyme was in the pH interval of 3.5–4.5 (poorly resolved bands). After neuraminidase treatment only one band with pI of 4.5 was obtained. This activator generated the formation of factor X_a in the presence of Ca^{2+} ions. (The activation without Ca^{2+} was about 25%). It was inactive on synthetic substrates, on casein, prothrombin, and fibrinogen. The insulin B-chain was weakly cleaved at the positions Ala14-Leu15 and Tyr16-Leu17, and this process was inhibited by EDTA. The cleavage site in factor X is not yet determined (Samel and Siigur, 1995).

Thus the Viperidae venoms contain a wide variety of coagulating enzymes. Some of them are more, some less studied, but the final effect of all of these enzymes is generation of fibrin clot in the victim's blood due to activation of either factors V, X, or prothrombin, or converting fibrinogen. In addition to the individually isolated preparations there are many references on such kinds of activities in different crude venoms (Stocker *et al.*, 1986b; Stocker, 1990a; Kornalik, 1990).

3 Anticoagulant Action

3.1 Fibrin(ogen)olytic enzymes

Fibrin(ogen)olytic enzymes have been isolated from the venoms of a number of snake species, including also members of Viperidae. Snake venom proteinases that cleave peptide bonds in the fibrinogen can be devided into three groups (Kornalik, 1990).

- **Thrombin-like enzymes (thrombic proteases): fibrinogen-coagulant proteases.** These enzymes are serine proteinases that, like thrombin, catalyze the release of fibrinopeptide A or B or both from fibrinogen at the NDS knot of Aα or Bβchains of the molecule. Thrombin-like proteinases act *in vitro* as procoagulants converting fibrinogen to fibrin.
- **Fibrin(ogen)olytic enzymes.** These enzymes directly split off fragments from the C terminals of α, β and γ chains of fibrinogen, rendering it unclottable by thrombin. α(β)-Fibrinogenases are metalloproteinases that cleave mainly α-chain of fibrinogen. β-Fibrinogenases are mainly thermostable serine proteinases. γ-Fibrinogenases so far found are metalloproteinases.
- **Enzymes activating plasminogen and acting as indirect fibrinolytic enzymes.** Plasminogen-activating enzymes have not been separated from Viperidae snake venom so far.

The direct-acting fibrin(ogen)olytic enzymes from snake venom have been reviewed by several authors (Kornalik, 1990; Markland, 1991; Seegers and Ouyang, 1979; Siigur and Siigur, 1992; Stocker, 1990; Tu, 1977).

Although there are numerous reports of fibrin(ogen)olytic activities in venoms of the Viperidae family, there are relatively few investigations on the isolation and detailed characterization of these enzymes. Fibrinolytic activity has been detected in the venoms of *Vipera russellii*, *V. aspis*, *Bitis gabonica*, *Cerastes cerastes*, *V. lebetina*, *Echis carinatus* (for reviews see Markland, 1991; Siigur and Siigur, 1992).

A fibrin(ogen)olytic enzyme was purified from *V. aspis* venom. The enzyme hydrolyzed casein, bovine and human fibrinogen, and fibrin. It was inhibited by EDTA, but not by DFP (Boffa and Boffa, 1974). Viljoen and co-workers

(1979) fractionated the venom of *B. gabonica* and found the fraction that possessed fibrinolytic activity. However, Bajwa *et al.* (1982) were unable to detect fibrinolytic activity in *B.gabonica* venom.

Fibrin(ogen)olytic enzyme, cerastase F4, was isolated from the venom of *Cerastes cerastes* (Egyptian sand viper). Cerastase is a single-chain protein with a molecular weight of 22,500, and it hydrolyzes the Aα-chain of fibrinogen followed by the hydrolysis of the Bβ-chain; it also degrades the three chains of fibrin at different rates (Daoud *et al.*, 1986a,b). Cerastase F4 is a metalloenzyme that contains 1 mol of Ca^{2+} and mol of Zn^{2+} per mole of protein, it is completely inhibited by EDTA or EGTA (Daoud *et al.*, 1987).

Siigur and co-workers (1991a,b) isolated two fibrinogenases from the venom of *Vipera lebetina*. A fibrinogenase with molecular weight of 52,500 and isoelectric point at pH ~3 is inactivated by DFP and PMSF but not by EDTA, suggesting that this enzyme is a serine proteinase. Activity of the enzyme is directed preferentially towards the Bβ-chain of fibrinogen, but on prolonged incubation with fibrinogen, there was partial digestion of the Aα-chain also. Bβ-fibrinogenase is a glycoprotein containing 23% of neutral sugars. It hydrolyzes BAEE with K_m 7.7 × 10^{-5} M, k_{cat} 43.8 sec^{-1}, TAME with K_m 3.6 × 10^{-4} M and k_{cat} 39.8 sec^{-1}, and BAPNA with K_m 1.8 × 10^{-4}M and k_{cat} 0.94 sec^{-1}. Lysine esters are not hydrolyzed. The enzyme has weak caseinolytic activity and hydrolyzes glucagon at the sites Lys^{12}-Tyr^{13}, Arg^{17}-Arg^{18} and Arg^{18}-Ala^{19}(Siigur *et al.*, 1991).

The Bβ fibrinogenase from *V.lebetina* venom showed extreme heat stability: heating for 20 min at 95°C did not affect its esterolytic activity. The other fibrinogenolytic enzyme isolated from *V. lebetina* venom (named lebetase) is a metalloproteinase that is inhibited by EDTA and DTT and has a molecular weight of about 23,700. Lebetase readily hydrolyzes the Aα-chain and more slowly the Bβ-chain of fibrinogen. In fibrin the same chains are attacked. Thus the enzyme is an Aα,Bβ-fibrinogenase. The fibrinolytic activity of lebetase is direct, without any plasminogen activation. Lebetase also hydrolyzes casein, asocasein, and oxidized insulin B-chain in the positions Ala^{14}–Leu^{15} and Tyr^{16}–Leu^{17}. It consists of several isoforms in the pH range 4.6–5.4, all having fibrinolytic activity (Siigur and Siigur, 1991).

Immunological properties of lebetase have been investigated by Siigur *et al.* (1996b). The polyclonal antibodies against two fractions of fibrinolytic enzyme in *V. lebetina* venom were produced and purified by protein A-affinity chromatography. The antibodies reacted only with the fibrinolytic enzyme in *V. lebetina* venom as demonstrated by Western immunoblotting. Immunodot assays, ELISA, and Western immunoblotting revealed that 11 snake venoms including species of Viperidae and Crotalidae but not Elapidae cross-react with lebetase antibodies to varying degrees. The molecular weights of cross-reacting components show that most probably antibodies react with fibrin(ogen)olytic components in the other snake venoms.

Recently (Siigur *et al.*, 1996a) the complete amino acid sequence of lebetase was deduced from the nucleotide sequence of a cDNA clone isolated by screening a venomous gland cDNA library of Central Asian *V. lebetina* snake. The complementary DNA (cDNA) sequence with 2011 bp encodes an open reading frame of 478 amino acids, which includes an 18-amino acid signal peptide, plus a 175-amino acid segment of zymogen-like propeptide, a mature protein of 204 amino acids, a spacer of 18 amino acids, and a disintegrin-like peptide of 63 amino acids. In the disintegrin-like domain of protein, the RGD sequence is replaced by VGD. The mature protein lebetase as isolated from the crude venom has the molecular weight of approximately 23,700 and, thus, lebetase as well as several other snake venom metalloproteinases is translated as a precursor protein, which may be processed posttranslationally. The metalloproteinase domain has a typical zinc-chelating sequence (HEXXHXXGXXH). A fibrinogenase (M_r 26,000) was isolated from *V. lebetina* venom that is different from lebetase (Gasmi *et al.*, 1991). It hydrolyzes more effectively the Bß-chain of fibinogen and fibrin.

A fibrinogenolysin (α-fibrinogenase) was isolated from *E. carinatus* venom by use of hydrophobic interaction chromatography. After further purification by ion-exchange and reversed-phase HPLC (C_{18}), 14 amino acid residues of it's N-terminal were determined (Tseng *et al.*, 1989). N-terminal sequence of fibrinogenolysin from *E. carinatus* venom is similar to that of lebetase, but in the case of lebetase the N-terminal is blocked.

3.2 Coagulation factor IX/factor X-binding proteins

A new coagulation factor IX/factor X-binding protein (ECLV IX/X-bp) from *E. carinatus leucogaster* venom has been purified. ECLV IX/X-bp binds factor IX and X in a Ca^{2+}-dependent manner and is devoid of thrombin-inhibitory and platelet-aggregating activities. It prolonged the coagulation time *in vivo* by inhibiting prothrombin activation reversibly. It might be a useful anticoagulant agent for pharmaceutical use. Direct binding of IX/X-bp to factor IX and X could also be detected by far-Western blotting, and results of the experiment ruled out the lectin-like mechanism of ECLV IX/X-bp. The complete amino acid sequence and the disulfide pattern of ECLV IX/X-bp was deduced by enzymatic hydrolysis and automated sequencing of the S-pyridylethylated protein. The venom protein is a heterodimer with one subunit of 131 amino acid residues and another of 125 residues. Both subunits are homologous to each other and to other snake venom proteins of the C-type lectin superfamily (Chen and Tsai, 1996).

3.3 Hemorrhagic proteinases

Hemorrhagic proteins from Crotalid venoms have been highly purified and characterized (for review see Bjarnason and Fox, 1994). There are relatively few data about purified hemorrhagic proteins from viperid venoms.

The potency of the hemorrhagic toxins varies, and the minimum hemorrhagic dose (MHD) has been used to determine the potency of hemorrhagic toxins. The MHD is defined as the amount of toxin in micrograms which produces a hemorrhagic spot 10 mm in diameter at a specified time, usually 3 or 6 h, after intradermal or subcutaneous injection into rabbits or mice.In order to determine whether the hemorrhagic toxins were potentially capable of interfering with the blood coagulation system, they were assayed for their proteolytic effect on fibrinogen and fibrin.

With respect to fibrinogen, the hemorrhagic toxin from *Vipera berus berus* venom was able to digest fibrinogen. Within a 30-min digestion, the Aα-chain was lost. The Bβ-chain is more resistant to hydrolysis, and the γ-chain was not digested. The hemorrhagic toxin had very weak proteolytic activity towards fibrin at incubation times of up to 24 h (Samel and Siigur, 1990). The proteolytic specificity was investigated by using the oxidized B-chain of bovine insulin. The *V. berus* venom enzyme cleaved the Ala^{14}–Leu^{15}, Tyr^{16}–Leu^{17} and His^{10}–Leu^{11} bonds.

Although the molecular masses of the hemorrhagic toxins vary widely (22–90,000), they all appear to be metalloproteinases and thus share a somewhat conserved proteinase domain. Two hemorrhagic principles, *Bitis arietans* hemorrhagin a (BHRa) and b (BHRb), were purified from the venom of the viperous snake *B. arietans* (puff adder). They belong to a group of high molecular weight hemorrhagic metalloproteinases; molecular weights were 68,000 for BHRa and 75,000 for BHRb. (Omori-Satoh *et al.*, 1995).

The hemorrhagins hydrolyzed gelatins prepared from types I, II, III and IV collagen. On the other hand, among native collagens, only type IV was hydrolyzed. The hemorrhagins hydrolyzed angiotensin I and luteinizing hormone-releasing hormone. The insulin B-chain is cleaved by BHRa and BHRb at positions 11 and 10, respectively (Yamakawa *et al.*, 1995). There are no data about the fibrin(ogen)olytic activity of these hemorrhagins. *B. arietans* venom does not contain low molecular weight hemorrhagins.

4. Action on Platelets

4.1 Disintegrins

Snake venom components affect platelet function in various ways. Some venom proteins induce release reaction and platelet aggregation, while others inhibit these reactions. The most effective antiplatelet constituents are Arg-Gly-Asp

(RGD)–containing polypeptides that have been identified as specific fibrinogen receptors [i.e., glycoprotein (GP) IIb/IIIa complex ($\alpha_{II}b\beta_3$)] antagonists and have been termed disintegrins (Gould *et al.*, 1990). Disintegrins are a family of RGD-containing nonenzymatic proteins isolated from crotalid and viperid venoms that inhibit the integrin-mediated cell-cell and cell-matrix interaction (Gould *et al.*, 1990; Dennis *et al.*, 1990; Kini and Evans, 1992). Disintegrins have been classified into three families based on their polypeptide chain length (Gould *et al.*, 1990): short-chain disintegrins, 47–51 amino acid residues and four disulfide bridges; medium-chain disintegrins, 68–75 amino acid residues and six disulfide bridges; and long-chain disintegrins, 83–84 amino acid residues and seven disulfide bridges.

All disintegrins contain the Arg-Gly-Asp (RGD) or Lys-Gly-Asp (KGD) tripeptide sequence common to ligands for the GP IIb/IIIa receptor. The RGD motif is involved in the binding of fibrinogen, vitronectin, fibronectin, von Willebrand factor, and thrombospondin to their receptor. In all of the disintegrins for which the three-dimensional structures have been determined, the RGD sequence is presented to the receptor at the end of a loop structure in the carboxy-terminal portion of disintegrin.

The amino-terminal portion of disintegrin may not determine the platelet aggregation inhibitory activity. Removal of segments from this region does not significantly affect integrin inhibitory activity. Thus, despite the differences in sizes, both echistatin and carinatin from *E. carinatus* venom show similar inhibitory potency (Kini and Evans, 1992).

In vitro studies have shown that disintegrins are potent inhibitors of fibrinogen binding to its receptor on platelets. This inhibitory activity of disintegrin is about 1000–30,000 times more potent than that of the tetrapeptide Arg-Gly-Asp-Ser.

Echistatin.A 49-residue protein, echistatin α1, which inhibits platelet aggregation, was purified from the venom of the saw-scaled viper *Echis carinatus*. Echistatin is a single-chain polypeptide with a molecular weight of 5400 and a native isoelectric point of 8.3, and contains eight cysteines. Echistatin contains the sequence arginine-glycine-aspartic acid (RGD). Echistatin inhibits fibrinogen-dependent platelet aggregation initiated by ADP, thrombin, epinephrine, collagen, or platelet-activating factor (Gan *et al.*, 1988). The variant echistatin α2, which is two residues shorter than α1 at the C terminus, was also found (Dennis *et al.*, 1990). The structure of echistatin has been determined by 1H NMR, distance geometry calculations, and restrained molecular dynamic simulations (Atkinson *et al.*, 1994).

Echistatin has been successfully produced by solid-phase peptide synthesis (Garsky *et al.*, 1989) and by recombinant DNA expression in nonmammalian cells such as yeast (Jacobson *et al*, 1989). The products folded correctly and were indistinguishable from the native peptides. Met at position 28 of the native echistatin was replaced by Leu in the recombinant peptide. Leu 28 echistatin was identical to the native one in inhibiting platelet aggregation (Gan *et al.*, 1989).

Two new variants of short disintegrins were purified from the venom of *E. carinatus leakeyi* and named echistatin β and γ. These proteins were found to be about 85% similar in amino acid sequence to echistatin α. They all contain the adhesive recognition sequence Arg-Gly-Asp (RGD) but inhibit the aggregation of platelets from human and other mammals with different potencies (Chen *et al.*, 1995). Recently structural analysis of a biologically active echistatin analogue des (46–49)-[Ala 8,37]-echistatin with three disulfide bonds was performed by 2D-NMR and computer graphics (Chuang *et al.*, 1996).

A mutant of echistatin was constructed by substituting C̲RGDC̲ for A̲RGDD̲ in the Arg-Gly-Asp (RGD) region. The mutant more strongly inhibited the binding of human fibrinogen to its receptor, integrin $\alpha_{II}b\beta_3$. (The mutant is twice as potent as the native form.) The two new Cys residues are linked to each other in the mutant (Yamada and Kidera, 1996). These results indicate that it is possible to increase the affinity to the integrins by introducing a conformational constraint into an RGD-containing peptide.

Multisquamatin. A small platelet aggregation inhibitor, multisquamatin (M_r = 5700), was purified from *E. multisquamatus* venom. Multisquamatin inhibits ADP-induced human, canine, and rabbit platelet aggregation using platelet-rich plasma. It has an IC_{50} of 97 nm, 281 nM, and 333 nM for human, canine, and rabbit PRP, respectively. In a competitive binding assay using 125I-7E3 (a monoclonal antibody to GPIIb/IIIa that inhibits platelet aggregation) multisquamatin demonstrated GPIIb/IIIa-specific binding to human and canine platelets. The IC_{50} for multisquamatin displacement of 7E3 binding to human and canine GPIIb/IIIa is 3 nM and 63 nM, respectively. These results indicate that multisquamatin inhibits platelet aggregation by binding with high-affinity to the integrin receptor GPIIb/IIIa (Trikha *et al.*, 1994).

Gabonin. An antiplatelet peptide, gabonin, was purified from the venom of *Bitis gabonica*. Gabonin is a disulfide-linked dimer (monomer 11,000 Da, 84 amino acid residues). Gabonin dose-dependently inhibited human platelet aggregation stimulated by ADP, collagen, U46619, or thrombin in preparations of platelet-rich plasma and platelet suspension. Gabonin inhibits platelet aggregation mainly through the blockade of fibrinogen binding toward fibrinogen receptors of the activated platelets (Huang *et al.*, 1992).

Bitistatin. A platelet aggregation inhibitory protein, bitistatin was isolated from the venom of the puff adder *Bitis arietans*. This is a single-chain peptide containing 83 amino acids and seven disulfide bonds (Shebuski *et al.*, 1989). Bitistatin shows a high degree of homology to echistatin. Bitistatin is a potent inhibitor of platelet aggregation assessed *in vitro* and *in vivo* in a canine model of repetitive thrombus formation (Shebuski *et al.*, 1990).

Arietin (M_r 8,500), an RGD-containing protein, was separated from the venom of *Bitis arietans* (Huang *et al.*, 1991a,b).

Eristostatin (eristocophin). The platelet aggregation inhibitor eristostatin (eristocophin) has been characterized from leaf-nosed viper *Eristocophis macmahoni* venom. Eristocophin was more like the venom peptide echistatin (Scarborough *et al.*, 1991).

Mapping eristostatin peptides by proteolytic digestion suggested that it has the same alignment of S-S bridges as echistatin. Synthetic echistatin D27W resembled eristostatin since it had increased platelet aggregation inhibitory activity (McLane *et al.*, 1996). Echistatin and eristostatin share 62% sequence identity, including eight cysteines. However, they show significant difference in their biological activities. Eristostatin is the most potent inhibitor of ADP-induced platelet aggregation among disintegrins (McLane *et al.*, 1996). RGD-containing peptides may be valuable antithrombotic agents.

4.2 Novel antiplatelet aggregation polypeptides

Novel antiplatelet aggregation polypeptides Lebetins 1 and Lebetins 2 were isolated from *Vipera lebetina* venom by gel filtration and reverse-phase chromatography. Lebetins 1 contain 13 and 12 amino acid residues; lebetins 2 contain 38 and 37 residues, each with one disulfide bond. Sequence analysis revealed neither RGD sequence nor homology with other polypeptides. Lebetin 1 sequences were identical to the N-terminus of Lebetin 2. Lebetins inhibited platelet aggregation induced by thrombin, collagen, and PAF-acether. Both groups displayed strong *in vitro* antiplatelet activity and were able to prevent *in vivo* collagen-induced thrombocytopenia in rats (Barbouche *et al.*, 1996). As these peptides lack any known antiplatelet aggregation-active sequence, they might help to delineate still unknown mechanisms involved in platelet aggregation. Lebetins could have the potency to be used as antithrombotic agents.

4.3 GPIb-binding proteins

There exists a class of venom anticoagulants that are heterodimers covalently linked by one intersubunit disulfide bond. These include echicetin from *E. carinatus* venom.

Echicetin is a heterodimeric protein consisting of 16- and 14-kDa subunits and inhibiting platelet aggregation and secretion induced by low concentrations of thrombin (<0.2 U/ml), by binding to platelet glycoprotein (Peng *et al.*, 1993; Peng *et al.*, 1995).

Peng *et al.* (1994) demonstrated that the reduced and S-alkylated β-subunit of echicetin retains its inhibitory activity on vWF binding to GPIb, although with less potency than that of the intact protein. However, the structural features and entire amino acid sequence of the GPIb-binding protein echicetin have not yet been reported.

Bitiscetin: a von Willebrand factor (vWF) modulator protein. Recently bitiscetin was purified from the snake venom of *Bitis arietans*. Bitiscetin showed a heterodimeric structure composed of disulfide-linked α (16 kDa) and β (13 kDa) subunits and had a pI value of 9.1. Bitiscetin-induced platelet agglutination was dependent on vWF and platelet membrane glycoprotein GPIb, but not on Ca and GPIIb/IIIa. Bitiscetin directly interacts with vWF and requires the protein conformation for its interaction. Further characterization of bitiscetin including the determination of binding sites with vWF is necessary to understand the modulation mechanism of bitiscetin (Hamako *et al.*, 1996).

5 Methods

5.1 Materials and chemicals

Buffers and reagents

- **Citrated human plasma**
- **Solution 1**
  10 mM Tris-HCl, pH 7.4
- **Solution 2**
  50 mM Tris-HCl, pH 7.5
  150 mM NaCl
  5 mM $CaCl_2$
- **Solution 3**
  200 mM Tris-HCl, pH 8.3
  150 mM NaCl
  10 mM EGTA
- **Solution 4**
  5 mM Tris-HCl, pH 7.5
  10 mM NaCl
- **Solution 5**
  50 mM phosphate buffer, pH 7.5
- **Solution 6**
  50 mM Tris-HCl, pH 7.4
  100 mM NaCl
- **25 mM $CaCl_2$ in Solution 1**
- **100 mM $CaCl_2$** (in H_2O)
- **Denaturing solution (Solution 7)**
  10 M urea
  4% SDS
  4% 2-mercaptoethanol
- **Equilibrating buffer (Solution 8)**
  0.05 M Tris-H_3PO_4 buffer (pH 7.5)
  1 mM benzamidine hydrochloride

Chemicals

APTT reagent (cephalin from rabbit brain)	Sigma	A1801
Bz-Ile-Glu(Pip)-Gly-Arg-pNA (S-2337)	Chromogenix	
Bovine factor X	Chromogenix	
Fibrinogen type IV bovine	Sigma	F4753
Thrombin from bovine plasma	Sigma	T6634
Urea	Sigma	U0631
Na-dodecyl sulphate (SDS), laurylsulphate	Sigma	L4390
2-mercaptoethanol	Sigma	M3148

Equipment

- Water-bath
- Gel electrophoresis apparatus
- Spectrophotometer (UV-VIS)
- Aggregometer

5.2 Assays

Assay of factor X-activating enzyme

The assay of factor X activator is based on factor X activation and determines the activity of generated factor X_a:

$$FX \xrightarrow[\text{RVV}-\text{X}]{\text{VBXAE}} FX_a$$

$$\text{Bz-Ile-Glu(Pip)-Gly-Arg-pNA} \xrightarrow{FX_a} \text{Bz-Ile-Glu(Pip)-Gly-Arg} + \underline{\text{pNA}}$$

The activation of purified factor X was followed by measuring the amidolytic activity of the formed factor X_a (Hofmann and Bon, 1987).

Protocol 1 Continuous spectrophotometric assay

1. Bovine factor X (10–50 µl, 5 U/ml) was incubated at 25°C in **Solution 2** and an appropriate concentration of factor X activator.
2. Aliquots (1–50 µl) were removed at various times, and their amidolytic activity was immediately determined in 950 µl of **Solution 3** and 250 mM S-2337 by spectrophotometrically recording the formation of p-nitroaniline at 405 nM.

Protocol 2 Assay by plasma prothrombin time (Furie and Furie, 1976)

1. A test sample is diluted with **Solution 1** to give a clotting time between 10 and 19 s.
2. An 0.1-ml aliquot of the diluted sample is incubated at room temperature for 1 min with 0.1 ml of plasma and 0.1 ml of APTT reagent in glass tubes.
3. The reaction is initiated with the addition of 0.1 ml of 25 mM $CaCl_2$, and the clotting time is determined.

Assay of β-fibrinogenase

Protocol 3 Assay of β-fibrinogenase

1. 2% fibrinogen solution in **Solution 4** was mixed with 25 μg of purified enzyme and incubated at 37°C.
2. At various times, 300-μl aliquots were taken and added to 100 μl of thrombin (5 U/ml).
3. The coagulation time of the mixture was recorded.
4. The clotting time of fibrinogen by thrombin was taken as zero.

Fibrinogenolysis

(both for α- and β-fibrinogenases) (Ouyang and Teng, 1976)

Protocol 4 Fibrinogenolysis

1. 0.5 ml of a 1% fibrinogen solution was incubated with 0.5 ml of enzyme (10–100 μg/ml) at 37°C in **Solution 6**.
2. At various time intervals (0, 5, 15, 30, 60, 120 min), 0.1-ml aliquots were withdrawn and added to 0.1 ml of denaturing solution **(Solution 7)**.
3. The samples were reduced and denatured overnight at 37°C before being electrophoresed by SDS-PAGE (10% gels) (Laemmli, 1970).

Fibrinolysis assays

Protocol 5 Fibrin-plate method (Astrup and Müllertz, 1952)

1. 10 ml of 1% fibrinogen in **Solution 5** was clotted with 0.6 ml of thrombin (5 U/ml) in a petri dish.
2. After 30–60 min, 20 μl of enzyme solution was placed as a small drop on the surface of the fibrin plate.

Notes:
The area of digested fibrin is a quantitative measure of the enzyme activity. This is measured as follows:

3. After incubation for 18–20 h at 37°C, the product of two perpendicular diameters is recorded (in square millimeters), and the mean value of the three calculations is calculated.

Protocol 6 Fibrin degradation assay (Willis and Tu, 1988)

1. 0.1 ml of 1% fibrinogen in **Solution 6** was clotted with 0.1 ml of thrombin (5 U/ml in **Solution 6**).
2. The fibrin clot was allowed to form for 1 h at room temperature.
3. Then, 0.1 ml of enzyme (10 μg/ml) was added, and the mixtures were incubated at 37°C for various time (0, 15, 30, 60, 120 min, 22–24 h).
4. After incubation, 0.3 ml of denaturing solution was added, and the mixture was incubated overnight at 37°C before SDS-PAGE.

Assay for platelet aggregation

Protocol 7 Assay for platelet aggregation

1. Platelet aggregation was measured at 25°C in a Chrono-log whole blood aggregometer (Chrono-log Corp., Havertown, PA) performed in human PRP prepared from blood of human donors who had not taken aspirin within the previous 10 days.
2. 51 ml of whole blood was drawn into 9 ml of sodium citrate (3.8%) and centrifuged at 160 × g for 20 min, allowed to stand for 5 min, and PRP was decanted into a clean tube.
3. Platelet-poor plasma (PPP) was prepared from the PRP by further centrifugation at 2000 × g for 10 min, removed with a pipette, and used in the reference cell of the aggregometer
4. Test substances dissolved in water or buffer were added to PRP to a final volume of 0.5 ml, allowed to incubate for 1 min followed by the addition of ADP (20 μM final concentration) to initiate aggregation.
5. Platelet aggregation was quantitated by measuring the total amplitude on a chart recorder.
6. IC_{50} values were determined from dose-response curves generated from various concentrations of the purified peptides added to PRP (Scarborough *et al.*, 1993).

Notes:
In detecting biological activities of snake venom components several facts must be considered:

- The action of the clotting enzymes *in vitro* and *in vivo* must be distinguished. Some of the *in vitro* procoagulant enzymes, depending on the dose administered, cause defibrination *in vivo* and thus act as anticoagulants.
- Highly purified protein from snake venom can react with different components of the blood coagulation system.
- The use of the blood coagulation system components as substrates does not fully explain the influence of the snake venom component on the whole process of coagulation, since one of many reactions is watched.

Purification of enzymes from Viperidae venoms

Traditionally, isolation of biologically active proteins from snake venoms has been achieved by gel filtration and ion-exchange chromatography. More recently affinity chromatography, FPLC and HPLC were used for purification of snake venom proteins. Using reversed-phase HPLC has often resulted in denaturation of the venom proteins by the organic mobile phase with concomitant loss of biological activity. The reversed-phase HPLC was effective for purification of disintegrins from snake venoms. These components are stable in conditions of purification.

The purification of snake venom components cannot be standardized due to large differences in venom composition and between the components needed. We give some examples for isolation of factor X activators and fibrino(geno)lytic enzymes from the venoms of *Vipera russellii*, *V. berus berus* and *V. lebetina*. These are only **examples** for possible procedures which could be used in fractionation of snake venoms.

Protocol 8 Isolation of factor X activator (RVV-X) from Russell's viper venom (Kisiel *et al.*, 1976)

1. Lyophilized venom (1 g) was dissolved in 5 ml of 0.05 M sodium acetate buffer (pH 5.5) containing 0.5 M NaCl and 5 mM benzamidine-HCl.
2. Insoluble material was removed by centrifugation for 5 min at 1000 × g at 20°C.
3. The supernatant was applied to a Sephadex G-150 column (2.6 × 96 cm) previously equilibrated with 0.05 M sodium acetate buffer (pH 5.5) containing 0.5 M NaCl-5 mM benzamidine hydrochloride.
4. Chromatography was performed at room temperature with a flow rate of approximately 0.5 ml/min.

5. Those fractions containing coagulant activity were pooled and dialyzed at 4°C against 4 l of 0.05 M Tris-H_3PO_4 buffer (pH 7.5) containing 1 mM benzamidine hydrochloride.
6. The dialyzed sample was applied to a QAE-Sephadex A-50 column (1.6 × 30 cm) previously equilibrated at 4°C with **Solution 8**.
7. The column was then washed with approximately 30 ml of equilibrating buffer **(Solution 8)** and further washed with 1–2 column vol of 0.05 M Tris-H_3PO_4 buffer (pH 6.0) containing 1 mM benzamidine hydrochloride.
8. Coagulant activity was then eluted from the column with a linear gradient of NaCl prepared from 100 ml of 0.05 M Tris-H_3PO_4 buffer (pH 6.0) containing 1 mM benzamidine hydrochloride and 100 ml of 0.05 M Tris-H_3PO_4 buffer (pH 6.0) containing 0.5 M NaCl and 1 mM benzamidine hydrochloride.

Protocol 9 Isolation of factor X activator (VBXAE) from *V. berus* venom (Samel and Siigur, 1995)

The fractionation steps were performed at 4°C.

1. Crude *V. berus berus* venom (5.0 g) was dissolved in 45 ml of 0.2 M ammonium acetate, pH 6.7.
2. Insoluble material was removed by centrifugation (5000 × g for 15 min), and the supernatant was applied to the column (5 × 130 cm) of Sephadex G-100 superfine equilibrated with 0.2 M ammonium acetate.
3. The elution was carried out with the same solution at a flow rate 9.6 ml/h, and fractions were collected at 1-h intervals.
4. The column of bacitracin-agarose (1.8 × 10 cm) was equilibrated with 0.1 M ammonium acetate, pH 8.0.
5. Lyophilized fraction III (0.3 g) from gel filtration was dissolved in 20 ml of 0.1 M ammonium acetate and applied to the column.
6. Nonadsorbed material was washed with the equilibration solution (0.1 M ammonium acetate, pH 8.0).
7. The fraction containing factor X-activating protein was eluted with 1 M NaCl in 0.1 M ammonium acetate, pH 8.0. The flow rate was 12 ml/h; fractions of 6 ml were collected.

Protocol 10 Purification of a β-fibrinogenase from *V. lebetina* venom (Siigur *et al.*, 1991)

Fractionation of crude venom: gel filtration

1. *V. lebetina* venom (1.5 g) was dissolved in 15 ml of 0.2 M ammonium acetate (pH 6.7), and the insoluble material was removed by centrifugation (5000 × g for 15 min).
2. The supernatant was applied to a column (2.8 × 120 cm) of Sephadex G-100 superfine equilibrated at 4°C with 0.2 M ammonium acetate.
3. The elution of venom components was at the rate of 6.8 ml/h, and 6.8-ml aliquots were collected. Fractions from each peak were pooled and lyophilized.

Affinity chromatography

4. The column of p-aminobenzamidine-succinylhexamethylenediamine Sepharose 4B (1.1 × 15 cm) was equilibrated with 0.05 M Tris-HCl buffer, pH 7.5, containing 0.5 M KCl and 0.05 M $CaCl_2$.
5. The material from Sephadex G-100 was divided into portions (40–50 mg each) and separately chromatographed.
6. Unbound material was eluted with the equilibration buffer at a flow rate of 2.4 ml/h.
7. The arginine esterolytic activity was eluted with 0.01 M HCl-0.5 M KCl-0.05 M $CaCl_2$ at a flow rate of 4.8 ml/h.
8. The enzyme fraction was concentrated by ultrafiltration (PM-10 membrane) and applied to a column of Sephadex G-25 equilibrated with 0.01 M Tris-HCl buffer, pH 8.0. The column was run with the same buffer at a flow rate of 16 ml/h.

DEAE-cellulose chromatography

9. The arginine esterase from the previous step was further fractionated by ion-exchange chromatography on a column of DE-52 cellulose (1.5 × 12.5 cm) equilibrated with 0.01 M Tris-HCl buffer, pH 8.0.
10. The column was developed using a linear salt gradient of 0–0.2 M NaCl in the starting buffer. The flow rate was 12 ml/h, and fractions of 4 ml were collected.

Protocol 11 Purification of lebetase, a fibrinolytic enzyme from *V. lebetina* venom (Siigur and Siigur, 1991)

Gel filtration

1. Crude *V. lebetina* venom (4.1 g) was dissolved in 25 ml of 0.2 M ammonium acetate (pH 6.7), and the insoluble material was removed by centrifugation (5000 xg for 15 min).
2. The supernatant was applied to a column (4.8 × 130 cm) of Sephadex G-100 superfine equilibrated with 0.2 M ammonium acetate.
3. The elution of venom components was effected at the rate of 5.5 ml/h, and 8.2 ml aliquots were collected.
4. The fractions were tested for caseinolytic and fibrinolytic activities.
5. Fibrinolytic fractions were combined and concentrated by ultrafiltration on PM-10 membranes.

DEAE-cellulose chromatography

6. DEAE-cellulose chromatography was performed on a column of Whatman DE-52 cellulose (1.5 × 12.5 cm) equilibrated with 0.08 M ammonium acetate (pH 8.3).
7. Aliquots from the concentrated fibrinolytic fraction from Sephadex G-100 were diluted to 0.08 M of ammonium acetate, and the pH brought to 8.3 with 25% ammonia.
8. The column was washed with equilibration buffer to elute the nonadsorbed protein.
9. Bound proteins were eluted using a linear gradient from 0.08 M to 0.15 M of ammonium acetate (pH 8.3). The flow rate was 16 ml/h; fractions of 4 ml were collected.

6 Conclusions

We can conclude that Viperidae venoms contain a wide variety of components with different specificities affecting hemostasis at various stages. Many of them could be used in medicine either for therapy or for diagnostics, for example, cerastobin for defibrinogenation, lebetase for dissolving thrombi and preventing their formation. Disintegrins are powerful platelet aggregation antagonists. Factor X activator RVVX is widely used for detection of factor X in blood. Ecarin is being used in patients with hepatic diseases or those being treated with vitamin K antagonists. Preparation of recombinant snake venom components will open new perspectives for designing more effective substances for clinical medicine.

Acknowledgements

The authors express gratitude to Mrs. Ene Valdmann for typing the manuscript.

References

Astrup T, Müllertz S (1952) The fibrin plate method for estimating fibrinolytic activity. Archs. Biochem. Biophys **40** 346–51

Atkinson RA, Saudek V, Pelton JT (1994): Echistatin: the refined structure of a disintegrin in solution by 1H NMR and restrained molecular dynamics. In J. Pept. Protein Res. **43**:563–72

Atoda H, Hyuga M, Morita T (1991): The primary structure of coagulation factor IX/factor X-binding protein isolated from the venom of Trimeresurus flavoviridis. In J. Biol. Chem. **266**:14903–11

Bajwa SS, Kirakossian H, Reddy KNN, Markland FS (1982): Thrombin-like and fibrinolytic enzymes in the venoms from the gaboon viper (Bitis gabonica), eastern cottonmouth mocassin (Agkistrodon p. piscivorus) and southern copperhead (Agkistrodon c. contortrix) snakes. In Toxicon **20**:427–32

Barbouche R, Marrakchi N, Mansuelle P, Krifi M, Fenouiller E, Rochat H, El Ayeb M (1996): Novel anti-platelet aggregation polypeptides from Vipera lebetina venom: isolation and characterization. In FEBS Lett **392**:6–10

Bjarnason JB, Fox JW (1994): Hemorrhagic metalloproteinases from snake venoms. In Pharmac. Ther. **62**:325–72

Boffa MC, Boffa GA (1974): Correlations between the enzymatic activities and the factors active on blood coagulation and platelet aggregation from the venom of Vipera aspis. In Biochim. Biophys. Acta **354**:275–90

Chen YL, Huang TF, Chen SW, Tsai IH (1995): Determination of the structure of two novel echistatin variants and comparison of the ability of echistatin variants to inhibit aggregation of platelets from different species. In Biochem. J. **305**:513–20

Chen YL, Tsai IH (1996): Functional and sequence characterization of coagulation factor IX/factor X-binding protein from the venom of Echis carinatus leucogaster. In Biochemistry **35**:5264–71

Chuang LC, Chen PY, Chen C, Huang TH, Wang KT, Chiou SH, Wu H (1996): Structural analysis of a biologically active echistatin analogue des(46–49)-[Ala8,37]-echistatin gamma with three disulfide bonds by 2D-NMR and computer graphics. In Biochem. Biophys. Res. Commun. **220**:246–54

Daoud E, Tu AT, El-Asmar MF (1986a): Mechanism of the anticoagulant, cerastase F-4, isolated from Cerastes cerastes (Egyptian sand viper) venom. In Thromb. Res. **41**:791–9

Daoud E, Tu AT, El-Asmar MF (1986b): Isolation and characterization of an anticoagulant proteinase, cerastase F-4, from Cerastes cerastes (Egyptian sand viper) venom. In Thromb. Res. **42**:55–62

Daoud EW, Halim HY, Shaban EA, El-Asmar MF (1987): Further characterization of the anticoagulant proteinase, cerastase F-4 from Cerastes cerastes (Egyptian sand viper) venom. In Toxicon **25**:891–7

Dennis MS, Henzel WJ, Pitti RM, Lipari MT, Napier MA, Deisher TA, Bunting S, Lazarus RA (1990): Platelet glycoprotein IIb-IIIa protein antagonists from snake venoms: Evidence for a family of platelet-aggregation inhibitors. In Biochemistry **87**:2471–5

Durkee KH, Roh BH, Doellgast GJ (1993): Immunoaffinity chromatographic purification of Russell's viper venom factor X activator using elution in high concentrations of magnesium chloride. In Protein Expression and Purification **4**:405–11

Farid TM, Tu AT (1989): Characterization of cerastobin, a thrombin-like enzyme from the venom of Cerastes vipera (Sahara sand viper). In Biochemistry **28**:371–7

Farid TM, Tu AT, El Asmar MF (1990): Effect of cerastobin, a thrombinlike enzyme from Cerastes vipera (Egyptian sand snake) venom, on human platelets. In Haemostasis **20**:296–304

Farid T, Nasser H, Zaki K, El-Asmar MF (1993): Low molecular weight factor X activator from Cerastes vipera (Sahara sand viper) venom. In Toxicon **31**:1007–17

Fortová F, Dyr JE, Vodrázka Z (1983): Isolation of the prothrombin-converting enzyme from fibrinogenolytic enzymes of Echis carinatus venom by chromatographic and electrophoretic methods. In J. Chromatog. **259**:473–9

Franssen JHL, Janssen-Claessen T, van Dieijen G (1983): Purification and properties of an activating enzyme of blood clotting factor X from the venom of Cerastes cerastes. In Biochim. Biophys. Acta **747**:186–90

Fujikawa K, Titani K, Davie EW (1975): Activation of bovine factor X (Stuart factor): conversion of factor $X_{a\alpha}$ to factor $X_{a\beta}$. In Proc. Natl. Acad. Sci. USA **72**:3359–63

Furie BC, Furie B (1975): Factor X activating enzyme from Russell's viper venom: isolation and characterization. In J. Biol. Chem. **250**:601–8

Furie BC, Furie B (1976): Coagulant protein of Russell's viper venom. In Methods Enzymol vol 45 (Lorand L, ed), pp 191–205: New York: Academic Press

Furukawa Y, Matsunaga Y, Hayashi K (1976): Purification and characterization of a coagulant protein from the venom of Russell's viper. In Biochim. Biophys. Acta **453**:48–61

Gan ZR, Condra JH, Gould RJ et al. (1989): High-level expression in Escherichia coli of a chemically synthesized gene for [Leu-28] echistatin. Gene 79: 159–66

Gan ZR, Gould RJ, Jacobs JW, Friedman PA, Polokoff MA (1988): Echistatin. A potent platelet aggregation inhibitor from the venom of the viper, Echis carinatus. In J. Biol. Chem. **263**:19827–32

Garsky VM, Lumma PK, Freidinger RM, Pitzenberger SM, Randall WC, Veber DF, Gould RJ, Friedman PA (1989): Chemical synthesis of echistatin, a potent inhibitor of platelet aggregation from Echis carinatus: Synthesis and biological activity of selected analogs. In Proc. Natl. Acad. Sci. **86**:4022–6

Gasmi A, Karoui M, Benlastar Z, Karoui H, El Ayeb M, Dellaga K (1991): Purification and characterization of a fibrinogenase from Vipera lebetina (desert adder) venom. In Toxicon **29**:827–36

Gould RJ, Polokoff MA, Friedman PA, Huang TF, Holt JC, Cook JJ, Niewiarowski S (1990): Disintegrins: A family of integrin inhibitory proteins from viper venoms. In Proc. Soc. Exp. Biol. Med. **195**:168–71

Gowda DC, Jackson CM, Hensley P, Davidson EA (1994): Factor X-activating glycoprotein of Russell's viper venom. Polypeptide composition and characterization of the carbohydrate moieties. In J. Biol. Chem. **269**:10644–50

Gowda DC, Jackson CM, Kurzban GP, McPhil P, Davidson EA (1996): Core sugar residues of the N-linked oligosaccharides of Russell's viper venom factor X-activator maintain functionally active polypeptide structure. In Biochemistry **35**:5833–7

Hamako J, Matsui T, Suzuki M, Ito M, Makita K, Fujimura Y, Ozeki Y, Titani K (1996): Purification and characterization of bitiscetin, a novel von Willebrand factor modulator protein from Bitis arietans snake venom. In Biochem. Biophys. Res. Commun. **226**:273–9

Hofmann H, Bon C (1987): Blood coagulation induced by the venom of Bothrops atrox. 2. Identification, purification, and properties of two factor X activators. In Biochemistry **26**:780–7

Huang TF, Wang WJ, Teng MC, Liu CS, Ouyang C (1991a): Purification and characterization of an antiplatelet peptide, arietin, from Bitis arietans venom. In Biochim. Biophys. Acta **1074**:136–43

Huang TF, Wang WJ, Teng MC, Ouyang C (1991b): Mechanism of action of the antiplatelet peptide, arietin, from Bitis arietans venom. In Biochim. Biophys. Acta **1074**:144–50

Huang TF, Peng HC, Peng IS, Teng MC, Ouyang C (1992): An antiplatelet peptide, gabonin, from Bitis gabonica snake venom. In Arch. Biochem. Biophys. **298**:13–20

Jacobson MA, Forma FM, Buenaga RF, Hofmann KJ, Schultz LD, Gould RJ, Friedman PA (1989): Expression and secretion of biologically active echistatin in Saccharomyces cerevisiae. In Gene **85**:511–6

Jayanthi GP, Gowda TV (1990): Synergistic interaction of a protease and protease inhibitors from Russell's viper (Vipera russelli) venom. In Toxicon **28**:65–74

Kini RM, Evans HJ (1992): Structural domains in venom proteins: evidence that metalloproteinases and nonenzymatic

platelet aggregation inhibitors (disintegrins) from snake venoms are derived by proteolysis from a common precursor. In Toxicon **30**:265–93
Kisiel W, Hermodson MA, Davie EW (1976): Interaction of Lanthanide ions with bovine factor X and their use in the affinity chromatography of the venom coagulant protein of Vipera russellii. In Biochemistry **15**:4901–6
Kisiel W (1979): Molecular properties of the Factor V-activating enzyme from Russell's viper venom. In J. Biol. Chem. **254**:12,230–4
Kisiel W (1991): Effect of snake venoms on factor V. In Handbook of Natural Toxins, vol. 5: Reptile Venoms and Toxins (Tu AT, ed) pp 253–64: New York: Marcel Dekker
Komori Y, Nikai T, Sugihara H (1990): Isolation and characterization of factor X activator from the venom of Vipera aspis aspis. In Int. J. Biochem. **22**:1053–60
Komori Y, Nikai T, Sugihara H (1993): Isolation and characterization of procoagulant from the venom of Vipera aspis aspis. In Int. J. Biochem. **25**:761–7
Kornalik F (1985): The influence of snake venom enzymes on blood coagulation. In Pharmac. Ther. **29**:353–405
Kornalik F (1990): Toxins affecting blood coagulation and fibrinolysis. In Handbook of Toxinology (Shier WT, Mebs D, eds) p 683: New York: Marcel Dekker
Laemmli UK (1970): Cleavage of structural proteins during the assembly of the head of bacteriophage T_4. In Nature **227**:680–5
Laraba-Djebari F, Martin-Eauclaire MF, Marchot P (1992): A fibrinogen-clotting serine proteinase from Cerastes cerastes (Horned viper) venom with arginine-esterase and amidase activities. Purification, characterization and kinetic parameter determination. In Toxicon **30**:1399–410
Laraba-Djebari F, Martin-Eauclaire MF, Mauco G, Marchot P (1995): Afaácytin, an αβ-fibrinogenase from Cerastes cerastes (Horned Viper) venom activates purified factor X and induces serotonin release from human platelets. In Eur. J. Biochem. **233**:756–65
Markland FS (1991): Inventory of α- and β-fibrinogenases from snake venoms. In Thromb. Haemostasis **65**:438–43
McLane MA, Vijay-Kumar S, Marcinkiewicz C, Calvere JJ, Niewiarowski S (1996): Importance of the structure of the RGD-containing loop in the disintegrins echistatin and eristostatin for recognition of alpha IIb beta 3 and alpha v beta 3 integrins. In FEBS Lett. **391**:139–43
Morita T, Iwanaga S (1978): Purification and properties of prothrombin activator from the venom of Echis carinatus. In J. Biochem. **83**:559–70
Nishida S, Fujita T, Kohno N, Atoda H, Morita T, Takeya H, Kido I, Paine MJI, Kawabata S, Iwanaga S (1995): cDNA cloning and deduced amino acid sequence of prothrombin activator (Ecarin) from Kenyan Echis carinatus venom. In Biochemistry **34**:1771–8
Omori-Satoh T, Yamakawa Y, Nagaoka Y, Mebs D. (1995): Hemorrhagic principles in the venom of Bitis arietans, a viperous snake. I. Purification and characterization. In Biochim. Biophys. Acta **1246**:61–6
Ouyang C, Teng CM (1976): Fibrinogenolytic enzymes of Trimeresurus mucrosquamatus venom. In Biochim. Biophys. Acta **420**:298–308
Ouyang C, Teng CM, Huang TF (1992): Characterization of snake venom components acting on blood coagulation and platelet function. In Toxicon **30**:945–66
Peng M, Lu W, Beviglia L, Niewiarowski S, Kirby EP (1993): Echicetin: a snake venom protein that inhibits binding of von Willebrand factor and alboaggregins to platelet glycoprotein Ib. In Blood **81**:2321–8
Peng M, Holt JC, Niewiarowski S (1994): Isolation, characterization and amino acid sequence of echicetin beta subunit, a specific inhibitor of von Willebrand factor and thrombin interaction with glycoprotein Ib. In Biochem. Biophys. Res. Commun. **205**:68–72
Peng M, Emig FA, Mao A, Lu W, Kirby EP, Niewiarowski S, Kowalska MA (1995): Interaction of echicetin with a high-affinity thrombin binding site on platelet glycoprotein GPIb. **74**:954–7
Pirkle H, Theodor I, Miyada D, Simmons G (1986): Thrombin-like enzyme from venom of Bitis gabonica. Purification, properties and coagulant actions. In J. Biol. Chem. **261**:8830–5

Pukrittayakamee S, Esnouf MP, McMichael AJ (1983): Purification and inactivation of the factor X activator of Russell's viper venom with monoclonal antibodies. In Mol. Biol. Med. **1**:123–35
Rhee MJ, Morris S, Kosow DP (1982): Role of meizothrombin and meizothrombin-(desF1) in the conversion of prothrombin to thrombin by the Echis carinatus venom coagulant. In Biochemistry **21**: 3437–43
Rosing J, Tans G (1992): Structural and functional properties of snake venom prothrombin activators. In Toxicon **30**:1515–27
Samel M, Siigur J (1990): Isolation and characterization of hemorrhagic metalloproteinase from Vipera berus berus (common viper) venom. In Comp. Biochem. Physiol. **97C**:209–14
Samel M, Siigur J (1995): Medium molecular weight factor X activating enzyme from Vipera berus berus venom. In Toxicon **33**:41–52
Scarborough RM, Rose JW, Hsu MA, Phillips DR, Fried VA, Campbell AM, Nannizzi L, Charo IF (1991): Barbourin. A GPIIb-IIIa-specific integrin antagonist from the venom of Sistrurus m. barbouri. In J. Biol. Chem. **266**:9359–62
Scarborough RM, Rose JW, Naughton MA, Phillips DR, Nannizzi L, Arfsten A, Campbell AM, Charo IF (1993): Characterization of the integrin specificities of disintegrins isolated from American pit viper venoms. In J. Biol. Chem. **268**:1058–65
Schieck A, Kornalik F, Habermann E (1972): The prothrombin-activating principle from Echis carinatus venom. I. Preparation and biochemical properties. In Naunyn-Schmiedeberg's Arch. Pharmacol. **272**:402–16
Seegers WH, Ouyang C (1979): Snake venoms and blood coagulation. In Snake Venoms, Handbook of Experimental Pharmacology, vol 52 (Lee CY, ed) pp 684–750, Berlin: Springer
Shebuski RJ, Ramjit DR, Bencen GH, Polokoff MA (1989): Effect of bitistatin, a potent RGD-containing peptide from the venom of Bitis arietans, on platelet aggregation and bleeding time in the dog (Abstract). In Circulation 80 (Suppl II):422
Shebuski RJ, Ramjit DR, Sitko GR, Lumma PK, Garsky VM (1990): Prevention of canine coronary artery thrombosis with echistatin, a potent inhibitor of platelet aggregation from the venom of the viper, Echis carinatus. In Thromb. Haemostas. **64**:576–81
Siigur E, Mähar A, Siigur J (1991): β-Fibrinogenase from the venom of Vipera lebetina. In Toxicon **29**:107–18
Siigur E, Siigur J (1991): Purification and characterization of lebetase, a fibrinolytic enzyme from Vipera lebetina (snake) venom. In Biochim. Biophys. Acta **1074**:223–9
Siigur E, Aaspõllu A, Tu AT, Siigur J (1996a): cDNA cloning and deduced amino acid sequence of fibrinolytic enzyme (lebetase) from Vipera lebetina snake venom. In Biochem. Biophys. Res. Commun. **224**:229–36
Siigur J, Siigur E (1992): The direct acting α-fibrin(ogen)olytic enzymes from snake venoms. In J. Toxicol.–Toxin Reviews **11**:91–113
Siigur J, Tõnismägi K, Tu AT, Siigur E (1996b): Cross-reactivities of polyclonal antibodies against lebetase, fibrinolytic enzyme of Levantine viper (Vipera lebetina) venom. In Toxicon **34**:608–13
Smith CM, Hanahan DJ (1976): The activation of factor V by factor Xa or α-chymotrypsin and comparison with thrombin and RVV-V action. An improved factor V isolation procedure. In Biochemistry **15**:1830–8
Solovyov DA, Platonova TN, Ugarova TP (1996): Purification and characterization of ecamulin–a prothrombin activator from the venom of many-scaled viper Echis multisquamatus. In Biokhimiya **61**:1094–104 (Russ.)
Stocker K (1980): Snake venom proteases affecting blood coagulation and fibrinolysis. In Nat. Toxins:111–23
Stocker K, Fischer H, Brogli M (1986a): Chromogenic assay for the prothrombin activator ecarin from the venom of the saw-scaled viper (Echis carinatus). In Toxicon **24**:81–9
Stocker K, Fischer H, Brogli M (1986b): Determination of factor X activator in the venom of the saw-scaled viper (Echis carinatus). In Toxicon **24**:313–5
Stocker K (1990a): Snake venom proteins affecting hemostasis and fibrinolysis. In Medical Use of Snake Venom Proteins (Stocker K, ed) pp 97–160: Boca Raton, Florida: CRC Press

Stocker KF (1990b): Application of snake venom proteins in the diagnosis of hemostatic disorders. In Medical Use of Snake Venom Proteins (Stocker KE, ed) pp 213–52: Boca Raton, Florida: CRC Press

Takeya H, Nishida S, Miyata T, Kawada SI, Saisaka Y, Morita T, Iwanaga S (1992): Coagulation factor X activating enzyme from Russell's viper venom (RVV-X). In J. Biol. Chem. **267**:14109–17

Teng CM, Chen YH, Ouyang C (1984): Purification and properties of the main coagulant and anticoagulant principles of Vipera russellii snake venom. In Biochim. Biophys. Acta **786**:204–12

Tokunaga F, Nagasawa K, Tamura S, Miyata T, Iwanaga S, Kisiel W (1988): The factor V-activating enzyme (RVV-V) from Russell's viper venom. Identification of isoproteins RVV-Vα, -Vβ and Vγ and their complete amino acid sequences. In J. Biol. Chem. **263**:17471–81

Trikha M, Rote WE, Manley PJ, Lucchesi BR, Markland FS (1994): Purification and characterization of platelet aggregation inhibitors from snake venoms. In Thromb. Res. **73**:39–52

Tseng A, Berry SL, Cotton B, Exner T, Sheumack DD, Howden MEH (1989): Isolation and purification of a fibrinogenolysin from the venom of the saw-scaled viper (Echis carinatus) by high performance liquid chromatography. In J. Chromatog. **474**:424–9

Tu A T (1977): Blood coagulation. In Venoms: Chemistry and Molecular Biology, pp 329–360: New York: Wiley

Viljoen CC, Meehan CM, Botes DP (1979): Separation of Bitis gabonica (Gaboon adder) venom arginine esterases into kinin-releasing, clotting, and fibrinolytic factors. In Toxicon **17**:145–54

Willis TW, Tu AT (1988): Purification and biochemical characterization of atroxase, a nonhemorrhagic fibrinolytic protease from Western diamondback rattlesnake venom. In Biochemistry **27**:4769–77

Yamada T, Kidera A (1996): Tailoring echistatin to possess higher affinity for integrin $\alpha_{II}b\beta_3$. In FEBS Lett **387**:11–15

Yamada D, Sekiya F, Morita T (1996): Isolation and characterization of carinactivase, a novel prothrombin activator in Echis carinatus venom with a unique catalytic mechanism. In J. Biol. Chem. **271**:5200–7

Yamakawa Y, Omori-Satoh T, Mebs D (1995): Hemorrhagic principles in the venom of Bitis arietans, a viperous snake. II. Enzymatic properties with special reference to substrate specificity. Biochim. Biophys. Acta 1247: 17–23

Guide to Solutions

Guide to Protocols

Troubleshooting Guide

Index

Index nach Kapitel

Kapitel 1

Kapitel 2

Kapitel 3

Kapitel 8

Kapitel 9

Kapitel 10

Kapitel 11

Kapitel 12

Kapitel 13

Kapitel 18

Kapitel 19

Kapitel 20

GPSR Compliance

The European Union's (EU) General Product Safety Regulation (GPSR) is a set of rules that requires consumer products to be safe and our obligations to ensure this.

If you have any concerns about our products, you can contact us on ProductSafety@springernature.com

In case Publisher is established outside the EU, the EU authorized representative is:

Springer Nature Customer Service Center GmbH
Europaplatz 3
69115 Heidelberg, Germany

Batch number: 10578397

Printed by Printforce, the Netherlands